Molecular Biotechnology: Therapeutic Applications and Strategies

Molecular Biotechnology: Therapeutic Applications and Strategies

Contributors

Volker F. Wendisch, Steffen N. Lindner et al.

www.aurisreference.com

Molecular Biotechnology: Therapeutic Applications and Strategies

Contributors: Volker F. Wendisch, Steffen N. Lindner et al.

Published by Auris Reference Limited
www.aurisreference.com

United Kingdom

Copyright 2016
Printed in 2017 for Sale in the Indian Subcontinent

The information in this book has been obtained from highly regarded resources. The copyrights for individual articles remain with the authors, as indicated. All chapters are distributed under the terms of the Creative Commons Attribution License, which permit unrestricted use, distribution, and reproduction in any medium, provided the original author and source are credited.

Notice

Contributors, whose names have been given on the book cover, are not associated with the Publisher. The editors and the Publisher have attempted to trace the copyright holders of all material reproduced in this publication and apologise to copyright holders if permission has not been obtained. If any copyright holder has not been acknowledged, please write to us so we may rectify.

Reasonable efforts have been made to publish reliable data. The views articulated in the chapters are those of the individual contributors, and not necessarily those of the editors or the Publisher. Editors and/or the Publisher are not responsible for the accuracy of the information in the published chapters or consequences from their use. The Publisher accepts no responsibility for any damage or grievance to individual(s) or property arising out of the use of any material(s), instruction(s), methods or thoughts in the book.

Molecular Biotechnology: Therapeutic Applications and Strategies

ISBN: 978-1-78154-963-6

British Library Cataloguing in Publication Data
A CIP record for this book is available from the British Library

Printed in the United Kingdom

Exclusively distributed by CBS Publishers & Distributors Pvt. Ltd.

Sales & Distribution Rights only for India, Pakistan, Bangladesh, Sri Lanka, Nepal and Bhutan. This book is not to be sold outside these territories.

Contents

List of Abbreviations vii
List of Contributors..ix
Preface.. xv

Chapter 1 Use of Glycerol in Biotechnological Applications 1
          Volker F. Wendisch, Steffen N. Lindner, and Tobias M. Meiswinkel

Chapter 2 Phage Display and Synthetic Peptides as Promising
          Biotechnological Tools for the Serological Diagnosis of Leprosy 53
          Silvana Maria Alban, Juliana Ferreira de Moura, Vanete
          Thomaz-Soccol, Samira Bührer Sékula, Larissa Magalhães
          Alvarenga, Marcelo Távora Mira, Carlos Chávez Olortegui,
          and João Carlos Minozzo

Chapter 3 DNA Repair and Chemotherapy 69
          Seiya Sato and Hiroaki Itamochi

Chapter 4 Therapeutic Strategies Based on Polymeric Microparticles 93
          C. Vilos and L. A. Velasquez

Chapter 5 Mammalian Models of Duchenne Muscular Dystrophy:
          Pathological Characteristics and Therapeutic Applications 115
          Akinori Nakamura and Shin'ichi Takeda

Chapter 6 Advanced Nanomaterials in Multimodal Imaging: Design,
          Functionalization, and Biomedical Applications 135
          Zhe Liu, Fabian Kiessling, and Jessica Gätjens

Chapter 7 Nanophotonics for Molecular Diagnostics and Therapy
          Applications... 171
          João Conde, João Rosa, João C. Lima, and Pedro V. Baptista

Chapter 8 Neural Stem Cells: Ready for Therapeutic Applications? 197
          Simona Casarosa, Yuri Bozzi, and Luciano Conti

Chapter 9 Strategies of Mucosal Immunotherapy for Allergic Diseases 233
          Yi-Ling Ye, Ya-Hui Chuang, and Bor-Luen Chiang

Chapter 10	**Topical Application of Recombinant Type VII Collagen Incorporates Into the Dermal–Epidermal Junction and Promotes Wound Closure** 265	
	Xinyi Wang, Pedram Ghasri, Mahsa Amir, Brian Hwang, Yingpin Hou, Michael Khilili, Andrew Lin, Douglas Keene, Jouni Uitto, David T Woodley, and Mei Chen	
	Citations 297	
	Index 299	

List of Abbreviations

AAV	Adeno-Associated Virus
AF	Anchoring Fibril
ALS	Amyotrophic Lateral Sclerosis
AO	Antisense Oligonucleotide
ATM	Ataxia Telangiectasia Mutated
BER	Base Excision Repair
BMD	Becker Muscular Dystrophy
BP	Basal Progenitor
BSA	Bovine Serum Albumin
CDK	Cyclin-Dependent Kinase
CK	Creatine Kinase
CLIO	Cross-Linked Iron Oxide
CNS	Central Nervous System
CT	Computed Tomography
CTGF	Connective Tissue Growth Factor
DC	Dendritic Cell
DEJ	Dermal–Epidermal Junction
DGC	Dystrophin-Glycoprotein Complex
DMD	Duchenne Muscular Dystrophy
DSB	Double-Strand Break
EC	Endemic Controls
EGFR	Epidermal Growth Factor Receptor
ESC	Embryonic Stem Cell
FDA	Food and Drug Administration
FRET	Fluorescence Resonance Energy Transfer
GF	Growth Factor
GMP	Good Medical Practice
HC	Household Contacts
HD	Huntington's Disease
HDM	House Dust Mite
HR	Homologous Recombination
ICM	Inner Cell Mass
LSPR	Localized Surface Plasmon Resonance
MCK	Muscle-Specific Creatine Kinase
MRI	Magnetic Resonance Imaging
MS	Multiple Sclerosis
NDA	New Drug Application
NER	Nucleotide Excision Repair
NHEJ	Non-Homologous End Joining
NMR	Nuclear Magnetic Resonance
NNI	National Nanotechnology Initiative

NSC	Neural Stem Cell
OCT	Optical Coherence Tomography
OI	Optical Imaging
PBS	Phosphate-Buffered Saline
PD	Parkinson's Disease
PET	Positron Emission Tomography
PRP	Plasmon Resonant Particles
PSMA	Prostate Specific Membrane Antigen
QD	Quantum Dots
RDEB	Recessive Dystrophic Epidermolysis Bullosa
RG	Radial Glia
RMS	Rostral Migratory Stream
ROS	Reactive Oxygen Species
SERS	Surface-Enhanced Raman Scattering
SI	Signal Intensity
SSB	Single-Strand Break
WHO	World Health Organization

List of Contributors

Volker F. Wendisch
Chair of Genetics of Prokaryotes, Faculty of Biology & CeBiTec, Bielefeld University, Germany

Steffen N. Lindner
Chair of Genetics of Prokaryotes, Faculty of Biology & CeBiTec, Bielefeld University, Germany

Tobias M. Meiswinkel
Chair of Genetics of Prokaryotes, Faculty of Biology & CeBiTec, Bielefeld University, Germany

Silvana Maria Alban
Department of Biotechnology and Bioprocess Engineering, Federal University of Parana, Curitiba, Parana, Brazil

Juliana Ferreira de Moura
Basic Pathology Department, Federal University of Parana, Curitiba, Parana, Brazil

Vanete Thomaz-Soccol
Department of Biotechnology and Bioprocess Engineering, Federal University of Parana, Curitiba, Parana, Brazil

Samira Bührer Sékula
Immunology Department, Tropical Pathology and Public Health Institute, Federal University of Goias, Goiania, Goias, Brazil

Larissa Magalhães Alvarenga
Immunology Department, Tropical Pathology and Public Health Institute, Federal University of Goias, Goiania, Goias, Brazil

Marcelo Távora Mira
Health Sciences Postgraduate Program, School of Medicine, Pontifical Catholic University of Parana, Curitiba, Parana, Brazil

Carlos Chávez Olortegui
Department of Biochemistry and Immunology, Federal University of Minas Gerais, Belo Horizonte, Minas Gerais, Brazil

João Carlos Minozzo
Center for Production and Research of Immunobiological Products, Parana State Department of Health, Piraquara, Parana, Brazil

Seiya Sato
Department of Obstetrics and Gynecology, Tottori University School of Medicine, Nishicho, Yonago-City, Tottori, Japan

Hiroaki Itamochi
Department of Obstetrics and Gynecology, Tottori University School of Medicine, Nishicho, Yonago-City, Tottori, Japan

C. Vilos
Center for Integrative Medicine and Innovative Science (CIMIS), Facultad de Medicina, Universidad Andrés Bello, Santiago, Echaurren 183, 8370071 Santiago, Chile
Center for the Development of Nanoscience and Nanotechnology (CEDENNA), Avenida Ecuador 3493, 9170124 Santiago, Chile

L. A. Velasquez
Center for Integrative Medicine and Innovative Science (CIMIS), Facultad de Medicina, Universidad Andrés Bello, Santiago, Echaurren 183, 8370071 Santiago, Chile
Center for the Development of Nanoscience and Nanotechnology (CEDENNA), Avenida Ecuador 3493, 9170124 Santiago, Chile

Akinori Nakamura
Department of Medicine (Neurology and Rheumatology), School of Medicine Shinshu University, 3-1-1 Ahahi, Matsumoto 390-8621, Japan

Shin'ichi Takeda
Department of Molecular Therapy, National Institute of Neuroscience, National Center of Neurology and Psychiatry, 4-1-1 Ogawa-higashi, Kodaira, Tokyo 187-8502, Japan

Zhe Liu
Department of Experimental Molecular Imaging (ExMI), Helmholtz Institute for Biomedical Engineering, Medical Faculty, RWTH Aachen University, Pauwelsstraße 20, 52074 Aachen, Germany

Fabian Kiessling
Department of Experimental Molecular Imaging (ExMI), Helmholtz Institute

for Biomedical Engineering, Medical Faculty, RWTH Aachen University, Pauwelsstraße 20, 52074 Aachen, Germany

Jessica Gätjens
Department of Experimental Molecular Imaging (ExMI), Helmholtz Institute for Biomedical Engineering, Medical Faculty, RWTH Aachen University, Pauwelsstraße 20, 52074 Aachen, Germany

João Conde
CIGMH, Departamento de Ciências da Vida, Faculdade de Ciências e Tecnologia, Universidade Nova de Lisboa, Campus de Caparica, 2829-516 Caparica, Portugal
Instituto de Nanociencia de Aragón, Universidad de Zaragoza, Campus Río Ebro, Edifício I+D, Mariano Esquillor s/n, 50018 Zaragoza, Spain

João Rosa
CIGMH, Departamento de Ciências da Vida, Faculdade de Ciências e Tecnologia, Universidade Nova de Lisboa, Campus de Caparica, 2829-516 Caparica, Portugal
REQUIMTE, Departamento de Química, Faculdade de Ciências e Tecnologia, Universidade Nova de Lisboa, Campus de Caparica, 2829-516 Caparica, Portugal

João C. Lima
REQUIMTE, Departamento de Química, Faculdade de Ciências e Tecnologia, Universidade Nova de Lisboa, Campus de Caparica, 2829-516 Caparica, Portugal

Pedro V. Baptista
CIGMH, Departamento de Ciências da Vida, Faculdade de Ciências e Tecnologia, Universidade Nova de Lisboa, Campus de Caparica, 2829-516 Caparica, Portugal

Simona Casarosa
Center for Integrative Biology, Università degli Studi di Trento

Yuri Bozzi
Center for Integrative Biology, Università degli Studi di Trento

Luciano Conti
Center for Integrative Biology, Università degli Studi di Trento

Yi-Ling Ye
Department of Biotechnology, National Formosa University, Yunlin, Taiwan

Ya-Hui Chuang
Department of Clinical Laboratory Sciences and Medical Biotechnology, National Taiwan University, Taipei, Taiwan

Bor-Luen Chiang
Graduate Institute of Clinical Medicine, National Taiwan University, Taipei, Taiwan

Xinyi Wang
Department of Dermatology, University of Southern California, Los Angeles, California, USA

Pedram Ghasri
Department of Dermatology, University of Southern California, Los Angeles, California, USA

Mahsa Amir
Department of Dermatology, University of Southern California, Los Angeles, California, USA

Brian Hwang
Department of Dermatology, University of Southern California, Los Angeles, California, USA

Yingpin Hou
Department of Dermatology, University of Southern California, Los Angeles, California, USA

Michael Khilili
Department of Dermatology, University of Southern California, Los Angeles, California, USA

Andrew Lin
Department of Dermatology, University of Southern California, Los Angeles, California, USA

Douglas Keene
Department of Molecular and Medical Genetics, Shriners Hospital for Children, Portland, Oregon, USA

Jouni Uitto
Department of Dermatology and Cutaneous Biology, Jefferson Medical College, Philadelphia, Pennsylvania, USA

David T Woodley
Department of Dermatology, University of Southern California, Los Angeles, California, USA

Mei Chen
Department of Dermatology, University of Southern California, Los Angeles, California, USA

Preface

Molecular biotechnology is the use of laboratory techniques to study and modify nucleic acids and proteins for applications in areas such as human and animal health, agriculture, and the environment. Molecular biotechnology results from the convergence of many areas of research, such as molecular biology, microbiology, biochemistry, immunology, genetics, and cell biology. The text *Molecular Biotechnology: Therapeutic Applications and Strategies* provides a thorough introduction to modern biotechnology detailing specific aspects such as gene therapy, transgenic animals, drug design, and the human genome project. The use of glycerol in biotechnological applications has been discussed in first chapter. Second chapter proposes the use of the phage-display technique as a tool to identify new reagents that may be effectively used in immunological assays. In third chapter, we provide an overview of major DNA repair pathways and describe recent advances in anticancer therapy with a focus on DNA repair in cancer. Fourth chapter presents the foundations of polymeric microparticles based on their formulation, mechanisms of drug release and some of their innovative therapeutic strategies to board multiple diseases. In fifth chapter, we review the pathological features of duchenne muscular dystrophy (DMD) and therapeutic applications, especially of exon skipping using antisense oligonucleotides and gene therapies using viral vectors in murine and canine models of DMD. Sixth chapter focuses on advanced nanomaterials in multimodal imaging. In seventh chapter, we focus on the uses of nanobiophotonics for molecular diagnostics involving specific sequence characterization of nucleic acids and for gene delivery systems of relevance for therapy strategies. Eighth chapter deals with the biological properties of neural stem cells (NSCs) and discusses how these cells may be exploited to provide effective therapies for neurological disorders. Ninth chapter reviews recent progress in mucosal immunotherapy for allergic diseases. Topical application of recombinant type VII collagen incorporates into the dermal–epidermal junction and promotes wound closure has been presented in tenth chapter.

Chapter 1

USE OF GLYCEROL IN BIOTECHNOLOGICAL APPLICATIONS

Volker F. Wendisch, Steffen N. Lindner, and Tobias M. Meiswinkel

Chair of Genetics of Prokaryotes, Faculty of Biology & CeBiTec, Bielefeld University, Germany

INTRODUCTION

Since decades the limited access to petroleum oil is a major concern and substitutions for fossil fuels are needed. One promising substitute is biodiesel, which is widely produced from vegetable oils, e.g. from rape seeds, soybeans, sunflower seeds or animal fats. In the synthesis of biodiesel, oils and fats are transesterified to fatty acid methyl ester in the presence of sodium hydroxide or potassium hydroxide. In this process, glycerol is generated as stoichiometric byproduct with a ratio of 10% (w/w) with respect to biodiesel produced. In 2009, the biodiesel production of the world reached 16 million tons (Licht, 2010), with the lion's share produced by the European Union with 9 million tons (EBB, European Biodiesel Board 2010), followed by the United States with a production of 2.7 million tons (Licht, 2010). Hence, 1.6 million tons of glycerol were produced as obligatory by-product. Glycerol finds applications as an ingredient of various products, such as creams, food, feed, and pharmaceuticals, but the demand for glycerol in these processes is limited. Integrated conversion of raw glycerol from the biodiesel process to value-added products is a driver towards higher cost efficiency of biodiesel production.

Glycerol is a good source of carbon and energy for growth of several microorganisms and may be suitable for the biotechnological production of a number of chemicals in fermentative processes. To date several microbiological productions have been adjusted to glycerol as carbon and energy source or, if glycerol is close to the desired product, are based on glycerol as substrate anyway. For instance, the biotechnological production of 1,3- propanediol and dihydroxyacetone has predominantly been carried out from glycerol, since these processes are catalyzed in a two and one step reaction, respectively. Bacterial 1,3- propanediol production from glycerol is involving two enzymes. First glycerol is dehydrated by glycerol dehydratase to 3-hydroxypropionaldehyde, which is subsequently converted to 1,3-propanediol by 1,3-propanediol dehydrogenase. Predominantly, 1,3- propanediol production has been approved with Clostridium strains, Klebsiella pneumoniae, or Escherichia coli (Zhu et al. 2002, Biebl et al. 1998, Forsberg, 1987). 1,3-propanediol finds application in the production of the polyester polytrimethylene terephthalate. Strains of Gluconobacter oxydans are used for producing dihydroxyacetone (glycerone) from glycerol (Flickinger & Perlman, 1977). Dihydroxyacetone is used as a building block in organic chemistry and as a skin tanning agent in cosmetics.

Besides products that can be derived from glycerol in one or two reactions further products requiring more complex conversions have been investigated, e.g. succinic acid production with Anaerobiospirillum succiniciproducens (Lee et al. 2001) or Escherichia coli (Blankschien et al., 2010, Zhang et al., 2010), citrate production with Yarrowia lipolytica (Rywinska & Rymowicz, 2010), ethanol production with Escherichia coli, Saccharomyces cerevisiae, or Hansenula polymorpha (Hong et al., 2010, Shams Yazdani & Gonzalez, 2008, Yu et al., 2010), production of amino acids with Corynebacterium glutamicum (Rittmann et al., 2008), or propionate production with Propionibacteria (Himmi et al., 2000).

Crude glycerol preparations from biodiesel factories differ considerably from the pure chemical glycerol, e.g. by their salt content. The applicability of crude glycerol from biodiesel production plants has already been demonstrated for production strains of Clostridium and Klebsiella in 1,3-propanediol production (Gonzalez-Pajuelo et al., 2004, Mu et al., 2006), Yarrowia lipolytica in citrate production (Papanikolaou et al., 2002), Kluyvera cryocrescens and Klebsiella pneumoniae in ethanol production (Choi et al., 2011, Oh et al., 2011), as well as Basfia succiniciproducens in succinic acid production (Scholten et al., 2009). This chapter will summarize state-of-the-art of glycerol-based biotechnological processes and will discuss future developments.

USE OF GLYCEROL AS A CARBON SOURCE IN BIOTECHNOLOGICAL APPLICATIONS

Glycerol has many applications, it is used for the production of food, cosmetics, paints, pharmaceutics, paper, textiles, leather and for the production of various chemicals (Wang et al., 2001). It can be used as a stabilizing agent for storage of cells and proteins. Physiologically, glycerol is essential for the biosynthesis of membranes, since it is the backbone of glycerolipids. And for its function as a component of lipids and fats it is an abundant source of carbon and energy in nature.

Formerly, glycerol was a valuable product derived from glucose via dihydroxyacetonephosphate and glycerol-3-phosphate by glycerol-3-phosphate dehydrogenase (EC 1.1.1.94) and glycerol-3-phosphatase (EC 3.1.3.21). The yeast Saccharomyces cerevisiae which uses glycerol as an osmolyte under osmotic stress conditions was engineered for efficient glycerol production from glucose. S. cerevisiae possesses two isozymes of each, glycerol-3-phosphate forming glycerol-3-phosphate dehydrogenases and glycerol-3-phosphatases (Larsson et al., 1993, Pahlman et al., 2001). Sulphite treatment of yeasts enabled glycerol production (Petrovska et al., 1999), as did metabolic engineering on the glycerol production pathway, e.g. deletion of the triosephosphate isomerase gene (Overkamp et al., 2002), deletion of the alcohol dehydrogenase gene, overexpression of a glycerol-3-phosphate dehydrogenase gene (Navarro-Avino et al., 1999), or overexpression of a glycerol exporter gene (Cordier et al., 2007). Nowadays, glycerol arises from the biodiesel production. In 2010 glycerol formed as byproduct in the biodiesel process amounted to 1.6 million tons (Licht, 2010), which is extending the glycerol demand by far, thus, making microbial glycerol production unprofitable.

Glycerol as Carbon Source for Growth

Glycerol can be used as a source of carbon and energy by many organisms. The initial step of glycerol utilization is its uptake into the cell. Albeit the small and uncharged molecule can diffuse through membranes without a transport system, many organisms possess glycerol transporters. In Escherichia coli, glycerol transport is mediated by the glycerol facilitator (Heller et al., 1980). In the wine bacterium Pediococcus pentosaceus active glycerol uptake has been reported (Pasteris & Strasser de Saad, 2008).

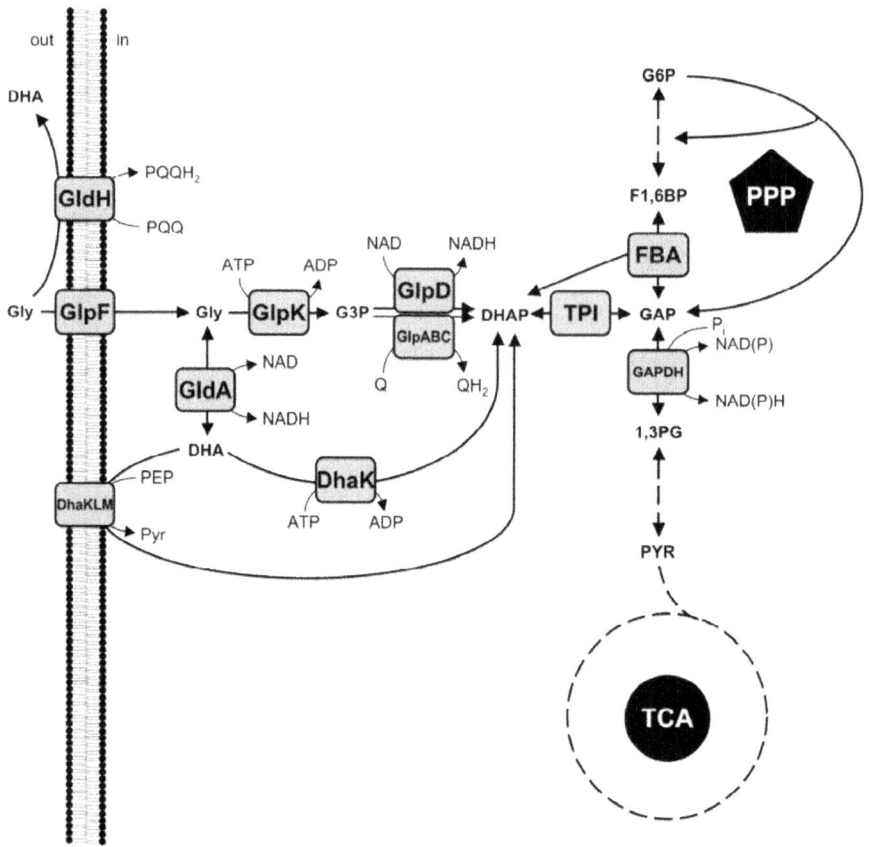

Figure 1. Pathways of glycerol utilization. Abbreviations: 1,3-PG 1,3-phosphoglycerate, DHA dihydroxyacetone, DhaK ATP dependent dihydroxyacetone kinase, DhaKLM PEP dependent dihydroxyacetone kinase, DHAP dihydroxyacetone-phosphate, F1,6BP fructose- 1,6-bisphosphate, G3P glycerol-3-phosphate, G6P glucose-6-phosphate, GAP glyceraldehyde-3-phosphate, GAPDH glyceraldehyde-3-phosphate dehydrogenase, GldA soluble glycerol dehydrogenase, GldH membrane bound glycerol dehydrogenase, GlpABC quinone dependent glycerol-3-phosphate dehydrogenase, GlpF glycerol facilitator, GlpK glycerol kinase, GlpD glycerol-3-phosphate dehydrogenase, Gly glycerol, PEP phosphoenolpyruvate, Pi inorganic phosphate, PPP pentose phosphate pathway, Pyr pyruvate, TCA tricarboxylic acid cycle, TPI triosephosphate isomerase.

Active glycerol transport has been described in yeasts, such as sodium dependent symport for Debaryomyces hansenii (Lucas et al., 1990) and proton dependent symport for Pichia sorbitophila (Lages & Lucas, 1995) and

Saccharomyces cerevisiae (Lages & Lucas, 1997, Ferreira et al., 2005). In yeasts, active glycerol transport is mostly regarded of importance with respect to the use of glycerol as an osmolyte under osmotic stress conditions.

The imported glycerol can enter the metabolism at the level of the glycolytic intermediate dihydroxyacetone-phosphate. Two routes for the formation of dihydroxyacetone-phosphate from glycerol have been reported. In the first, ATP-dependent glycerol kinase (EC 2.7.1.30) phosphorylates glycerol to glycerol-3-phosphate, which is subsequently oxidized to dihydroxyacetone-phosphate by glycerol-3-phosphate dehydrogenase, which are either quinone or FAD-dependent (EC 1.1.5.3) or NAD-dependent (EC 1.1.1.8). In the second pathway, glycerol is oxidized to dihydroxyacetone by glycerol dehydrogenase (EC 1.1.1.6) before being phosphorylated to dihydroxyacetone-phosphate by ATP- or phosphoenolpyruvate-dependent dihydroxyacetone kinase (EC 2.7.1.29).

The glycerol-3-phosphate pathway is active in e.g. Gluconobacter oxydans and Clostridium acetobutylicum (Claret et al., 1994, Gonzalez-Pajuelo et al., 2006), whereas Clostridium butyricum uses the dihydroxyacetone pathway (Gonzalez-Pajuelo et al., 2006). In E. coli, Klebsiella pneumoniae, Saccharomyces cerevisiae and other yeasts, both pathways are present (Gonzalez et al., 2008, Wang et al., 2001, Ruch et al., 1974, Norbeck & Blomberg, 1997). In S. cerevisiae, glycerol is mainly utilized via glycerol-3-phosphate while the pathway via dihydroxyacetone is suggested to play a role during hyperosmotic stress for regulation of the intracellular glycerol concentration (Blomberg, 2000). In E. coli, the pathway via dihydroxyacetone operates under certain anaerobic conditions only (Gonzalez et al., 2008), but the major pathway is via glycerol-3-phosphate. In K. pneumoniae, the glycerol-3-phosphate pathway is active under aerobic and the dihydroxyaceton pathway is active under anaerobic conditions (Ruch et al., 1974).

Dihydroxyacetone

Dihydroxyacetone (DHA; 1,3-Dihydroxypropan-2-one, Glycerone) is produced biotechnologically with a global market of 2,000 tons per year (Pagliaro et al., 2007). DHA is produced biotechnologically, since its chemical synthesis is expensive and requires safety measures to cope with hazardous reactants (Hekmat et al., 2003). The most popular use of DHA is as coloring agent in sunless tanning products, such as creams and lotions (Levy, 1992). The tanning effect depends on Maillard-like reactions of DHA with the amino acids of the outer skin layer. Historically, also applications for medical treatments of glycogenesis, a glycogen storage disease, and diabetes mellitus have been described. Currently, the use of DHA as building block for chemical synthesis appears to have the highest potential for a production process based on biodiesel-derived glycerol (Enders et al., 2005, Zheng et al., 2008, Hekmat et al., 2003).

Gluconobacter oxydans is used for the production of DHA from glycerol. This bacterium belongs to the family of Acetobacteraceae (acetic acid bacteria), which are able to oxidize many carbohydrates and alcohols incompletely. To this end, G. oxydans possesses a variety of membrane-bound dehydrogenases. The membrane-bound pyrroloquinoline quinone (PQQ)-dependent glycerol dehydrogenase (EC 1.1.99.22) catalyzes oxidation of the secondary hydroxy group of glycerol to DHA (Matsushita et al., 1994). The enzyme is the protein product of sldA and sldB (Prust et al., 2005) and its localization allows extracellular oxidization of glycerol to DHA in the periplasma with a concurrent reduction of the membrane-localized PQQ. Besides membrane-bound glycerol dehydrogenase (Fig. 2), G. oxydans also possesses an intracellular catabolic pathway for the use of glycerol as a carbon source for growth (Fig. 1), in which glycerol is phosphorylated by ATP-dependent glycerol kinase to yield glycerol-3-phosphate which in turn is oxidized to dihydroxyacetonephosphate by NAD-dependent glycerol-3-phosphate dehydrogenase (Claret et al., 1994). Since a functional glycolysis pathway is missing in G. oxydans and since its tricarboxylic acid cycle is incomplete, dihydroxyacetone-phosphate is metabolized via the pentose phosphate pathway (Greenfield & Claus, 1972).

Problems have occurred in the process of DHA production by G. oxydans, most importantly inhibition of the biotransformation process by the substrate glycerol and the product DHA as both inhibit growth and DHA production (Claret et al., 1992, Claret et al., 1993, Bauer et al., 2005). These problems have been met by optimizing production conditions including immobilization of G. oxydans cells to a polyvinyl alcohol matrix, which resulted in cells active for 14 days while maintaining glycerol dehydrogenase activity above 90 % (Wei et al., 2007).

Also fed-batch cultivations were shown to be supportive for DHA production compared to batch fermentations, as higher total glycerol amounts could be converted into DHA by avoiding inhibitory concentrations of glycerol, however, in this case production yields are limited with respect to DHA accumulation (Bories et al., 1991). Further enhancements of DHA production were achieved when using repeated fed-batch cultivations, in which DHA concentrations were kept below the inhibitory concentration. By this method immobilized cells were reused for up to 100 fed-batch cycles reducing costs and time for cleaning, sterilization, and inoculation (Hekmat et al., 2003). Other studies focused on optimizing culturing conditions for DHA production focused on cultivation media, aeration, or pH (Wethmar & Deckwer, 1999, Svitel & Sturdik, 1994, Tkac et al., 2001, Holst et al., 1985).

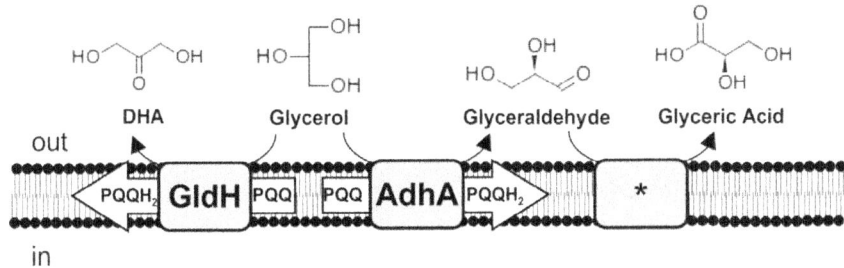

Figure 2. Extracellular glycerol oxidation in Gluconobacter oxydans. Abbreviations: DHA dihydroxyacetone, PQQ oxidized pyrroloquinoline quinone, PQQH$_2$ reduced pyrroloquinoline quinone, AdhA alcohol dehydrogenase, * unidentified glyceric acid forming dehydrogenase.

After the genome sequence of G. oxydans became public (Prust et al., 2005), metabolic engineering of G. oxydans has been reported for optimizing DHA production. Overexpression of the genes encoding the glycerol dehydrogenase, sldAB, led to increased biomass formation and higher DHA yields from glycerol (Herrmann et al., 2007). As the formation of glyceric acid as byproduct interferes with DHA production, disruption of adhA, the gene for the PQQ-dependent alcohol dehydrogenase, which is involved in the oxidation of glycerol to glyceric acid was shown to improve the product yield by abolishing glyceraldehyde and glyceric acid formation (Habe et al., 2010a). In addition, the mutant lacking adhA was less inhibited by high initial glycerol concentrations than the parent strain. In line with the notion that glyceric acid inhibits growth and DHA production by G. oxydans, it was found that addition of glyceric acid to the medium reduced growth and DHA production (Habe et al., 2010a) and that AdhA activity strongly increased when very high glycerol concentrations were used (Habe et al., 2009d). In the next step, the combination of adhA disruption and sldAB overexpression resulted in a strain with very high productivity and strongly increased tolerance towards glycerol (Li et al., 2010b). Since the production of DHA by the obligate aerobic G. oxydans is characterized by a very high demand for oxygen to oxidize reduced PQQ (Hekmat et al., 2003, Svitel & Sturdik, 1994), the gene encoding Vitreoscilla hemoglobin, an oxygen transporting protein, was shown to be beneficial (Li et al., 2010a).

Besides G. oxydans strains other acetic acid bacteria have also been reported for DHA production from glycerol, e.g. Gluconobacter melanogenus (Flickinger & Perlman, 1977) and Acetobacter xylinum (Nabe et al., 1979).

Until today, production of DHA from biodieselderived raw glycerol has not yet been reported.

1,2-Propanediol

1,2-propanediol (propylenglycol, 1,2-PDO) is a commodity chemical with a wide range of applications, including polyester resins, plastics, antifreeze agents, de-icing products, detergents, or paints. The global demand for 1,2-PDO is estimated to be up to 1.6 million tons per year (Shelley, 2007). Chemically, 1,2-PDO is produced from propylene.

A variety of microorganisms have been reported as natural producers of 1,2-PDO, such as the bacteria E. coli (Hacking & Lin, 1976, Gonzalez et al., 2008), Thermoanaerobacterium thermosaccharolyticum (Cameron & Cooney, 1986), Bacteroides ruminicola (Turner & Roberton, 1979), Salmonella typhimurium, and Klebsiella pneumoniae (Badia et al., 1985), but also yeasts have been shown to produce 1,2-PDO (Suzuki & Onishi, 1968).

Two main routes for the microbial production of 1,2-PDO exist. First, 1,2-PDO may be formed from lactaldehyde, an intermediate of dissimilatory desoxy sugar (e.g. fucose, rhamnose) utilization generated either by fuculose^{-1}-phosphate aldolase (EC 4.1.2.17) or by rhamnulose^{-1}-phosphate aldolase (EC 4.1.2.19). Lactaldehyde is reduced to 1,2-PDO by NADH-dependent lactaldehyde reductase (FucO, EC 1.1.1.77), as shown e.g. for Salmonella typhimurium (Suzuki & Onishi, 1968). Due to high prices of fucose and rhamnose this pathway is not applicable.

The second route to 1,2-PDO diverts from glycolysis and, thus, 1,2-PDO production from glucose is feasible. The glycolytic intermediate dihydroxyacetone-phosphate is funneled into the methylglyoxal pathway by methylglyoxal synthase (EC 4.2.3.3) forming methylglyoxal. Methylglyoxal synthases from e.g. Escherichia coli, Clostridium acetobutylicum, and Saccharomyces cerevisiae have been purified and characterized (Cooper & Anderson, 1970, Freedberg et al., 1971, Hopper & Cooper, 1972, Huang et al., 1999, Murata et al., 1985), all of which were inhibited by phosphate. Methylglyoxal in turn is reduced to 1,2-PDO in two subsequent NADH/NADPH-dependent reactions. Two variants of methylglyoxal reduction to 1,2-PDO are known with either acetol or lactaldehyde as intermediate. Acetol a may arise from methylglyoxal e.g. by aldehyde dehydrogenase (EC 1.1.1.2) or alcohol dehydrogenases (EC 1.1.1.1) from E. coli (Misra et al., 1996), by methylglyoxal reductase from S. cerevisiae (EC 1.1.1.283)

(Nakamura et al., 1997) or acetol oxidoreductase from E. coli (Boronat & Aguilar, 1981). In the second step acetol is converted to 1,2-PDO by e.g. E. coli glycerol dehydrogenase (EC 1.1.1.6). The variant with lactaldehyde as intermediate involves e.g. glycol dehydrogenase from Enterobacter aerogenes (EC 1.1.1.185) (Carballo et al., 1993) or glycerol dehydrogenase from E. coli (EC 1.1.1.6) (Altaras & Cameron, 1999) for the first reduction step and e.g. E. coli 1,2-PDO reductase for the second reduction. Both ways necessitate two reduction equivalents per 1,2-PDO formed.

1,2-PDO Production by E. Coli

First approaches towards metabolic engineering of E. coli strains for production of 1,2-PDO from glucose by Cameron et al. involved overexpression of genes for either aldose reductase or glycerol dehydrogenase (Cameron et al., 1998). Later on, the same group coexpressed glycerol dehydrogenase genes from E. coli or K. pneumoniae together with the E. coli methylglyoxal synthase gene in E. coli and reached up to 0.7 g l^{-1}, an almost three-fold increase when compared to overexpression of E. coli or K. pneumoniae glycerol dehydrogenase gene alone (0.25 g l^{-1}) (Altaras & Cameron, 1999). Additional overexpression of yeast alcohol dehydrogenase or E. coli 1,2-PDO reductase further improved production performance and 1,2-PDO concentrations of 4.5 g l^{-1} and a yield of 0.19 g (g glucose) $^{-1}$ were achieved in a fed-batch fermentation process (Altaras & Cameron, 2000). Elimination of lactate dehydrogenase by gene deletion improved 1,2-PDO production by an E. coli strain overexpressing genes coding for methylglyoxal synthase from Clostridium acetobutylicum and glycerol dehydrogenase from E. coli (Berrios-Rivera et al., 2003).

Glycerol has a higher degree of reduction than glucose, thus, higher 1,2-PDO yields are theoretically possible using glycerol (0.72 g g^{-1}) as compared to glucose (0.63 g g^{-1}) (Clomburg & Gonzalez, 2011). Moreover, 1,2-PDO has been reported to be a natural product of anaerobic fermentation of glycerol in E. coli by Gonzalez et al. (Gonzalez et al., 2008).

Glycerol is converted to 1,2-PDO in a pathway consisting of glycerol dehydrogenase for oxidation of glycerol to dihydroxyacetone and phosphorylation of the latter to dihydroxyacetone-phosphate by phosphoenolypyruvate-dependent dihydroxyacetone kinase. Subsequently, dihydroxyacetone-phosphate is reduced to 1,2-PDO which requires two NADH. Thus, to generate NADH and ATP required in these reactions a portion of dihydroxyacetone-phosphate needs to be catabolized in glycolysis and onwards to ethanol (Gonzalez et al., 2008).

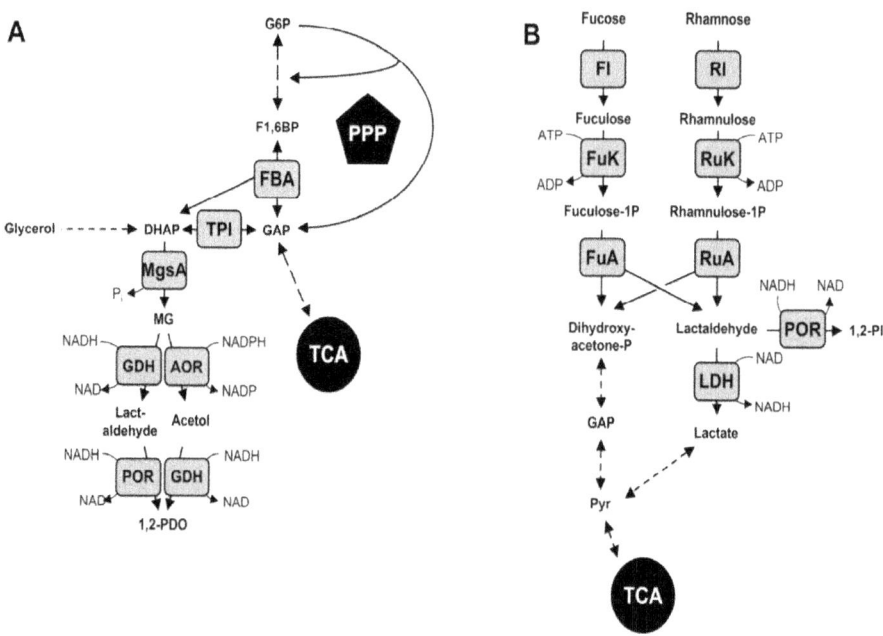

Figure 3. 1,2-propanediol production pathways. A, 1,2-propanediol production from glycolytic intermediate dihydroxyacetone-phosphate; Abbreviations: 1,2-PDO 1,2-propanediol, AOR, acetol oxidoreductase, DHAP, dihydroxyacetone-phosphate, F1,6BP, fructose-1,6- bisphosphate, FBA, fructose-1,6-bisphosphate aldolase, G6P, glucose-6-phosphate, GAP, glyceraldehyde-3-phosphate, GDH, glycerol dehydrogenase, MG, methylglyoxal, MgsA, methylglyoxal synthase, P_i inorganic phosphate, POR 1,2-propanediol-oxidoreductase, TCA tricarboxylic acid cycle, TPI triosephosphate isomerase; B, 1,2-propanediol production from fucose and rhamnose; Abbreviations: FI fucose isomerase, FuA fuculose-1-phosphate aldolase, FuK fuculose kinase, LDH lactate dehydrogenase, RI rhamnose isomerase, RuK rhamnulose kinase, RuA rhamnulose-1-phosphate aldolase.

Recently, Clomburg et al. rationally engineered E. coli for effective 1,2-PDO production from glycerol. They introduced a 1,2-PDO production pathway via methylglyoxal synthase, glycerol dehydrogenase, and aldehyde oxidoreductase. Replacement of the phosphoenolpyruvate-dependent dihydroxyacetone kinase by ATP-dependent dihydroxyacetone kinase from Citrobacter freudii elevated dihydroxyacetone-phosphate availability and enhanced 1,2-PDO production from 0.02 to 0.15 g g^{-1}. Moreover, the formation of byproducts succinate, acetate, ethanol, and formate increased but lactate was reduced. To eliminate byproduct formation several gene deletions, e.g. of genes coding for lactate dehydrogenase, acetate kinase, phosphate acetyltransferase, formate hydrogen lyase, fumarate reductase, alcohol dehydrogenase, and pyruvate dehydrogenase, were tested alone or in combination. Ethanol formation could not be abolished as deletion of the alcohol dehydrogenase gene strongly decreased 1,2-PDO production (0.02 compared to 0.12 g g^{-1}) and glycerol consumption, but increased formation of acetate and succinate as byproducts. However, deletion of the genes for acetate kinase, phosphate acetyltransferase, and lactate dehydrogenase resulted in an increased product yield of 0.21 g g^{-1}, but entailed increased ethanol, formate, and pyruvate formation. The use of raw glycerol by this strain reduced formate formation and increased the 1,2-PDO yield (0.24 g g^{-1}) (Clomburg & Gonzalez, 2011). Taken together, engineered E. coli strains allow for 1,2-PDO yields from glycerol that are comparable to yields from glucose, making glycerol a feasible substrate for microbial 1,2- PDO production.

1,2-PDO Production by other Microorganisms

Recombinant strains of S. cerevisiae and of C. glutamicum have been developed for production of 1,2-PDO, as well. S. cerevisiae carrying multiple genome-integrated copies of E. coli methylglyoxal synthase and glycerol dehydrogenase genes was able to produce 1,2- PDO (Lee & da Silva, 2006). Plasmid-borne expression of the latter genes combined with the deletion of the endogenous triosephosphate isomerase gene resulted in the production of 1.1 g l^{-1} 1,2-PDO (Jung et al., 2008). Expression of E. coli methylglyoxal synthase gene and a Corynebacterium glutamicum aldo-keto reductase gene enabled C. glutamicum for the production of 1,2-PDO from glucose (1.8 g l^{-1}) (Niimi et al., 2011).

1,3-Propanediol

1,3-propanediol (1,3-PDO) can be used in several chemical applications. It is a substrate in the formulation reactions for polyesters, polyethers, polyurethanes, adhesives, composites, laminates, coatings, and moldings. In addition, 1,3-PDO itself is used e.g. as solvent or antifreeze agent. Most importantly, the production of polytrimethylene terephthalate (PTT) from 1,3-PDO and terephthalic acid is the driver of the growing market for 1,3-PDO. The polymer PTT is a promising polyester with numerous applications, e.g. as compound in textile, carpet, upholstery, or specialty resins. Due to its properties, PTT might be favored over polymers such as nylon, polyethylene terephthalate and polybutylene terephthalate. PTT has environmental benefits since it is biodegradable (for recent reviews see (da Silva et al., 2009, Liu et al., 2010, Carole et al., 2004). The chemical producers Shell and DuPont produce PTT from 1,3-PDO which is commercialized under the names Sorona, Hytrel (DuPont) and Corterra (Shell).

1,3-PDO is a success story of a glycerol-based biotechnological process. When 1,3-PDO was first discovered to be a fermentative product of glycerol in 1881 (Werkman & Gillen, 1932), it received little interest until the development of PTT in 1941 (Whinfield & Dickinson, 1946).

However, while terephthalic acid was readily available the expense of 1,3-PDO production limited efficient PTT production. Nowadays, chemical production of 1,3-PDO from ethylene oxide or acrolein as well as its biotechnological production enable supply of 1,3-PDO at low cost. It has to be noted that chemical synthesis of 1,3-PDO suffers from high energy consumption, toxic intermediates, and expensive catalysts as major drawbacks.

Currently, the market for 1,3-PDO is estimated to be about 50,000 tons per year (Liu et al., 2010), but due to a growing production of PTT a market volume of 230,000 tons is foreseen for 2020 (Carole et al., 2004). An indication of the growing interests and the potential of biotechnological 1,3-PDO production are the current decisions of the joint venture of DuPont and Tate & Lyle to extend their 1,3-PDO biotech plant in Louden, TN, USA, by 35 %. The actual capacity of the plant is 45,000 tons per year (Greenwood, 2010). Also the French company Metabolic Explorer (Clermont-Ferrand, France) decided to build a plant for biotechnological 1,3-PDO production in Malaysia in partnership with Malaysian Bio-XCell. The plant with a capacity to produce 50,000 tons 1,3-PDO annually is expected to start with a production of 8,000 tons per year (Degalard, 2011). While the process of DuPont and Tate & Lyle is based on sugars from corn hydrolysates, the Metabolic Explorer process will be based on crude glycerol.

Biotechnological Production of 1,3-PDO

Several bacteria have been shown to naturally possess the ability of 1,3-PDO production, all of which belong either to enterobacteria or to firmicutes. Production of 1,3-PDO has been shown e.g. for strains of Klebsiella pneumoniae, Clostridium butyricum, Clostridium acetobutylicum, Clostridium pasteurianum, Clostridium butylicum, Clostridium beijerinckii, Clostridium kainantoi, Lactobacillus brevis and Lactobacillus buchneri, and Enterobacter agglomerans (Forage & Foster, 1982, Barbirato et al., 1996, Nakas et al., 1983, Schutz & Radler, 1984, Homann et al., 1990, Biebl, 1991, Biebl et al., 1992, Daniel et al., 1995).

Biosynthesis of 1,3-PDO from glycerol is catalyzed in a reducing pathway involving a cytosolic two-step process. First, glycerol dehydratase (EC 4.2.1.30) or diol dehydratase (EC 4.2.1.28) convert glycerol into 3-hydroxypropionaldehyde (3-HPA) (Schneider & Pawelkiewicz, 1966, Toraya et al., 1978). Glycerol dehydratase from K. pneumoniae has been reported to be inactivated by glycerol and necessitates a reactivating factor encoded by gdrAB (Tobimatsu et al., 1999). In the second step, 3-HPA is reduced to 1,3-PDO by NADPH- or NADH-dependent 1,3-PDO dehydrogenases (EC 1.1.1.202) e.g. by DhaT from K. pneumoniae or YqhD from E. coli (Johnson & Lin, 1987). Besides NADH-dependent DhaT, K. pneumoniae possesses a second NADPH dependent enzyme active as a 1,3-propanediol dehydrogenase. This enzyme was found, since a dhaT deletion mutant was still able to produce 1,3-PDO and 3-hydroxypropionic acid from glycerol albeit with reduced efficiency (Ashok et al., 2011). The sought enzyme was later identified as a homolog to E. coli YqhD and overexpression of the respective gene was shown to restore 1,3-PDO production by the dhaT deletion mutant (Seo et al., 2010).

The regeneration of reduction equivalents for 1,3-PDO production from glycerol is ensured in K. pneumoniae by simultaneous operation of the oxidative pathway of glycerol utilization. Here, NADH is generated by glycerol dehydrogenase during oxidation of glycerol to dihydroxyacetone and during catabolism of dihydroxyacetone phosphate, which is formed from dihydroxyacetone by dihydroxyacetone kinase (Seo et al., 2009). Thus, the demand of NADH limits the theoretical yield of 1,3-PDO production from glycerol. The requirement of some glycerol being oxidized and catabolized in glycolysis and the tricarboxylic acid cycle entails the formation of unwanted byproducts such as acetic acid, lactic acid, formic acid, succinic acid, butyric acid, 2,3-butanediol, and ethanol. The main byproducts of K. pneumoniae strains are 2,3-butanediol, acetic acid, ethanol, and lactic acid (Menzel et al., 1997, Zhang et al., 2006, Kretschmann et al., 1993), whereas C. butyricum and metabolically engineered C. acetobutylicum strains mainly accumulate byproducts acetic acid and butyric acid during 1,3-PDO production (Papanikolaou et al., 2000, Papanikolaou et al., 2004, Saintamans et al., 1994, Gonzalez-Pajuelo et al., 2005, Soucaille, 2008, Sarcabal et al., 2007). With C. pasteurianum, butanol but not 1,3-PDO is the main fermentation product from glycerol, and ethanol, acetic acid, butyric acid and lactic acid are formed as further byproducts (Biebl, 2001). Metabolically engineered E. coli were reported to accumulate formic acid, acetic acid, lactic acid, and pyruvic acid as the major byproducts (Tong et al., 1991, Skraly et al., 1998) and accumulate growth inhibiting metabolites glycerol-3-phosphate and methylglyoxylate (Tkac et al., 2001, Zhu et al., 2001).

A major concern in 1,3-PDO production is the fact that the substrate glycerol, the intermediate 3-HPA, the product 1,3-PDO, and several byproducts inhibit growth and production. In K. pneumoniae, 1,3-PDO yields decrease with increasing glycerol concentrations and metabolic flux analyses revealed a higher carbon flux via the oxidative glycerol utilization pathway to the loss of 1,3-PDO (Xiu et al., 2011). In C. butyricum, growth is completely inhibited at 1,3-PDO concentrations higher than 60 g l^{-1}. Also the byproducts acetic acid (27 g l^{-1}) and butyric acid (19 g l^{-1}) abolished growth of this bacterium as did glycerol concentrations of 80 g l^{-1} or more (Biebl, 1991, Colin et al., 2000). Growth of K. pneumoniae is inhibited at glycerol concentrations above 110 g l^{-1} under aerobic and above 133 g l^{-1} under anaerobic conditions. Also the byproducts acetic acid (15 g l^{-1}), lactic acid (19 g l^{-1}), and ethanol (26 g l^{-1}) (15, 19, 26 g l^{-1} under anaerobic and 24, 26, and 17 g l^{-1} under aerobic conditions) inhibit growth of K. pneumoniae (Cheng et al., 2005). The accumulation of 3-HPA, the intermediate product of 1,3-PDO production has a toxic effect on growth and 1,3-PDO fermentation in K. pneumoniae. Both, glycerol dehydratase and 1,3-propanediol dehydrogenase are sensitive to 3-HPA. 1,3-propanediol dehydrogenase activity decreased as 3-HPA accumulated, leading to a further increase in 3-HPA concentrations (Hao et al., 2008a). Purified 1,3-propanediol dehydrogenase from E. agglomerans CNCM 1210 was shown to be inhibited by NAD$^+$ (K_i 0.29 mM) and 1,3-PDO (Ki 13.7 mM) and therefore might be limiting production yields of 1,3-PDO (Barbirato et al., 1997). Also the glycerol dehydratases from C. freundii and metagenome samples were shown to be inhibited by 1,3- PDO (Knietsch et al., 2003), moreover glycerol dehydratases from K. pneumoniae and C. freundii are inhibited by deactivation by glycerol (Tobimatsu et al., 1999, Tobimatsu et al., 2000, Kajiura et al., 2001, Seifert et al., 2001).

To overcome production limitations by e.g. substrate and product inhibition or byproduct inhibition, several approaches to optimize cultivation conditions were followed. Because the oxidative glycerol utilization pathway is necessitated for NADH regeneration but also leads to the formation of unwanted byproducts the cultivations of K. pneumoniae is preferably carried out under micro-aerobic conditions. Chen et al. compared cultivation conditions during 1,3-PDO production with K. pneumoniae. They found that final 1,3-PDO concentrations and yields were increased in batch fermentations under micro-aerobic conditions. Productivity increased from 0.8 to 1.57 under anaerobic and micro-aerobic conditions, respectively, and ethanol was reduced as well (Chen et al., 2003).

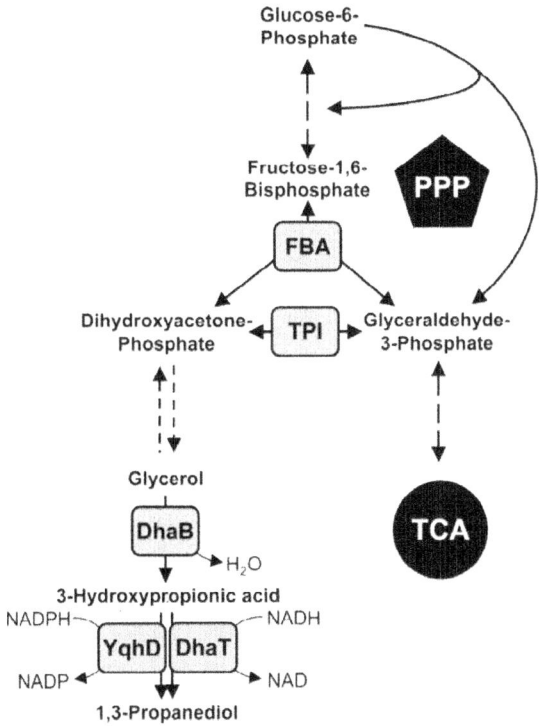

Figure 4. 1,3-propanediol production pathway. Abbreviations: DhaB glycerol dehydratase, DhaT NADH dependent 1,3-propanediol dehydrogenase, FBA fructose-1,6-bisphosphate aldolase, PPP pentose phosphate pathway, TCA tricarboxylic acid cycle, TPI triosephosphate isomerase, YqhD NADPH dependent 1,3-propanediol dehydrogenase.

Hao et al. postulated that the use of K. pneumoniae in a fed batch fermentation process using initial glycerol concentration of 30 g l^{-1} and subsequently keeping it to 7-8 g l^{-1} during exponential growth phase avoids toxic concentrations of 3-HPA (Hao et al., 2008a). For Citrobacter freundii, Pflugmacher et al. could show that cell immobilization to polyurethane carrier particles supports productivity to 8.2 g l^{-1} h^{-1} (Pflugmacher & Gottschalk, 1994). Gungormusler et al. reported 1,3-PDO production from raw glycerol with use Clostridium beijerinckii cells immobilized to ceramic rings and pumice stones in combination with a hydraulic retention time system. They could show an increase in productivity and yield for 1,3-PDO production and predicted a maximal 1,3-PDO production rate of 30 g l^{-1} h^{-1} (Gungormusler et al., 2011).

Metabolic Engineering of K. Pneumoniae, C. Acetobutylicum and E. Coli for 1,3-PDO Production

Several metabolic engineering approaches were made for the optimization of 1,3-PDO production and reduction of byproduct formation with K. pneumoniae and C. acetobutylicum. E. coli, not a natural producer of 1,3-PDO, was engineered for 1,3-PDO production.

K. Pneumonia

K. pneumoniae was genetically manipulated to gain higher 1,3-PDO titers and production rates and to reduce process competing byproducts. When overexpressing the gene encoding the first enzyme of 1,3-PDO production glycerol dehydratase (dhaB) Zhao et al. (2009) found no effect on 1,3-PDO yields but reported decreased formation of the byproducts ethanol and 2,3-butanediol and an increase of acetic acid production (Ma et al., 2009). The overexpression of the 1,3-PDO dehydrogenase gene dhaT was shown to positively affect 1,3-PDO production in many studies and to reduce the formation of the toxic intermediate and substrate of 1,3-PDO dehydrogenase 3-HPA (Rao et al., 2010, Ma et al., 2009, Zhuge et al., 2010, Hao et al., 2008b). While aiming to reduce concentrations of the inhibitory intermediate 3-HPA Hao et al. (2008) overexpressed the 1,3-PDO dehydrogenase gene (dhaT) in K. pneumoniae TUAC01. During fermentation with 30 or 50 g l^{-1} glycerol 3-HPA accumulation was significantly reduced to 1.49 and 2.02 mM, respectively, compared to the parental strain which produced 7.55 and 12.57 mM, respectively (Hao et al., 2008b). Accordingly, Ma et al. (2010) overexpressed the 1,3-PDO dehydrogenase gene in K. pneumoniae and found a 3-fold decrease of 3-HPA production and an increase in 1,3-PDO production by 16.5% (Ma et al., 2010b). Similarly, Zhao et al. (2009) reported a strong decrease of the toxic intermediate 3-HPA by overexpression of the 1,3-PDO dehydrogenase gene leading to an increased molar yield of 0.64 mol mol^{-1} compared to 0.51 mol mol^{-1} of the parental strain and decreased lactic acid, ethanol and succinic acid concentrations in fed-batch fermentations (Ma et al., 2009). Zhuge et al. (2010) combined overexpression of dhaT and yqhD (encoding NADH- and NADPHdependent 1,3-PDO dehydrogenases from K. pneumoniae and E. coli, respectively) in K. pneumoniae. The recombinant strain had a slightly elevated product titer (18.3 g l^{-1} compared to 17.1 g l^{-1}) and increased molar 1,3-PDO yield (0.51 to 0.57 mol mol^{-1}) in batch fermentations. Furthermore, the byproducts 3-HPA, succinic acid, lactic acid, acetic acid and ethanol were significantly reduced (Zhuge et al., 2010). When analyzing a K. pneumoniae mutant strain defective in the genes for NADH-dependent 1,3-PDO dehydrogenase and the oxydative glycerol utilization pathway Seo et al. (2009) reported that the mutant surprisingly retained the ability of 1,3-PDO production. 1,3-PDO yields were low but a strongly reduced byproduct formation was reported (Seo et al., 2009). Later, Seo et al. (2010) published the identification of a second 1,3-PDO dehydrogenase, which possesses high homology to E. coli YqhD and is, in contrast to the dhaT encoded enzyme, NADPH-dependent. Overproduction of the NADPH-dependent 1,3-PDO dehydrogenase from K. pneumoniae resulted in restoration of 1,3-PDO production in the mutant defective in the genes for NADH-dependent 1,3-PDO dehydrogenase and the oxydative glycerol utilization pathway and moreover byproduct formation remained low in the recombinant strain, as the oxidative pathway of glycerol utilization was absent. Although the 1,3-PDO concentrations were lower as compared to the parental strain (4.7 compared to 7.9 g l^{-1}), the molar yield of 0.54 mol mol^{-1} of the recombinant strain was higher compared to 0.48 mol mol^{-1} (Seo et al., 2010). To reduce byproduct formation Horng et al. (2010) constructed a K. pneumoniae mutant lacking the genes for glycerol dehydrogenase and dihydroxyacetone kinase. This mutant was reported to have ceased lactate, 2,3-butanediol, and ethanol byproduct formation and furthermore showed higher 1,3-PDO productivity when glycerol dehydratase and 1,3-PDO dehydrogenase were overexpressed (Horng et al., 2010). A lactate dehydrogenase mutant was constructed by Xu et al. (2009) in K. pneumoniae HR526. The accumulation of lactate was reduced from 40 g l^{-1} in the parental strain to less than 3 g l^{-1} in the lactate dehydrogenase mutant, which showed very low lactate dehydrogenase activity. The mutant furthermore produced higher 1,3-PDO concentrations (95 g l^{-1} as compared to 102 g l^{-1}) with a higher yield (0.48 mol mol^{-1} to 0.52 mol mol^{-1}) and higher production rate (1.98 g l^{-1} h^{-1} to 2.13 g l^{-1} h^{-1}). Reduced lactate production increased NADH availability for 1,2-PDO production in K. pneumoniae (Xu et al., 2009) and K. oxytoca (Yang et al., 2007). Similarly, inactivation of the NADH-dependent aldehyde dehydrogenase gene in K. pneumoniae abolished ethanol formation, improved NADH availability and resulted in a 1,3-PDO production rate of 1.07 g l^{-1} h^{-1} and a yield of 0.70 mol mol^{-1} (Zhang et al., 2006).

E. Coli

Wild-type E. coli is unable to convert glycerol to 1,3-PDO (Tong et al., 1991), but already in 1991 Tong et al. constructed a recombinant E. coli strain for the production of 1,3-PDO in an early application of metabolic engineering. A K. pneumoniae ATCC 25955 genomic library in E. coli was screened for anaerobic growth on glycerol and dihydroxyacetone and 1,3-PDO production. The selected recombinant possessed glycerol dehydratase, 1,3-PDO dehydrogenase, glycerol dehydrogenase, and dihydroxyacetone kinase from K. pneumoniae and produced 1 g l^{-1} 1,3-PDO with a yield of 0.46 mol mol^{-1} (Tong et al., 1991). Cofermentation of glycerol with glucose or xylose increased yields from 0.46 mol mol^{-1} to 0.63 mol mol^{-1} in the presence of glucose and to 0.55 mol mol^{-1} when cofermented with xylose (Tong & Cameron, 1992). Skraly et al. (1998) further optimized 1,3-PDO production with E. coli as they constructed an artificial operon containing the K. pneumoniae genes of 1,3- PDO production. They could show that the recombinant E. coli strain yielded 0.82 mol mol^{-1} (glycerol only) in a fed-batch process cofermenting glycerol and glucose, but only 9.3 g l^{-1} of glycerol where converted (Skraly et al., 1998). The additional expression of the glycerol dehydratase reactivating factor genes gdrAB together with the expression of K. pneumoniae genes encoding glycerol dehydratase and 1,3-PDO dehydrogenase yielded 8.6 g l^{-1} 1,3-PDO in a fed-batch fermentation (Wang et al., 2007). A higher final 1,3-PDO concentration of 13.2 g l^{-1} was obtained when they substituted K. pneumoniae 1,3-PDO dehydrogenase with the E. coli YqhD, which possesses NADPH-dependent 1,3-PDO dehydrogenase activity (Tan et al., 2007). YqhD from E. coli was also used in addition to the vitamin B12-independent glycerol dehydratase from C. butyricum in a two-stage fermentation process (aerobic biomass production from glucose followed by anaerobic 1,3-PDO production from glycerol). By this process 1,3-PDO concentrations of 104.4 g l^{-1} were reached, with a production rate of 2.62 g l- 1 h^{-1} and a molar yield of 1.09 mol mol^{-1} (90.2% g g^{-1}) (referred to glycerol only) (Tang et al., 2009). Some efforts have also been made in the abolishment of toxic intermediate metabolites in E. coli during 1,3-PDO production. In E. coli high glycerol concentrations inhibit growth and 1,3-PDO production due to intracellular accumulation of glycerol-3- phosphate (Cozzarelli et al., 1965), the product of glycerol kinase, first enzyme in the oxidative pathway of glycerol utilization in E. coli. Zhu et al. (2002) could show that the glycerol-3-phosphate concentration increased when glycerol concentrations were elevated. The glycerol-3-phosphate accumulation was due to inefficient expression of the glycerol-3- phosphate dehydrogenase gene, and was overcome by usage of a glycerol kinase mutant, which showed 2.5-fold increased 1,3-PDO production (Zhu et al., 2002). The reduction of methylglyoxal formation by expression of the Pseudomonas putida glyoxalase I resulted in increased 1,3-PDO production by 50% (Zhu et al., 2001).

C. Acetobutylicum

C. butyricum is a promising candidate for efficient 1,3-PDO production because it produces high concentrations of 1,3-PDO and possesses a vitamin

B12-independent glycerol dehydratase circumventing the addition of expensive vitamin B12. Because tools for genetic manipulations of C. butyricum are unavailable, Gonzalez-Pajuelo et al. (2005) introduced the 1,3-PDO pathway from C. butyricum into C. acetobutylicum. The recombinant C. acetobutylicum strain (DG1(pSPD5)) accumulated 84 g l^{-1} 1,3-PDO in a fed-batch culture and reached a high production rate of 3 g l^{-1} h^{-1} (Gonzalez-Pajuelo et al., 2005).

Raw Glycerol

Investigations of raw glycerol use for the production of 1,3-PDO have been made with Clostridium and Klebsiella strains. K. pneumoniae produced 1,3-PDO concentrations from raw glycerol from soybean oil biodiesel production close to that from pure glycerol (51.3 g l^{-1}) with a productivity of 1.7 g l^{-1} h^{-1} (compared to 2 g l^{-1} h^{-1}) (Mu et al., 2006). Similar concentrations (56 g l^{-1}) were reported by Hiremath et al. who used glycerol obtained from Jatropha seed biodiesel production (Hiremath et al., 2011). C. beijerinckii was also reported to produce 1,3-PDO from raw glycerol (Gungormusler et al., 2011). Notably, productivity was increased to 1.51 g l^{-1} h^{-1} when using raw glycerol as compared to 0.84 g l^{-1} h^{-1} when using pure glycerol (Um et al., 2010). Moon et al. compared 1,3-PDO production with Klebsiella and Clostridium strains from glycerol derived from biodiesel production from waste vegetable oil and soybean oil (Moon et al., 2010). Possibly due to inhibitory methanol concentrations, the use of soybean derived glycerol was better than use of raw glycerol derived from waste vegetable oil from different suppliers. In this study, a higher tolerance of Klebsiella strains to the different raw glycerols used compared to the Clostridium strains was observed (Moon et al., 2010). C. butyricum was also shown to utilize raw glycerol for 1,3- PDO production (Papanikolaou et al., 2000, Gonzalez-Pajuelo et al., 2004). Inhibitory effects of raw glycerol components on 1,3-PDO production with C. butyricum were found not to be due to NaCl or methanol, but rather due to oleic acid (Chatzifragkou et al., 2010).

Ethanol

Ethanol as a bio-fuel is mainly gained from sugarcane in Brazil, from corn in the USA and from sugar beets in the EU (da Silva et al., 2009). Since ethanol production is already done in a tens of billions scale, production deriving from crude glycerol may contribute only a small fraction. Nevertheless there has been considerable research on this topic in order to use crude glycerol efficiently to produce ethanol (Licht, 2010).

Unfortunately, the well known ethanol producer Saccharomyces cerevisiae grows very slow on glycerol and, thus, the growth had to be considerably improved, e.g. by selecting S. cerevisiae strain CBS8066-FL20 which grows much faster (0.2 h^{-1} rather than at 0.01 h^{-1}) (OchoaEstopier et al., 2011). Ethanol accumulation of the yeast Hansenula polymorpha was improved from 0.83 g l^{-1} to 2.74 g l^{-1} by expression of genes encoding pyruvate decarboxylase (pdc) and aldehyde dehydrogenase II (adhB) from Zymomonas mobilis. Combined with the expression of glycerol dehydrogenase (dhaD) and dehydroxyacetone kinase (dhaKLM) genes from Klebsiella pneumonia even more ethanol (3.1 g l^{-1}) was produced (Hong et al., 2010). Elementary mode analysis and metabolic evolution of E. coli mutants led to conversion of 40 g l^{-1} glycerol to ethanol reaching 90% of the theoretical yield (Trinh & Srienc, 2009).

E. coli strains have been engineered to produce ethanol and H_2 or ethanol and formate from crude glycerol. Due to overexpression of genes for glycerol dehydrogenase (gldA) and dihydroxyacetonekinase (dhaKLM) 95% of the theoretical maximum yield and specific production rates of 15-30 mmol (g cell)$^{-1}$ h^{-1} could be obtained (Shams Yazdani & Gonzalez, 2008). A newly isolated bacterium, Kluyvera cryocrescens S26, was able to produce 27 g l^{-1} ethanol with yield of 0.8 mol mol^{-1} and a productivity of 0.61 g l^{-1} h^{-1} (Choi et al., 2011).

Succinate

Succinic acid (succinate) is a so called platform chemical based on which a variety of other chemicals are produced, e.g. tetrahydrofuran, γ-butyrolactone, adipic acid, 1,4-butanediol and n-methyl-pyrrolidone. Based on this spectrum of products there are several markets succinic acid is involved in, such as pharmaceuticals, chemistry of biodegradable polymers, surfactants and detergents (Zeikus et al., 1999). Various microorganisms have been engineered for succinate production (Wendisch et al., 2006).

Succinic acid is an intermediate of the TCA cycle with four carbon atoms and two carboxylic groups. Today most of the produced succinic acid derives from the petrochemical industry and only a small part already comes from biotechnological processes. For the chemical synthesis the nonrenewable fossil fuel butane is the starting point leading through maleic anhydride to succinic acid (Zeikus et al., 1999).

Production of succinate from glycerol is interesting because both share the same level of reduction, thus, when produced from glycerol and CO_2 no further electron source is necessary. Various attempts have been made to use bacteria to efficiently produce succinate using natural succinic acid producers as well as metabolically engineered strains (Zeikus et al., 2007, Ingram et al., 2008, Lin et al., 2005, Samuelov et al., 1991, Okino et al., 2005, Singh et al., 2009, Van der Werf et al., 1997, Zhang et al., 2009). Among the natural producers,

Anaerobiospirillum succiniciproducens was shown to use glycerol as sole or combined carbon source to efficiently produce succinic acid (Lee et al., 2004). They found a high succinic acid yield of 133% (or 160% when yeast extract was additionally fed) for glycerol concentrations of 6.5 g l^{-1}. Glycerol entailed less formation of acetic acid as byproduct and, thus, an easier downstream processing (Lee et al., 2001). Succinate production by the related bacterium Basfia succiniciproducens from crude glycerol in a continuous cultivation process allowed for product yields of about 1 g g^{-1} (Scholten et al., 2009).

Succinate production from glycerol by E. coli appeared to be favored under aerobic conditions as indicated by elementary mode analysis (Chen et al., 2010). Using microaerobic conditions, a recombinant producing pyruvate carboxylase from Lactococcus lactis and lacking pathways to byproducts showed succinic acid yields on glycerol comparable to those on glucose (Blankschien et al., 2010). About 80% of the maximum theoretical yield could be achieved by inserting three key mutations affecting phosphoenolpyruvate carboxykinase (pck), part of the phosphotransferase system (ptsI) and the pyruvate formate lyase (pflB) (Zhang et al., 2010).

Citrate

Citric acid is produced by fermentation at a scale of about 1,600,000 t/a (Papanikolaou et al., 2002, Berovic & Legisa, 2007), and it is sold for about 0.8 €/kg (Weusthuis et al., 2011). The main markets are the food and the pharmaceutical industries as well as applications in cosmetics, detergents and cleaning products. Citric acid is produced almost exclusively by Aspergillus niger in a submerged fermentation process using starch- or sucrose-based media like molasses (Soccol et al., 2006).

Different strains of the yeast Yarrowia lipolytica have been investigated for citric acid production using glycerol as a carbon source. For Yarrowia lipolytica Wratislavia AWG7 a maximal yield of 0.67 g g^{-1} was reported in continuous culture using glycerol as carbon source with only low contamination by the common byproduct isocitric acid (Rywinska et al., 2011, Rywinska & Rymowicz, 2010).

In 2002, Papanikolaou et al. reported about the Y. lipolytica strain LGAM S(7)1 being capable of growing on crude glycerol as carbon source and producing up to 35 g l^{-1} citric acid (Papanikolaou et al., 2002) . In 2010 and 2011, much higher citric acid concentrations of 112 g l -1 and a yield of 0.6 g g^{-1} using crude glycerol and the acetate-negative mutant strain Y. lipolytica A-101-1.22 (Rymowicz et al., 2010) or the acetate-negative strain Y. lipolytica N15 could be achieved (Kamzolova et al., 2011).

Amino Acids

Amino acids are a multi-billion dollar business (Wendisch, 2007). They are used as flavor enhancers (L-glutamate), feed additives (L-lysine, L-methionine, L-threonine, Ltryptophane), to produce sweeteners such as aspartam (L-aspartate, L-phenylalanine), and in various pharmaceutical applications. The biggest products are L-glutamate (2,160,000 tons per year) and L-lysine (1,330,000 tons per year) (Ajinomoto, 2010a, Ajinomoto, 2010b). Pathways for amino acid synthesis start from intermediates of glycolysis (e.g. L-serine from 3-phosphoglycerate, L-valine from pyruvate), glycolysis and pentose phosphate pathway (aromatic amino acids L-tyrosine, L-phenylalanine, and Ltryptophane, from phosphoenolpyruvate (PEP) and erythrose-4-phosphate), and tricarboxylic acid cycle intermediates (L-glutamate, L-glutamine from 2-oxoglutarate, L-lysine, L-aspartate from oxaloacetate) (Schneider & Wendisch, 2001, Wendisch, 2007, Gopinath et al., 2011). In principle, production of amino acids from glycerol should be possible. In the following section the biosynthesis of L-glutamate, L-lysine, and L-phenylalanine are described, since strains for production of these from glycerol have been described.

Corynebacterium glutamicum is a natural L-glutamate producer (Eggeling & Bott, 2005), but the excretion of L-glutamate needs to be "triggered", e.g. by limitation of biotin (Shiio et al., 1962). Biotin is essential for the activity of acetyl-CoA carboxylase, necessary for fatty acid synthesis, and hence for membrane precursors, thus effect of biotin limitation on Lglutamate production is thought to be due to a higher permeability of the cell membrane (Shimizu & Hirasawa, 2007). Also addition of detergents like Tween 40 (Takinami et al., 1965), antibiotics like penicillin (Nara et al., 1964), and ethambutol (Radmacher et al., 2005, Stansen et al., 2005) trigger L-glutamate production. In C. glutamicum L-glutamate is mainly synthesized by NADPH dependent glutamate dehydrogenase from the tricarboxylic acid cycle intermediate 2-oxoglutarate (Bormann et al., 1992). This holds true for high ammonia concentrations, when ammonia is low L-glutamate is synthesized via L-glutamine by glutamine synthetase and glutamine-2-oxoglutarate aminotransferase. Crucial for Lglutamate production is the anaplerosis of the tricarboxylic acid cycle by either pyruvate carboxylase or PEP carboxylase. Pyruvate carboxylase has been shown to be indispensable under detergent triggered production conditions (Peters-Wendisch et al., 2001) and vice versa under biotin limiting conditions PEPcarboxylase is responsible for anaplerosis (Sato et al., 2008, Delaunay et al., 2004, Lapujade et al., 1999). C. glutamicum was engineered for glycerol utilization by expression of the genes for glycerol facilitator, glycerol kinase, and glycerol-3-phosphate dehydrogenase from

E. coli (Rittmann et al., 2008). Under ethambutol triggered L-glutamate production conditions recombinant C. glutamicum showed reduced Lglutamate yields from glycerol compared to glucose, 0.11 g g^{-1} compared to 0.20 g g^{-1}, respectively (Rittmann et al., 2008).

Production of L-lysine, which is used as a feed additive, is also carried out with C. glutamicum (Wendisch, 2007, Eggeling & Bott, 2005). The precursors of L-lysine production are the tricarboxylic acid cycle intermediate oxaloacetate and the glycolytic intermediate pyruvate. Deregulation of the L-lysine production pathway by introduction of feedback resistant variants of the key enzyme aspartate kinase, which usually is inhibited by L-lysine and L-threonine (Kalinowski et al., 1991) enables C. glutamicum for L-lysine production. Further increases were made by overexpression of the gene for pyruvate carboxylase (Peters-Wendisch et al., 2001), which provides L-lysine precursor oxaloacetate by the anaplerotic reaction from pyruvate (Peters-Wendisch et al., 1998). The anaplerotic reaction from PEP was shown to be dispensable for L-lysine production from glucose, however, it might play an important role if glucose is phosphorylated by use of ATP or polyphosphate and not PEP, which was shown to enhance L-lysine production and might elevate PEP availability (Lindner et al., 2011). Vice versa also an inactivation of the gene for PEP carboxykinase, catalyzing decarboxylation of oxaloacetate to PEP, entailed increased L-lysine production (Riedel et al., 2001). NADPH supply is very important for L-lysine production, as four molecules of NADPH are needed for one molecule L-lysine. The main path of NADPH generation is the oxidative branch of the pentose phosphate pathway (Marx et al., 1996), where NADP is reduced to NADPH by glucose-6-phosphate dehydrogenase and 6-phosphogluconate dehydrogenase, thus to enhance L-lysine production numerous attempts have been made towards increasing the pentose phosphate pathway flux, hence NADPH availability, hence L-lysine production. A deletion of the phosphoglucose isomerase gene drives the complete flux from glucose- 6-phosphate into the pentose phosphate pathway and was shown to increase L-lysine production but to the cost of reduced growth (Marx et al., 2003). Redirection of the glycolytic flux towards the entry of the pentose phosphate pathway was furthermore achieved by overexpression of the fructose-bisphosphatase gene (Becker et al., 2005, Georgi et al., 2005) as well as use of feedback resistant variants of glucose-6-phosphate dehydrogenase and 6-phosphogluconate dehydrogenase (Becker et al., 2007, Ohnishi et al., 2005). Also the increase of NADP availability by overexpression of a NAD kinase gene resulted in increased L-lysine production (Lindner et al., 2010). To establish L-lysine production from glycerol Rittmann et al. introduced the Escherichia coli glycerol utilization genes in a metabolic engineered C. glutamicum L-lysine producing strain (deregulated Llysine pathway and higher anaplerotic from pyruvate to oxaloacetate). L-lysine yields were slightly lower from glycerol as glucose, 0.19 g g^{-1} compared to 0.26 g g^{-1}, respectively (Rittmann et al., 2008). Glycerol has also been used as a source of carbon for the production of the polymer of L-lysine e-Poly-L-lysine with Streptomyces sp.(Chen et al., 2011b, Chen et al., 2011a). e-Poly-L-lysine is an antimicrobial agent against bacteria, yeasts, and viruses (Shima et al., 1984) and therefore interesting for the pharmaceutical industry (Shih et al., 2004) and it is used as food preservative.

The main use of the aromatic amino acid phenylalanine is in production of the sweetener aspartam. Biosynthesis of aromatic amino acids from PEP and erythrose-4-phosphate involves the shikimic acid pathway and dedicated terminal biosynthesis pathways for tryptophan, tyrosine and phenylalanine (Sprenger, 2007). Biosynthesis of aromatic amino acids e.g. in Escherichia coli and C. glutamicum was engineered e.g. by gene deregulation (Berry, 1996, Herry & Dunican, 1993), by gene copy number increase (Chan et al., 1993), and by the use of feedback-resistant enzyme variants, e.g. variants of 3-deoxy-D-arabinoheptulosonate 7-phosphate synthase, the first enzyme of the shikimic acid pathway, variants of anthranilate synthase of the tryptophan pathway in E. coli (Tribe & Pittard, 1979) or anthranilate phosphoribosyltransferase of the tryptophan pathway in C. glutamicum (O'Gara & Dunican, 1995). In addition, strains were engineered for increased supply of the precursors PEP and erythrose-4-phosphate. In C. glutamicum, PEP availability was increased in PEP carboxylase mutants and erythrose-4-phosphate concentrations were elevated by overexpression of the transketolase gene (Ikeda & Katsumata, 1999, Ikeda et al., 1999, Katsumata & Kino, 1989). Similar approaches were made in E. coli, where PEP carboxylase or pyruvate kinase gene knock outs and overexpression of PEP carboxykinase increased PEP supply (Miller et al., 1987, Gosset et al., 1996, Chao & Liao, 1993, Backman, 1992). Furthermore, overexpression of genes encoding PEP synthase, PEP carboxykinase, and the use of an ATP-dependent glucose phosphorylation system instead of the PEP-dependent phosphotransferase system had positive effects on the availability of shikimic acid pathway precursor PEP (Patnaik et al., 1995, Liao, 1996, Gulevich et al., 2004). In E. coli, the availability of erythrose-4-phosphate could be increased by overproduction of transketolase and transaldolase or by phosphoglucose isomerase gene disruption (Draths & Frost, 1990, Draths et al., 1992, Lu & Liao, 1997, Mascarenhas et al., 1991, Frost, 1992). Up to now, only Lphenylalanine production from glycerol has been shown, but results might be transferable to the other aromatic amino acids. Similar final concentrations of L-phenylalanine were reported for an engineered E. coli strain regardless of the use of glycerol, glucose or sucrose as carbon source.

Notably, a higher yield was reported when glycerol was used (0.58 g g^{-1}) as compared to the use of sucrose (0.25 g g^{-1}) (Khamduang et al., 2009).

Polyamines may be derived from amino acids (Schneider & Wendisch, 2010). While strain development for sugar-based production of polyamines such as the diamine 1,4- diaminobutane, which is used e.g. in the polyamide market, has been successful (Schneider & Wendisch, 2010), glycerol-based production of poylamines has not yet been reported.

2,3-Butanediol

2,3-Butanediol (2,3-BDO) is used as a solvent, fuel, and for the production of polymers and chemicals (Perego et al., 2003, Saha & Bothast, 1999). Bacterial 2,3-BDO production has been shown e.g. with strains of Klebsiella pneumoniae, Klebsiella oxytoca, Enterobacter aerogenes, Bacillus polymyxa, and Bacillus licheniformis (Grover et al., 1990, Perego et al., 2000, De Mas et al., 1988, Nilegaonkar et al., 1996, Jansen et al., 1984). Biosynthesis of 2,3-BDO is funneled from pyruvate in three steps. First, acetolactate synthase (EC 2.2.1.6) catalyses the condensation of two pyruvate molecules to acetolactate with concomitant CO_2 liberation. Second, acetolactate is decarboxylated by acetolactate decarboxylase (EC 4.1.1.5) to acetoin. Third, acetoin is reduced to 2,3-butanediol by 2,3-BDO dehydrogenase (acetoin reductase; EC 1.1.1.4) (Juni, 1952). Thus for 2,3-BDO production all substrates first need to be converted to pyruvate, the intermediate of glycolysis.

2,3-BDO is a product of mixed acid fermentation and, thus, associated with byproduct formation. Byproduct reduction approaches were made with K. oxytoca mutants defective in genes encoding lactate dehydrogenase and phosphotransacetylase, reducing lactate and acetate byproduct formation by 88% and 92%, respectively, but increasing 2,3-BDO production only by 7.8% (Ji et al., 2008). Also formation of ethanol, a major byproduct of 2,3- BDO production with K. oxytoca, could be eliminated by insertion mutagenesis of the aldehyde dehydrogenase gene and 2,3-BDO production from glucose in a fed-batch process was improved to yield 130 g l^{-1} 2,3-BDO with a productivity of 1.63 g l^{-1} h^{-1} and a yield of 0.48 g g^{-1} (Ji et al., 2010).

Many substrates have been used for the production of 2,3-BDO. Use of starch as a substrate for 2,3-BDO production has been shown with K. pneumoniae by overexpression of a secretory a-amylase (Wei et al., 2008). With B. licheniformis corn starch hydrolysates were applied to 2,3-BDO production (Perego et al., 2003). With E. aerogenes, food industry wastes such as starch hydrolysates, raw and decoloured molasses, and whey permeate were used for the fermentation of 2,3-BDO (Perego et al., 2000). The use of lignocellulosic compounds for 2,3- BDO has also been reported, e.g. corncob hydrolysates were used in processes with K. oxytoca (Cheng et al., 2010) and K. pneumoniae (Ma et al., 2010a).

Glycerol was used for the production of 2,3-BDO as well. Because 1,3-propanediol production is preferably carried out from glycerol e.g. by K. pneumoniae and because 2,3- BDO is a known byproduct of this process (Biebl et al., 1998), glycerol might be a good substrate for 2,3-BDO production. Production of 2,3-BDO from glycerol by K. pneumoniae G31 resulted in final concentrations of 49.2 g l^{-1}. The medium pH had a large influence on 2,3-BDO fermentation from glycerol with 2,3-BDO production being favored at alkaline pH (Petrov & Petrova, 2009). In addition, intense aeration increased 2,3-BDO synthesis and reduced byproducts (Petrov & Petrova, 2010).

Hydrogen

Hydrogen production is highly desirable as a source of clean energy to be used, e.g. in fuel cells. Processes for the use of glycerol or crude glycerol respectively are under investigation. Besides microbial strategies to generate H_2 from crude glycerol there are also promising chemicals techniques such as steam reforming, partial oxidation, auto thermal reforming, aqueous-phase reforming, and supercritical water reforming (Xiaohu Fan, 2010). Currently, only low concentrations of glycerol can be used in microbial H_2 production process to avoid that other products like 1,3-propanediol or ethanol are produced along with H_2. Enterobacter aerogenes HU-101 showed hydrogen yields of 1.12 mol mol^{-1} using crude glycerol, but at relatively low glycerol concentrations of 1.7 g l^{-1} (Ito et al., 2005). Mixed cultures isolated from soil or wastewater converted crude glycerol to H_2 with a yield 0.31 mol mol^{-1} and to 1,3-Propanediol with a yield of 0.59 mol mol^{-1}. These values are lower than the ones on glucose but similar to the ones with pure glycerol, suggesting that inhibiting substances in crude glycerol may not be a problem in this process (Selembo et al., 2009). Production rates of 0.68 ± 0.16 mmol H_2 l^{-1} h^{-1} could be achieved by an evolved E. coli BW25113 frdC negative strain along with some ethanol production (Hu & Wood, 2010).

Glyceric Acid

Glyceric acid is a known byproduct of dihydroxyacetone production from glycerol with Gluconobacter oxydans. The path from glycerol to glyceric aid, which might be suitable for chemical applications (Habe et al., 2009a), occurs via two dehydrogenases. First, alcohol dehydrogenase oxidizes glycerol to glyceraldehyde which is subsequently oxidized further to glyceric acid by a so far unidentified enzyme (Habe et al., 2009d). In a screen of various acetic acid bacteria Acetobacter tropicalis was the best glyceric acid producing strain

(Habe et al., 2009b). A. tropicalis produced 101.8 g l^{-1} glyceric acid while Gluconobacter frateurii accumulated 136.5 g l^{-1} (Habe et al., 2009d). The involvement of a membrane-bound alcohol dehydrogenase in glyceric acid production was investigated with G. oxydans IFO12528. Gene disruption of the alcohol dehydrogenase entailed severely reduced glyceric acid concentrations, indicating a role of the alcohol dehydrogenase in glyceric acid production (Habe et al., 2009d). G. frateurii was engineered for glyceric acid production by disruption of the glycerol dehydrogenase gene sldA, thus, eliminating dihydroxyacetone production. The growth retardation of this strain on glycerol alone was overcome by addition of sorbitol to the medium. A higher glyceric acid concentration of 89.1 g l^{-1} was reached with the sldA mutant compared to the parental strain (54.7 g l^{-1}) as production dihydroxyacetone as byproduct was avoided (Habe et al., 2010b). Glyceric acid production from raw glycerol pretreated with activated charcoal by Gluconobacter sp. NBRC3259 reached comparable concentration of glyceric acid (45.9 g l^{-1} and 54.7 g l^{-1}) and of dihydroxyacetone (28.2 g l^{-1} and 33.7 g l^{-1}) as production from pure glycerol (Habe et al., 2009c).

Biosurfactants

Surfactants are used in numerous applications such as cleaners, emulsifiers, in coatings, laundry detergents, or in paints. The global surfactant market is predicted to reach 17.9 billion dollars by 2015 (Global Industry Analysts, 2010). Biosurfactants consist of a hydrophilic part and a hydrophobic/lipophilic part, making them amphiphiles/tensides. The hydrophilic part can consist of a sugar, peptide or protein, while the hydrophobic part contains fatty acids or fatty alcohols. The great advantage of biosurfactants over chemically produced tensides is that they are biodegradable, hence environmentally friendly, less toxic, and they can be produced from renewable resources. Natural biosurfactant producers are bacteria, yeast, and fungi (Mulligan, 2005). Biosurfactants are already used in many applications, e.g. in the food, agriculture, chemical, pharmaceutical, and cosmetic industries (for recent review see (Pacwa-Plociniczak et al., 2011)). Glycerol is the backbone of the lipid component of biosurfactants and use of pure and raw glycerol in biosurfactant production has been investigated with a variety of organisms. The yeast Pseudozyma antarctica produced 16.3 g l^{-1} glycolipid biosurfactants from glycerol (Morita et al., 2007). Glycolipid surfactants production was also shown with Candida bombicola and product concentrations of 12.7 g l^{-1} of sophorolipids could be obtained from glycerol and oleic acid (Ashby & Solaiman, 2010). When a biodiesel co-product stream consisting of 40% glycerol, 34% hexane-solubles, and 26% water was used production of sophorolipids by C. bombicola was strongly increased (from 9 g l^{-1} to 60 g l^{-1}) (Ashby et al., 2005). Production of glucosylmannosyl-glycerolipid with Microbacterium spec. was reported to be 1.5-fold higher when glycerol instead of glucose is used in media containing the complex medium compounds peptone and yeast extract (Lang et al., 2004, Wicke et al., 2000). When several carbon sources were analyzed for rhamnolipid production by Pseudomonas aeruginosa EM1, glycerol was the best carbon source besides glucose, yielding 4.9 and 7.5 g l^{-1} respectively (Wu et al., 2008). Glycerolbased biosurfactant production with Pseudomonas aeruginosa UCP0992 yielded 8.0 g l^{-1} biosurfactants (Silva et al., 2010), processes with Rhodococcus erythropolis yielded 1.7 g l^{-1} biosurfactants (Ciapina et al., 2006) and with Ustilago maydis 32.1 g l^{-1} glycolipid biosurfactants from 50 g l^{-1} glycerol could be produced (Liu et al., 2011).

CONCLUSIONS

Glycerol availability has increased tremendously as it arises as byproduct of the biodiesel process. Besides using glycerol as a chemical in creams and other small-scale applications, glycerol may be used as starting material for large-scale biotechnological processes. Several microbiological process are based on glycerol as substrate anyway, e.g. the biotechnological production of 1,3-propanediol and dihydroxyacetone. In addition, as glycerol is a good source of carbon and energy for growth of several microorganisms it may be suitable for the biotechnological production of a number of chemicals in fermentative processes. Microbial catalysts have been developed for the production of succinic acid, citric acid, glyceric acid, propionic acid, ethanol, 1,2-propanediol, 2,3-butanediol, biosurfactants and amino acids. Successful implementation of these processes critically depends on strain optimization, recalcitrance to inhibitors present in crude glycerol preparations from biodiesel factories and measures to reduce glycerol price volatility. If successful, coupling biodiesel production to the production of value-added products from the side-stream glycerol "on the spot" would represent an excellent example of applying the biorefinery concept to a large-scale process.

ACKNOWLEDGEMENTS

Work in the laboratory of the authors is supported in part by grants from the BMBF (0315589G, 0315598E, 316017A), ERA-IB (22009508B) and ESF (PAK529).

REFERENCES

1. Ajinomoto, (2010a) Feed-Use Amino Acids Business. Available from

World Wide Web: http://www.ajinomoto.com/ir/pdf/Feed-useAA-Oct2010.pdf. Cited 18 March 2011.
2. Ajinomoto, (2010b) Food Products Business. Available from World Wide Web: http://www.ajinomoto.com/ir/pdf/Food-Oct2010.pdf. Cited 18 March 2011.
3. Altaras, N. E. & D. C. Cameron, (1999) Metabolic engineering of a 1,2-propanediol pathway in Escherichia coli. Appl Environ Microbiol 65: 1180-1185.
4. Altaras, N. E. & D. C. Cameron, (2000) Enhanced production of (R)-1,2-propanediol by metabolically engineered Escherichia coli. Biotechnol Prog 16: 940-946.
5. Ashby, R. D., A. Nunez, D. K. Y. Solaiman & T. A. Foglia, (2005) Sophorolipid biosynthesis from a biodiesel co-product stream. J Am Oil Chem Soc 82: 625-630.
6. Ashby, R. D. & D. K. Solaiman, (2010) The influence of increasing media methanol concentration on sophorolipid biosynthesis from glycerol-based feedstocks. Biotechnol Lett 32: 1429-1437.
7. Ashok, S., S. M. Raj, C. Rathnasingh & S. Park, (2011) Development of recombinant Klebsiella pneumoniae ΔdhaT strain for the co-production of 3-hydroxypropionic acid and 1,3- propanediol from glycerol. Appl Microbiol Biotechnol 90: 1253-1265.
8. Backman, K., (1992) Method of biosynthesis of phenylalanine. U.S. Patent No. US 5169768.
9. Badia, J., J. Ros & J. Aguilar, (1985) Fermentation mechanism of fucose and rhamnose in Salmonella typhimurium and Klebsiella pneumoniae. J Bacteriol 161: 435-437.
10. Barbirato, F., J. P. Grivet, P. Soucaille & A. Bories, (1996) 3-Hydroxypropionaldehyde, an inhibitory metabolite of glycerol fermentation to 1,3-propanediol by enterobacterial species. Appl Environ Microbiol 62: 1448-1451.
11. Barbirato, F., A. Larguier, T. Conte, S. Astruc & A. Bories, (1997) Sensitivity to pH, product inhibition, and inhibition by NAD+ of 1,3-propanediol dehydrogenase purified from Enterobacter agglomerans CNCM 1210. Arch Microbiol 168: 160-163.
12. Bauer, R., N. Katsikis, S. Varga & D. Hekmat, (2005) Study of the inhibitory effect of the product dihydroxyacetone on Gluconobacter oxydans in a semi-continuous two-stage repeated-fed-batch process. Bioprocess Biosyst Eng 28: 37-43.

13. Becker, J., C. Klopprogge, A. Herold, O. Zelder, C. J. Bolten & C. Wittmann, (2007) Metabolic flux engineering of L-lysine production in Corynebacterium glutamicum--over expression and modification of G6P dehydrogenase. J Biotechnol 132: 99-109.
14. Becker, J., C. Klopprogge, O. Zelder, E. Heinzle & C. Wittmann, (2005) Amplified expression of fructose 1,6-bisphosphatase in Corynebacterium glutamicum increases in vivo flux through the pentose phosphate pathway and lysine production on different carbon sources. Appl Environ Microbiol 71: 8587-8596.
15. Berovic, M. & M. Legisa, (2007) Citric acid production. Biotechnol Annu Rev 13: 303-343.
16. Berrios-Rivera, S. J., K. Y. San & G. N. Bennett, (2003) The effect of carbon sources and lactate dehydrogenase deletion on 1,2-propanediol production in Escherichia coli. J Ind Microbiol Biotechnol 30: 34-40.
17. Berry, A., (1996) Improving production of aromatic compounds in Escherichia coli by metabolic engineering. Trends Biotechnol 14: 250-256.
18. Biebl, H., (1991) Glycerol Fermentation of 1,3-Propanediol by Clostridium butyricum - Measurement of Product Inhibition by Use of a Ph-Auxostat. Appl Microbiol Biot 35: 701-705.
19. Biebl, H., (2001) Fermentation of glycerol by Clostridium pasteurianum - batch and continuous culture studies. J Ind Microbiol Biot 27: 18-26.
20. Biebl, H., S. Marten, H. Hippe & W. D. Deckwer, (1992) Glycerol Conversion to 1,3- Propanediol by Newly Isolated Clostridia. Appl Microbiol Biot 36: 592-597.
21. Biebl, H., A. P. Zeng, K. Menzel & W. D. Deckwer, (1998) Fermentation of glycerol to 1,3- propanediol and 2,3-butanediol by Klebsiella pneumoniae. Appl Microbiol Biotechnol 50: 24-29.
22. Blankschien, M. D., J. M. Clomburg & R. Gonzalez, (2010) Metabolic engineering of Escherichia coli for the production of succinate from glycerol. Metab Eng 12: 409-419.
23. Blomberg, A., (2000) Metabolic surprises in Saccharomyces cerevisiae during adaptation to saline conditions: questions, some answers and a model. FEMS Microbiol Lett 182: 1-8.
24. Bories, A., C. Claret & P. Soucaille, (1991) Kinetic-Study and Optimization of the Production of Dihydroxyacetone from Glycerol Using Gluconobacter oxydans. Process Biochem 26: 243-248.
25. Bormann, E. R., B. J. Eikmanns & H. Sahm, (1992) Molecular analysis

of the Corynebacterium glutamicum gdh gene encoding glutamate dehydrogenase. Mol Microbiol 6: 317-326.

26. Boronat, A. & J. Aguilar, (1981) Experimental evolution of propanediol oxidoreductase in Escherichia coli. Comparative analysis of the wild-type and mutant enzymes. Biochim Biophys Acta 672: 98-107.

27. Cameron, D. C., N. E. Altaras, M. L. Hoffman & A. J. Shaw, (1998) Metabolic engineering of propanediol pathways. Biotechnol Prog 14: 116-125.

28. Cameron, D. C. & C. L. Cooney, (1986) A Novel Fermentation - the Production of R(-)-1,2- Propanediol and Acetol by Clostridium thermosaccharolyticum. Bio-Technol 4: 651-654.

29. Carballo, J., A. Bernardo, M. J. Prieto & R. M. Sarmiento, (1993) Kinetics of alphadicarbonyls reduction by L-glycol dehydrogenase (NAD+) from Enterobacter aerogenes. Ital J Biochem 42: 79-89.

30. Carole, T. M., J. Pellegrino & M. D. Paster, (2004) Opportunities in the industrial biobased products industry. Appl Biochem Biotechnol 113-116: 871-885.

31. Chan, E. C., H. L. Tsai, S. L. Chen & D. G. Mou, (1993) Amplification of the Tryptophan Operon Gene in Escherichia coli Chromosome to Increase L-Tryptophan Biosynthesis. Appl Microbiol Biot 40: 301-305.

32. Chao, Y. P. & J. C. Liao, (1993) Alteration of growth yield by overexpression of phosphoenolpyruvate carboxylase and phosphoenolpyruvate carboxykinase in Escherichia coli. Appl Environ Microbiol 59: 4261-4265.

33. Chatzifragkou, A., D. Dietz, M. Komaitis, A. P. Zeng & S. Papanikolaou, (2010) Effect of biodiesel-derived waste glycerol impurities on biomass and 1,3-propanediol production of Clostridium butyricum VPI 1718. Biotechnol Bioeng 107: 76-84.

34. Chen, X., L. Tang, S. Li, L. Liao, J. Zhang & Z. Mao, (2011a) Optimization of medium for enhancement of epsilon-poly-L-lysine production by Streptomyces sp. M-Z18 with glycerol as carbon source. Bioresour Technol 102: 1727-1732.

35. Chen, X., D. J. Zhang, W. T. Qi, S. J. Gao, Z. L. Xiu & P. Xu, (2003) Microbial fed-batch production of 1,3-propanediol by Klebsiella pneumoniae under micro-aerobic conditions. Appl Microbiol Biot 63: 143-146.

36. Chen, X. S., S. Li, L. J. Liao, X. D. Ren, F. Li, L. Tang, J. H. Zhang & Z. G. Mao, (2011b) Production of epsilon-poly-L: -lysine using a novel

two-stage pH control strategy by Streptomyces sp. M-Z18 from glycerol. Bioprocess Biosyst Eng 34: 561-567.
37. Chen, Z., H. Liu, J. Zhang & D. Liu, (2010) Elementary mode analysis for the rational design of efficient succinate conversion from glycerol by Escherichia coli. J Biomed Biotechnol 2010: 518743.
38. Cheng, K. K., H. J. Liu & D. H. Liu, (2005) Multiple growth inhibition of Klebsiella pneumoniae in 1,3-propanediol fermentation. Biotechnol Lett 27: 19-22.
39. Cheng, K. K., Q. Liu, J. A. Zhang, J. P. Li, J. M. Xu & G. H. Wang, (2010) Improved 2,3- butanediol production from corncob acid hydrolysate by fed-batch fermentation using Klebsiella oxytoca. Process Biochem 45: 613-616.
40. Choi, W. J., M. R. Hartono, W. H. Chan & S. S. Yeo, (2011) Ethanol production from biodiesel-derived crude glycerol by newly isolated Kluyvera cryocrescens. Appl Microbiol Biotechnol 89: 1255-1264.
41. Ciapina, E. M., W. C. Melo, L. M. Santa Anna, A. S. Santos, D. M. Freire & N. Pereira, Jr., (2006) Biosurfactant production by Rhodococcus erythropolis grown on glycerol as sole carbon source. Appl Biochem Biotechnol 131: 880-886.
42. Claret, C., A. Bories & P. Soucaille, (1992) Glycerol Inhibition of Growth and Dihydroxyacetone Production by Gluconobacter oxydans. Current Microbiology 25: 149-155.
43. Claret, C., A. Bories & P. Soucaille, (1993) Inhibitory Effect of Dihydroxyacetone on Gluconobacter oxydans - Kinetic Aspects and Expression by Mathematical Equations. J Ind Microbiol 11: 105-112.
44. Claret, C., J. M. Salmon, C. Romieu & A. Bories, (1994) Physiology of Gluconabacter oxydans during Dihydroxyacetone Production from Glycerol. Appl Microbiol Biot 41: 359-365.
45. Clomburg, J. M. & R. Gonzalez, (2011) Metabolic engineering of Escherichia coli for the production of 1,2-propanediol from glycerol. Biotechnol Bioeng 108: 867-879.
46. Colin, T., A. Bories & G. Moulin, (2000) Inhibition of Clostridium butyricum by 1,3- propanediol and diols during glycerol fermentation. Appl Microbiol Biotechnol 54: 201-205.
47. Cooper, R. A. & A. Anderson, (1970) Formation and Catabolism of Methylglyoxal during Glycolysis in Escherichia coli. Febs Letters 11: 273-&.
48. Cordier, H., F. Mendes, I. Vasconcelos & J. M. Francois, (2007) A

metabolic and genomic study of engineered Saccharomyces cerevisiae strains for high glycerol production. Metab Eng 9: 364-378.

49. Cozzarelli, N. R., J. P. Koch, S. Hayashi & E. C. Lin, (1965) Growth stasis by accumulated Lalpha-glycerophosphate in Escherichia coli. J Bacteriol 90: 1325-1329.

50. da Silva, G. P., M. Mack & J. Contiero, (2009) Glycerol: a promising and abundant carbon source for industrial microbiology. Biotechnol Adv 27: 30-39.

51. Daniel, R., K. Stuertz & G. Gottschalk, (1995) Biochemical and Molecular Characterization of the Oxidative Branch of Glycerol Utilization by Citrobacter freundii. Journal of Bacteriology 177: 4392-4401.

52. De Mas, C., N. B. Jansen & G. T. Tsao, (1988) Production of optically active 2,3-butanediol by Bacillus polymyxa. Biotechnol Bioeng 31: 366-377.

53. Degalard, P., (2011) METabolic EXplorer develops its first plant in Malaysia. Available from World Wide Web: http://www.ubifrance.com/my/Posts-1483-METabolicEXplorer-develops-its-first-plant-in-Malaysia. Cited 22 June 2011.

54. Delaunay, S., P. Daran-Lapujade, J. M. Engasser & J. L. Goergen, (2004) Glutamate as an inhibitor of phosphoenolpyruvate carboxylase activity in Corynebacterium glutamicum. J Ind Microbiol Biotechnol 31: 183-188.

55. Draths, K. M. & J. W. Frost, (1990) Synthesis Using Plasmid-Based Biocatalysis - Plasmid Assembly and 3-Deoxy-D-Arabino-Heptulosonate Production. Journal of the American Chemical Society 112: 1657-1659.

56. Draths, K. M., D. L. Pompliano, D. L. Conley, J. W. Frost, A. Berry, G. L. Disbrow, R. J. Staversky & J. C. Lievense, (1992) Biocatalytic Synthesis of Aromatics from DGlucose - the Role of Transketolase. Journal of the American Chemical Society 114: 3956-3962.

57. EBB, (European Biodiesel Board 2010) 2009-2010: EU biodiesel industry restained growth in challenging times. Annual biodiesel production statistics. Available in: http://www.ebb-eu.org/EBBpressreleases/EBB press release 2009 prod 2010_capacity FINAL.pdf.

58. Eggeling, L. & M. Bott, (2005) Handbook of Corynebacterium glutamicum. CRC Press, Boca Raton, USA.

59. Enders, D., M. Voith & A. Lenzen, (2005) The dihydroxyacetone unit--a versatile C(3) building block in organic synthesis. Angew Chem Int Ed Engl 44: 1304-1325.

60. Ferreira, C., F. van Voorst, A. Martins, L. Neves, R. Oliveira, M. C.

Kielland-Brandt, C. Lucas & A. Brandt, (2005) A member of the sugar transporter family, Stl1p is the glycerol/H+ symporter in Saccharomyces cerevisiae. Mol Biol Cell 16: 2068-2076.

61. Flickinger, M. C. & D. Perlman, (1977) Application of Oxygen-Enriched Aeration in the Conversion of Glycerol to Dihydroxyacetone by Gluconobacter melanogenus IFO 3293. Appl Environ Microbiol 33: 706-712.

62. Forage, R. G. & M. A. Foster, (1982) Glycerol fermentation in Klebsiella pneumoniae: functions of the coenzyme B12-dependent glycerol and diol dehydratases. J Bacteriol 149: 413-419.

63. Forsberg, C. W., (1987) Production of 1,3-Propanediol from Glycerol by Clostridium acetobutylicum and Other Clostridium Species. Appl Environ Microbiol 53: 639-643.

64. Freedberg, W. B., W. S. Kistler & E. C. Lin, (1971) Lethal synthesis of methylglyoxal by Escherichia coli during unregulated glycerol metabolism. J Bacteriol 108: 137-144.

65. Frost, J., (1992) Enhanced production of common aromatic pathway compounds. U.S. Patent No. US 5168056.

66. Georgi, T., D. Rittmann & V. F. Wendisch, (2005) Lysine and glutamate production by Corynebacterium glutamicum on glucose, fructose and sucrose: Roles of malic enzyme and fructose-1,6-bisphosphatase. Metab Eng 7: 291-301.

67. Global Industry Analysts, I., (2010) MCP-2146: Surface active agents - a global strategic business report. Available from World Wide Web: http://www.strategyr.com/pressMCP-2146.asp. Cited 22 June 2011.

68. Gonzalez-Pajuelo, M., J. C. Andrade & I. Vasconcelos, (2004) Production of 1,3-propanediol by Clostridium butyricum VPI 3266 using a synthetic medium and raw glycerol. J Ind Microbiol Biotechnol 31: 442-446.

69. Gonzalez-Pajuelo, M., I. Meynial-Salles, F. Mendes, J. C. Andrade, I. Vasconcelos & P. Soucaille, (2005) Metabolic engineering of Clostridium acetobutylicum for the industrial production of 1,3-propanediol from glycerol. Metab Eng 7: 329-336.

70. Gonzalez-Pajuelo, M., I. Meynial-Salles, F. Mendes, P. Soucaille & I. Vasconcelos, (2006) Microbial conversion of glycerol to 1,3-propanediol: physiological comparison of a natural producer, Clostridium butyricum VPI 3266, and an engineered strain, Clostridium acetobutylicum DG1(pSPD5). Appl Environ Microbiol 72: 96-101.

71. Gonzalez, R., A. Murarka, Y. Dharmadi & S. S. Yazdani, (2008) A new

model for the anaerobic fermentation of glycerol in enteric bacteria: trunk and auxiliary pathways in Escherichia coli. Metab Eng 10: 234-245.

72. Gopinath V, Meiswinkel TM, Wendisch VF, Nampoothiri KM (2011) Amino acid production from rice straw and wheat bran hydrolysates by recombinant pentose-utilizing Corynebacterium glutamicum. Appl Microbiol Biotechnol. doi:10.1007/s00253-011- 3478-x

73. Gosset, G., J. Yong-Xiao & A. Berry, (1996) A direct comparison of approaches for increasing carbon flow to aromatic biosynthesis in Escherichia coli. J Ind Microbiol 17: 47-52.

74. Greenfield, S. & G. W. Claus, (1972) Nonfunctional tricarboxylic acid cycle and the mechanism of glutamate biosynthesis in Acetobacter suboxydans. J Bacteriol 112: 1295-1301.

75. Greenwood, A., (2010) DuPont, Tate & Lyle JV expand US propandiol plant. Available from World Wide Web: http://www.icis.com/Articles/2010/05/04/9356156/duponttate-lyle-jv-expand-us-propandiol-plant.html. Cited 22 June 2011.

76. Grover, B. P., S. K. Garg & J. Verma, (1990) Production of 2,3-Butanediol from Wood Hydrolysate by Klebsiella-Pneumoniae. World J Microb Biot 6: 328-332.

77. Gulevich, A., I. Biryukova, D. Zimenkov, A. Skorokhodova, A. Kivero, A. Belareva & S. Mashko, (2004) Method for producing l-amino acid using bacterium having enhanced expression of pckA gene. World Patent No.WO 090125.

78. Gungormusler, M., C. Gonen & N. Azbar, (2011) Continuous production of 1,3-propanediol using raw glycerol with immobilized Clostridium beijerinckii NRRL B-593 in comparison to suspended culture. Bioprocess Biosyst Eng.

79. Habe, H., T. Fukuoka, D. Kitamoto & K. Sakaki, (2009a) Biotechnological production of Dglyceric acid and its application. Appl Microbiol Biotechnol 84: 445-452.

80. Habe, H., T. Fukuoka, D. Kitamoto & K. Sakaki, (2009b) Biotransformation of glycerol to Dglyceric acid by Acetobacter tropicalis. Appl Microbiol Biotechnol 81: 1033-1039.

81. Habe, H., T. Fukuoka, T. Morita, D. Kitamoto, T. Yakushi, K. Matsushita & K. Sakaki, (2010a) Disruption of the membrane-bound alcohol dehydrogenase-encoding gene improved glycerol use and dihydroxyacetone productivity in Gluconobacter oxydans. Biosci Biotechnol Biochem 74: 1391-1395.

82. Habe, H., Y. Shimada, T. Fukuoka, D. Kitamoto, M. Itagaki, K. Watanabe, H. Yanagishita & K. Sakaki, (2009c) Production of glyceric acid by Gluconobacter sp. NBRC3259 using raw glycerol. Biosci Biotechnol Biochem 73: 1799-1805.

83. Habe, H., Y. Shimada, T. Fukuoka, D. Kitamoto, M. Itagaki, K. Watanabe, H. Yanagishita, T. Yakushi, K. Matsushita & K. Sakaki, (2010b) Use of a Gluconobacter frateurii mutant to prevent dihydroxyacetone accumulation during glyceric acid production from glycerol. Biosci Biotechnol Biochem 74: 2330-2332.

84. Habe, H., Y. Shimada, T. Yakushi, H. Hattori, Y. Ano, T. Fukuoka, D. Kitamoto, M. Itagaki, K. Watanabe, H. Yanagishita, and others, (2009d) Microbial production of glyceric acid, an organic acid that can be mass produced from glycerol. Appl Environ Microbiol 75: 7760-7766.

85. Hacking, A. J. & E. C. Lin, (1976) Disruption of the fucose pathway as a consequence of genetic adaptation to propanediol as a carbon source in Escherichia coli. J Bacteriol 126: 1166-1172.

86. Hao, J., R. Lin, Z. Zheng, Y. Sun & D. Liu, (2008a) 3-Hydroxypropionaldehyde guided glycerol feeding strategy in aerobic 1,3-propanediol production by Klebsiella pneumoniae. J Ind Microbiol Biotechnol 35: 1615-1624.

87. Hao, J., W. Wang, J. Tian, J. Li & D. Liu, (2008b) Decrease of 3-hydroxypropionaldehyde accumulation in 1,3-propanediol production by over-expressing dhaT gene in Klebsiella pneumoniae TUAC01. J Ind Microbiol Biotechnol 35: 735-741.

88. Hekmat, D., R. Bauer & J. Fricke, (2003) Optimization of the microbial synthesis of dihydroxyacetone from glycerol with Gluconobacter oxydans. Bioprocess Biosyst Eng 26: 109-116.

89. Heller, K. B., E. C. Lin & T. H. Wilson, (1980) Substrate specificity and transport properties of the glycerol facilitator of Escherichia coli. J Bacteriol 144: 274-278.

90. Herrmann, U., C. Gatgens, U. Degner & S. Bringer-Meyer, (2007) Biotransformation of glycerol to dihydroxyacetone by recombinant Gluconobacter oxydans DSM 2343. Appl Microbiol Biot 76: 553-559.

91. Herry, D. M. & L. K. Dunican, (1993) Cloning of the trp gene cluster from a tryptophanhyperproducing strain of Corynebacterium glutamicum: identification of a mutation in the trp leader sequence. Appl Environ Microbiol 59: 791-799.

92. Himmi, E. H., A. Bories, A. Boussaid & L. Hassani, (2000) Propionic acid fermentation of glycerol and glucose by Propionibacterium

acidipropionici and Propionibacterium freudenreichii ssp. shermanii. Appl Microbiol Biotechnol 53: 435-440.

93. Hiremath, A., M. Kannabiran & V. Rangaswamy, (2011) 1,3-Propanediol production from crude glycerol from Jatropha biodiesel process. N Biotechnol 28: 19-23.

94. Holst, O., H. Lundback & B. Mattiasson, (1985) Hydrogen-Peroxide as an Oxygen Source for Immobilized Gluconobacter oxydans Converting Glycerol to Dihydroxyacetone. Appl Microbiol Biot 22: 383-388.

95. Homann, T., C. Tag, H. Biebl, W. D. Deckwer & B. Schink, (1990) Fermentation of Glycerol to 1,3-Propanediol by Klebsiella and Citrobacter Strains. Appl Microbiol Biot 33: 121-126.

96. Hong, W. K., C. H. Kim, S. Y. Heo, L. H. Luo, B. R. Oh & J. W. Seo, (2010) Enhanced production of ethanol from glycerol by engineered Hansenula polymorpha expressing pyruvate decarboxylase and aldehyde dehydrogenase genes from Zymomonas mobilis. Biotechnol Lett 32: 1077-1082.

97. Hopper, D. J. & R. A. Cooper, (1972) The purification and properties of Escherichia coli methylglyoxal synthase. Biochem J 128: 321-329.

98. Horng, Y. T., K. C. Chang, T. C. Chou, C. J. Yu, C. C. Chien, Y. H. Wei & P. C. Soo, (2010) Inactivation of dhaD and dhaK abolishes by-product accumulation during 1,3- propanediol production in Klebsiella pneumoniae. J Ind Microbiol Biotechnol 37: 707-716.

99. Hu, H. & T. K. Wood, (2010) An evolved Escherichia coli strain for producing hydrogen and ethanol from glycerol. Biochem Biophys Res Commun 391: 1033-1038.

100. Huang, K., F. B. Rudolph & G. N. Bennett, (1999) Characterization of methylglyoxal synthase from Clostridium acetobutylicum ATCC 824 and its use in the formation of 1, 2-propanediol. Appl Environ Microbiol 65: 3244-3247.

101. Ikeda, M. & R. Katsumata, (1999) Hyperproduction of tryptophan by Corynebacterium glutamicum with the modified pentose phosphate pathway. Appl Environ Microbiol 65: 2497-2502.

102. Ikeda, M., K. Okamoto & R. Katsumata, (1999) Cloning of the transketolase gene and the effect of its dosage on aromatic amino acid production in Corynebacterium glutamicum. Appl Microbiol Biotechnol 51: 201-206.

103. Ingram, L. O., K. Jantama, M. J. Haupt, S. A. Svoronos, X. L. Zhang, J. C. Moore & K. T. Shanmugam, (2008) Combining metabolic

engineering and metabolic evolution to develop nonrecombinant strains of Escherichia coli C that produce succinate and malate. Biotechnology and Bioengineering 99: 1140-1153.

104. Ito, T., Y. Nakashimada, K. Senba, T. Matsui & N. Nishio, (2005) Hydrogen and ethanol production from glycerol-containing wastes discharged after biodiesel manufacturing process. Journal of Bioscience and Bioengineering 100: 260-265.

105. Jansen, N. B., M. C. Flickinger & G. T. Tsao, (1984) Production of 2,3-butanediol from Dxylose by Klebsiella oxytoca ATCC 8724. Biotechnol Bioeng 26: 362-369.

106. Ji, X. J., H. Huang, S. Li, J. Du & M. Lian, (2008) Enhanced 2,3-butanediol production by altering the mixed acid fermentation pathway in Klebsiella oxytoca. Biotechnol Lett 30: 731-734.

107. Ji, X. J., H. Huang, J. G. Zhu, L. J. Ren, Z. K. Nie, J. Du & S. Li, (2010) Engineering Klebsiella oxytoca for efficient 2, 3-butanediol production through insertional inactivation of acetaldehyde dehydrogenase gene. Appl Microbiol Biotechnol 85: 1751-1758.

108. Johnson, E. A. & E. C. Lin, (1987) Klebsiella pneumoniae 1,3-propanediol:NAD+ oxidoreductase. J Bacteriol 169: 2050-2054.

109. Jung, J. Y., E. S. Choi & M. K. Oh, (2008) Enhanced production of 1,2-propanediol by tpi1 deletion in Saccharomyces cerevisiae. J Microbiol Biotechnol 18: 1797-1802.

110. Juni, E., (1952) Mechanisms of Formation of Acetoin by Bacteria. Journal of Biological Chemistry 195: 715-726.

111. Kajiura, H., K. Mori, T. Tobimatsu & T. Toraya, (2001) Characterization and mechanism of action of a reactivating factor for adenosylcobalamin-dependent glycerol dehydratase. Journal of Biological Chemistry 276: 36514-36519.

112. Kalinowski, J., J. Cremer, B. Bachmann, L. Eggeling, H. Sahm & A. Puhler, (1991) Genetic and biochemical analysis of the aspartokinase from Corynebacterium glutamicum. Mol Microbiol 5: 1197-1204.

113. Kamzolova, S. V., A. R. Fatykhova, E. G. Dedyukhina, S. G. Anastassiadis, N. P. Golovchenko & I. G. Morgunov, (2011) Citric Acid Production by Yeast Grown on Glycerol-Containing Waste from Biodiesel Industry. Food Technol Biotech 49: 65-74.

114. Katsumata, R. & K. Kino, (1989) Process for producing amino acids by fermentation. Japan Patent 01317395 A (P2578488).

115. Khamduang, M., K. Packdibamrung, J. Chutmanop, Y. Chisti & P.

Srinophakun, (2009) Production of L-phenylalanine from glycerol by a recombinant Escherichia coli. J Ind Microbiol Biotechnol 36: 1267-1274.

116. Knietsch, A., S. Bowien, G. Whited, G. Gottschalk & R. Daniel, (2003) Identification and characterization of coenzyme B12-dependent glycerol dehydratase- and diol dehydratase-encoding genes from metagenomic DNA libraries derived from enrichment cultures. Appl Environ Microbiol 69: 3048-3060.

117. Kretschmann, J., F.-J. Carduck, W.-D. Deckwer, C. Tag & B. H, (1993) Fermentive production of 1,3-propanediol. U.S. Patent No. US 5254467.

118. Lages, F. & C. Lucas, (1995) Characterization of a Glycerol H+ Symport in the Halotolerant Yeast Pichia sorbitophila. Yeast 11: 111-119.

119. Lages, F. & C. Lucas, (1997) Contribution to the physiological characterization of glycerol active uptake in Saccharomyces cerevisiae. Biochim Biophys Acta 1322: 8-18.

120. Lang, S., W. Beil, H. Tokuda, C. Wicke & V. Lurtz, (2004) Improved production of bioactive glucosylmannosyl-glycerolipid by sponge-associated Microbacterium species. Mar Biotechnol (NY) 6: 152-156.

121. Lapujade, P., J. L. Goergen & J. M. Engasser, (1999) Glutamate excretion as a major kinetic bottleneck for the thermally triggered production of glutamic acid by Corynebacterium glutamicum. Metab Eng 1: 255-261.

122. Larsson, K., R. Ansell, P. Eriksson & L. Adler, (1993) A gene encoding sn-glycerol 3- phosphate dehydrogenase (NAD+) complements an osmosensitive mutant of Saccharomyces cerevisiae. Mol Microbiol 10: 1101-1111.

123. Lee, P. C., W. G. Lee, S. Y. Lee & H. N. Chang, (2001) Succinic acid production with reduced by-product formation in the fermentation of Anaerobiospirillum succiniciproducens using glycerol as a carbon source. Biotechnol Bioeng 72: 41-48.

124. Lee, S. Y., S. H. Hong, S. H. Lee & S. J. Park, (2004) Fermentative production of chemicals that can be used for polymer synthesis. Macromol Biosci 4: 157-164.

125. Lee, W. & N. A. Dasilva, (2006) Application of sequential integration for metabolic engineering of 1,2-propanediol production in yeast. Metab Eng 8: 58-65.

126. Levy, S. B., (1992) Dihydroxyacetone-Containing Sunless or Self-Tanning Lotions. J Am Acad Dermatol 27: 989-993.

127. Li, M., J. Wu, J. Lin & D. Wei, (2010a) Expression of Vitreoscilla hemoglobin enhances cell growth and dihydroxyacetone production in

Gluconobacter oxydans. Curr Microbiol 61: 370-375.

128. Li, M. H., J. Wu, X. Liu, J. P. Lin, D. Z. Wei & H. Chen, (2010b) Enhanced production of dihydroxyacetone from glycerol by overexpression of glycerol dehydrogenase in an alcohol dehydrogenase-deficient mutant of Gluconobacter oxydans. Bioresour Technol 101: 8294-8299.

129. Liao, J., (1996) Microorganisms and methods for overproduction of DAHP by cloned pps gene. World Patent No. WO 9608567.

130. Licht, F. O., (2010) World Ethanol and Biofuels Report. 8.

131. Lin, H., G. N. Bennett & K. Y. San, (2005) Fed-batch culture of a metabolically engineered Escherichia coli strain designed for high-level succinate production and yield under aerobic conditions. Biotechnology and Bioengineering 90: 775-779.

132. Lindner, S. N., H. Niederholtmeyer, K. Schmitz, S. M. Schoberth & V. F. Wendisch, (2010) Polyphosphate/ATP-dependent NAD kinase of Corynebacterium glutamicum: biochemical properties and impact of ppnK overexpression on lysine production. Appl Microbiol Biotechnol 87: 583-593.

133. Lindner, S. N., G. M. Seibold, A. Henrich, R. Kramer & V. F. Wendisch, (2011) Phosphotransferase System-Independent Glucose Utilization in Corynebacterium glutamicum by Inositol Permeases and Glucokinases. Appl Environ Microbiol 77: 3571-3581.

134. Lindner SN, Seibold GM, Kramer R, Wendisch VF (2011) Impact of a new glucose utilization pathway in amino acid-producing Corynebacterium glutamicum. Bio engineered Bugs 2 (5)

135. Liu, H., Y. Xu, Z. Zheng & D. Liu, (2010) 1,3-Propanediol and its copolymers: research, development and industrialization. Biotechnol J 5: 1137-1148.

136. Liu, Y., C. M. Koh & L. Ji, (2011) Bioconversion of crude glycerol to glycolipids in Ustilago maydis. Bioresour Technol 102: 3927-3933.

137. Lu, J. L. & J. C. Liao, (1997) Metabolic engineering and control analysis for production of aromatics: Role of transaldolase. Biotechnol Bioeng 53: 132-138.

138. Lucas, C., M. Dacosta & N. Vanuden, (1990) Osmoregulatory Active Sodium-Glycerol Cotransport in the Halotolerant Yeast Debaryomyces-Hansenii. Yeast 6: 187-191.

139. Ma, C. Q., A. L. Wang, Y. Wang, T. Y. Jiang, L. X. Li & P. Xu, (2010a) Production of 2,3- butanediol from corncob molasses, a waste by-product in xylitol production. Appl Microbiol Biot 87: 965-970.

140. Ma, X. Y., L. Zhao, Y. Zheng & D. Z. Wei, (2009) Effects of over-expression of glycerol dehydrogenase and 1,3-propanediol oxidoreductase on bioconversion of glycerol into 1,3-propandediol by Klebsiella pneumoniae under micro-aerobic conditions. Bioproc Biosyst Eng 32: 313-320.

141. Ma, Z., Z. Rao, B. Zhuge, H. Fang, X. Liao & J. Zhuge, (2010b) Construction of a novel expression system in Klebsiella pneumoniae and its application for 1,3-propanediol production. Appl Biochem Biotechnol 162: 399-407.

142. Marx, A., A. A. de Graaf, W. Wiechert, L. Eggeling & H. Sahm, (1996) Determination of the fluxes in the central metabolism of Corynebacterium glutamicum by nuclear magnetic resonance spetroscopy combined with metabolite balancing. Biotechnol Bioeng 49: 111-129.

143. Marx, A., S. Hans, B. Mockel, B. Bathe & A. A. de Graaf, (2003) Metabolic phenotype of phosphoglucose isomerase mutants of Corynebacterium glutamicum. J Biotechnol 104: 185-197.

144. Mascarenhas, D., D. J. Ashworth & C. S. Chen, (1991) Deletion of pgi alters tryptophan biosynthesis in a genetically engineered strain of Escherichia coli. Appl Environ Microbiol 57: 2995-2999.

145. Matsushita, K., H. Toyama & O. Adachi, (1994) Respiratory chains and bioenergetics of acetic acid bacteria. Adv Microb Physiol 36: 247-301.

146. Menzel, K., A. P. Zeng & W. D. Deckwer, (1997) High concentration and productivity of 1,3- propanediol from continuous fermentation of glycerol by Klebsiella pneumoniae. Enzyme Microb Tech 20: 82-86.

147. Miller, J. E., K. C. Backman, M. J. Oconnor & R. T. Hatch, (1987) Production of Phenylalanine and Organic-Acids by Phosphoenolpyruvate Carboxylase-Deficient Mutants of Escherichia coli. J Ind Microbiol 2: 143-149.

148. Misra, K., A. B. Banerjee, S. Ray & M. Ray, (1996) Reduction of methylglyoxal in Escherichia coli K12 by an aldehyde reductase and alcohol dehydrogenase. Mol Cell Biochem 156: 117-124.

149. Moon, C., J. H. Ahn, S. W. Kim, B. I. Sang & Y. Um, (2010) Effect of biodiesel-derived raw glycerol on 1,3-propanediol production by different microorganisms. Appl Biochem Biotechnol 161: 502-510.

150. Morita, T., M. Konishi, T. Fukuoka, T. Imura & D. Kitamoto, (2007) Microbial conversion of glycerol into glycolipid biosurfactants, mannosylerythritol lipids, by a basidiomycete yeast, Pseudozyma antarctica JCM 10317(T). J Biosci Bioeng 104: 78-81.

151. Mu, Y., H. Teng, D. J. Zhang, W. Wang & Z. L. Xiu, (2006) Microbial production of 1,3- propanediol by Klebsiella pneumoniae using crude glycerol from biodiesel preparations. Biotechnol Lett 28: 1755-1759.
152. Mulligan, C. N., (2005) Environmental applications for biosurfactants. Environ Pollut 133: 183-198.
153. Murata, K., Y. Fukuda, K. Watanabe, T. Saikusa, M. Shimosaka & A. Kimura, (1985) Characterization of methylglyoxal synthase in Saccharomyces cerevisiae. Biochem Biophys Res Commun 131: 190-198.
154. Nabe, K., N. Izuo, S. Yamada & I. Chibata, (1979) Conversion of Glycerol to Dihydroxyacetone by Immobilized Whole Cells of Acetobacter xylinum. Appl Environ Microbiol 38: 1056-1060.
155. Nakamura, K., S. Kondo, Y. Kawai, N. Nakajima & A. Ohno, (1997) Amino acid sequence and characterization of aldo-keto reductase from bakers' yeast. Biosci Biotechnol Biochem 61: 375-377.
156. Nakas, J. P., M. Schaedle, C. M. Parkinson, C. E. Coonley & S. W. Tanenbaum, (1983) System development for linked-fermentation production of solvents from algal biomass. Appl Environ Microbiol 46: 1017-1023.
157. Nara, T., S. Kinoshita & H. Samejima, (1964) Effect of Penicillin on Amino Acid Fermentation. Agr Biol Chem Tokyo 28: 120-124.
158. Navarro-Avino, J. P., R. Prasad, V. J. Miralles, R. M. Benito & R. Serrano, (1999) A proposal for nomenclature of aldehyde dehydrogenases in Saccharomyces cerevisiae and characterization of the stress-inducible ALD2 and ALD3 genes. Yeast 15: 829-842.
159. Niimi, S., N. Suzuki, M. Inui & H. Yukawa, (2011) Metabolic engineering of 1,2-propanediol pathways in Corynebacterium glutamicum. Appl Microbiol Biotechnol 90: 1721-1729.
160. Nilegaonkar, S. S., S. B. Bhosale, C. N. Dandage & A. H. Kapadi, (1996) Potential of Bacillus licheniformis for the production of 2,3-butanediol. J Ferment Bioeng 82: 408-410.
161. Norbeck, J. & A. Blomberg, (1997) Metabolic and regulatory changes associated with growth of Saccharomyces cerevisiae in 1.4 M NaCl. Evidence for osmotic induction of glycerol dissimilation via the dihydroxyacetone pathway. J Biol Chem 272: 5544-5554.
162. O'Gara, J. P. & L. K. Dunican, (1995) Mutations in the trpD gene of Corynebacterium glutamicum confer 5-methyltryptophan resistance by encoding a feedback-resistant anthranilate phosphoribosyltransferase.

Appl Environ Microbiol 61: 4477-4479.

163. Ochoa-Estopier, A., J. Lesage, N. Gorret & S. E. Guillouet, (2011) Kinetic analysis of a Saccharomyces cerevisiae strain adapted for improved growth on glycerol: Implications for the development of yeast bioprocesses on glycerol. Bioresour Technol 102: 1521-1527.

164. Oh, B. R., J. W. Seo, S. Y. Heo, W. K. Hong, L. H. Luo, M. H. Joe, D. H. Park & C. H. Kim, (2011) Efficient production of ethanol from crude glycerol by a Klebsiella pneumoniae mutant strain. Bioresour Technol 102: 3918-3922.

165. Ohnishi, J., R. Katahira, S. Mitsuhashi, S. Kakita & M. Ikeda, (2005) A novel gnd mutation leading to increased L-lysine production in Corynebacterium glutamicum. FEMS Microbiol Lett 242: 265-274.

166. Okino, S., M. Inui & H. Yukawa, (2005) Production of organic acids by Corynebacterium glutamicum under oxygen deprivation. Appl Microbiol Biotechnol 68: 475-480.

167. Overkamp, K. M., B. M. Bakker, P. Kotter, M. A. Luttik, J. P. Van Dijken & J. T. Pronk, (2002) Metabolic engineering of glycerol production in Saccharomyces cerevisiae. Appl Environ Microbiol 68: 2814-2821.

168. Pacwa-Plociniczak, M., G. A. Plaza, Z. Piotrowska-Seget & S. S. Cameotra, (2011) Environmental applications of biosurfactants: recent advances. Int J Mol Sci 12: 633-654.

169. Pagliaro, M., R. Ciriminna, H. Kimura, M. Rossi & C. Della Pina, (2007) From glycerol to value-added products. Angew Chem Int Ed Engl 46: 4434-4440.

170. Pahlman, A. K., K. Granath, R. Ansell, S. Hohmann & L. Adler, (2001) The yeast glycerol 3- phosphatases Gpp1p and Gpp2p are required for glycerol biosynthesis and differentially involved in the cellular responses to osmotic, anaerobic, and oxidative stress. J Biol Chem 276: 3555-3563.

171. Papanikolaou, S., M. Fick & G. Aggelis, (2004) The effect of raw glycerol concentration on the production of 1,3-propanediol by Clostridium butyricum. J Chem Technol Biot 79: 1189-1196.

172. Papanikolaou, S., L. Muniglia, I. Chevalot, G. Aggelis & I. Marc, (2002) Yarrowia lipolytica as a potential producer of citric acid from raw glycerol. J Appl Microbiol 92: 737-744.

173. Papanikolaou, S., P. Ruiz-Sanchez, B. Pariset, F. Blanchard & M. Fick, (2000) High production of 1,3-propanediol from industrial glycerol by a newly isolated Clostridium butyricum strain. J Biotechnol 77: 191-208.

174. Pasteris, S. E. & A. M. Strasser de Saad, (2008) Transport of glycerol by

Pediococcus pentosaceus isolated from wine. Food Microbiol 25: 545-549.

175. Patnaik, R., R. G. Spitzer & J. C. Liao, (1995) Pathway engineering for production of aromatics in Escherichia coli: Confirmation of stoichiometric analysis by independent modulation of AroG, TktA, and Pps activities. Biotechnol Bioeng 46: 361-370.

176. Perego, P., A. Converti, A. Del Borghi & P. Canepa, (2000) 2,3-butanediol production by Enterobacter aerogenes: selection of the optimal conditions and application to food industry residues. Bioprocess Eng 23: 613-620.

177. Perego, P., A. Converti & M. Del Borghi, (2003) Effects of temperature, inoculum size and starch hydrolyzate concentration on butanediol production by Bacillus licheniformis. Bioresour Technol 89: 125-131.

178. Peters-Wendisch, P. G., C. Kreutzer, J. Kalinowski, M. Patek, H. Sahm & B. J. Eikmanns, (1998) Pyruvate carboxylase from Corynebacterium glutamicum: characterization, expression and inactivation of the pyc gene. Microbiology 144: 915-927.

179. Peters-Wendisch, P. G., B. Schiel, V. F. Wendisch, E. Katsoulidis, B. Mockel, H. Sahm & B. J. Eikmanns, (2001) Pyruvate carboxylase is a major bottleneck for glutamate and lysine production by Corynebacterium glutamicum. J Mol Microbiol Biotechnol 3: 295-300.

180. Petrov, K. & P. Petrova, (2009) High production of 2,3-butanediol from glycerol by Klebsiella pneumoniae G31. Appl Microbiol Biotechnol 84: 659-665.

181. Petrov, K. & P. Petrova, (2010) Enhanced production of 2,3-butanediol from glycerol by forced pH fluctuations. Appl Microbiol Biotechnol 87: 943-949.

182. Petrovska, B., E. Winkelhausen & S. Kuzmanova, (1999) Glycerol production by yeasts under osmotic and sulfite stress. Can J Microbiol 45: 695-699.

183. Pflugmacher, U. & G. Gottschalk, (1994) Development of an Immobilized Cell Reactor for the Production of 1,3-Propanediol by Citrobacter-Freundii. Appl Microbiol Biot 41: 313-316.

184. Prust, C., M. Hoffmeister, H. Liesegang, A. Wiezer, W. F. Fricke, A. Ehrenreich, G. Gottschalk & U. Deppenmeier, (2005) Complete genome sequence of the acetic acid bacterium Gluconobacter oxydans. Nat Biotechnol 23: 195-200.

185. Radmacher, E., K. C. Stansen, G. S. Besra, L. J. Alderwick, W. N. Maughan, G. Hollweg, H. Sahm, V. F. Wendisch & L. Eggeling, (2005)

Ethambutol, a cell wall inhibitor of Mycobacterium tuberculosis, elicits L-glutamate efflux of Corynebacterium glutamicum. Microbiology 151: 1359-1368.

186. Rao, Z. M., Z. Ma, B. Zhuge, H. Y. Fang, X. R. Liao & J. Zhuge, (2010) Construction of a Novel Expression System in Klebsiella pneumoniae and its Application for 1,3- Propanediol Production. Appl Biochem Biotech 162: 399-407.

187. Riedel, C., D. Rittmann, P. Dangel, B. Mockel, S. Petersen, H. Sahm & B. J. Eikmanns, (2001) Characterization of the phosphoenolpyruvate carboxykinase gene from Corynebacterium glutamicum and significance of the enzyme for growth and amino acid production. J Mol Microbiol Biotechnol 3: 573-583.

188. Rittmann, D., S. N. Lindner & V. F. Wendisch, (2008) Engineering of a glycerol utilization pathway for amino acid production by Corynebacterium glutamicum. Appl Environ Microbiol 74: 6216-6222.

189. Ruch, F. E., J. Lengeler & E. C. Lin, (1974) Regulation of glycerol catabolism in Klebsiella aerogenes. J Bacteriol 119: 50-56.

190. Rymowicz, W., A. R. Fatykhova, S. V. Kamzolova, A. Rywinska & I. G. Morgunov, (2010) Citric acid production from glycerol-containing waste of biodiesel industry by Yarrowia lipolytica in batch, repeated batch, and cell recycle regimes. Appl Microbiol Biotechnol 87: 971-979.

191. Rywinska, A., P. Juszczyk, M. Wojtatowicz & W. Rymowicz, (2011) Chemostat study of citric acid production from glycerol by Yarrowia lipolytica. J Biotechnol 152: 54-57.

192. Rywinska, A. & W. Rymowicz, (2010) High-yield production of citric acid by Yarrowia lipolytica on glycerol in repeated-batch bioreactors. J Ind Microbiol Biotechnol 37: 431-435.

193. Saha, B. C. & R. J. Bothast, (1999) Production of 2,3-butanediol by newly isolated Enterobacter cloacae. Appl Microbiol Biotechnol 52: 321-326.

194. Saintamans, S., P. Perlot, G. Goma & P. Soucaille, (1994) High Production of 1,3-Propanediol from Glycerol by Clostridium butyricum Vpi-3266 in a Simply Controlled Fed-Batch System. Biotechnol Lett 16: 831-836.

195. Samuelov, N. S., R. Lamed, S. Lowe & J. G. Zeikus, (1991) Influence of CO2-HCO3- Levels and Ph on Growth, Succinate Production, and Enzyme-Activities of Anaerobiospirillum succiniciproducens. Appl Environ Microb 57: 3013-3019.

196. Sarcabal, P., C. Croux & P. Soucaille, (2007) Method for preparing 1,3-propanediol by a recombinant micro-organism in the absence of

coenzyme b12 or one of its precursors. U.S. Patent No. US 7267972

197. Sato, H., K. Orishimo, T. Shirai, T. Hirasawa, K. Nagahisa, H. Shimizu & M. Wachi, (2008) Distinct roles of two anaplerotic pathways in glutamate production induced by biotin limitation in Corynebacterium glutamicum. J Biosci Bioeng 106: 51-58.

198. Schneider, J. & V. F. Wendisch, (2010) Putrescine production by engineered Corynebacterium glutamicum. Appl Microbiol Biotechnol. 88(4):859–868

199. Schneider J, Niermann K, Wendisch VF (2011) Production of the amino acids l-glutamate, llysine, l-ornithine and l-arginine from arabinose by recombinant Corynebacterium glutamicum. J Biotechnol 154 (2-3):191-198.

200. Schneider, Z. & J. Pawelkiewicz, (1966) The properties of glycerol dehydratase isolated from Aerobacter aerogenes, and the properties of the apoenzyme subunits. Acta Biochim Pol 13: 311-328.

201. Scholten, E., T. Renz & J. Thomas, (2009) Continuous cultivation approach for fermentative succinic acid production from crude glycerol by Basfia succiniciproducens DD1. Biotechnol Lett 31: 1947-1951.

202. Schutz, H. & F. Radler, (1984) Anaerobic Reduction of Glycerol to Propanediol-1.3 by Lactobacillus brevis and Lactobacillus buchneri. Syst Appl Microbiol 5: 169-178.

203. Seifert, C., S. Bowien, G. Gottschalk & R. Daniel, (2001) Identification and expression of the genes and purification and characterization of the gene products involved in reactivation of coenzyme B-12-dependent glycerol dehydratase of Citrobacter freundii. European Journal of Biochemistry 268: 2369-2378.

204. Selembo, P. A., J. M. Perez, W. A. Lloyd & B. E. Logan, (2009) Enhanced hydrogen and 1,3- propanediol production from glycerol by fermentation using mixed cultures. Biotechnol Bioeng 104: 1098-1106.

205. Seo, J. W., M. Y. Seo, B. R. Oh, S. Y. Heo, J. O. Baek, D. Rairakhwada, L. H. Luo, W. K. Hong & C. H. Kim, (2010) Identification and utilization of a 1,3-propanediol oxidoreductase isoenzyme for production of 1,3-propanediol from glycerol in Klebsiella pneumoniae. Appl Microbiol Biotechnol 85: 659-666.

206. Seo, M. Y., J. W. Seo, S. Y. Heo, J. O. Baek, D. Rairakhwada, B. R. Oh, P. S. Seo, M. H. Choi & C. H. Kim, (2009) Elimination of by-product formation during production of 1,3- propanediol in Klebsiella pneumoniae by inactivation of glycerol oxidative pathway. Appl Microbiol Biotechnol 84: 527-534.

207. Shams Yazdani, S. & R. Gonzalez, (2008) Engineering Escherichia coli for the efficient conversion of glycerol to ethanol and co-products. Metab Eng 10: 340-351.
208. Shelley, S., (2007) A renewable route to propylene glycol. Chem Eng Prog 103: 6-9.
209. Shih, I. L., Y. T. Van & M. H. Shen, (2004) Biomedical applications of chemically and microbiologically synthesized poly(glutamic acid) and poly(lysine). Mini Rev Med Chem 4: 179-188.
210. Shiio, I., S. I. Otsuka & M. Takahashi, (1962) Effect of biotin on the bacterial formation of glutamic acid. I. Glutamate formation and cellular premeability of amino acids. J Biochem 51: 56-62.
211. Shima, S., H. Matsuoka, T. Iwamoto & H. Sakai, (1984) Antimicrobial Action of EpsilonPoly-L-Lysine. J Antibiot 37: 1449-1455.
212. Shimizu, H. & T. Hirasawa, (2007) Production of Glutamate and Glutamate-Related Amino Acids: MolecularMechanism Analysis and Metabolic Engineering. In Amino Acid Biosynthesis – Pathways, Regulation and Metabolic Engineering (Wendisch V.F., ed), Springer, Heidelberg, Germany: DOI 10.1007/7171_2006_1064.
213. Silva, S. N., C. B. Farias, R. D. Rufino, J. M. Luna & L. A. Sarubbo, (2010) Glycerol as substrate for the production of biosurfactant by Pseudomonas aeruginosa UCP0992. Colloids Surf B Biointerfaces 79: 174-183.
214. Singh, A., M. D. Lynch & R. T. Gill, (2009) Genes restoring redox balance in fermentationdeficient E. coli NZN111. Metab Eng 11: 347-354.
215. Skraly, F. A., B. L. Lytle & D. C. Cameron, (1998) Construction and characterization of a 1,3- propanediol operon. Appl Environ Microbiol 64: 98-105.
216. Soccol, C. R., L. P. S. Vandenberghe, C. Rodrigues & A. Pandey, (2006) New perspectives for citric acid production and application. Food Technol Biotech 44: 141-149.
217. Soucaille, P., (2008) Process for the biological production of 1, 3-propanediol from glycerol with high yield. World Patent No. WO08052595
218. Sprenger, G., (2007) Aromatic Amino Acids In Amino Acid Biosynthesis - Pathways, Regulation and Metabolic Engineering (Wendisch, VF., ed), Springer, Berlin, Germany, pp. 93-128.
219. Stansen, C., D. Uy, S. Delaunay, L. Eggeling, J. L. Goergen & V. F. Wendisch, (2005) Characterization of a Corynebacterium glutamicum

lactate utilization operon induced during temperature-triggered glutamate production. Appl Environ Microbiol 71: 5920-5928.
220. Suzuki, T. & H. Onishi, (1968) Aerobic Dissimilation of L-Rhamnose and Production of LRhamnonic Acid and 1,2-Propanediol by Yeasts. Agr Biol Chem Tokyo 32: 888-&.
221. Svitel, J. & E. Sturdik, (1994) Product Yield and by-Product Formation in Glycerol Conversion to Dihydroxyacetone by Gluconobacter oxydans. J Ferment Bioeng 78: 351-355.
222. Takinami, K., H. Yoshii, H. Tsuri & H. Okada, (1965) Biochemical Effects of Fatty Acid and Its Derivatives on L-Glutamic Acid Fermentation .3. Biotin-Tween 60 Relationship in Accumulation of L-Glutamic Acid and Growth of Brevibacterium lactofermentum. Agr Biol Chem Tokyo 29: 351-&.
223. Tan, T. W., F. H. Wang, H. J. Qu, D. W. Zhang & P. F. Tian, (2007) Production of 1,3- propanediol from glycerol by recombinant E. coli using incompatible plasmids system. Mol Biotechnol 37: 112-119.
224. Tang, X., Y. Tan, H. Zhu, K. Zhao & W. Shen, (2009) Microbial conversion of glycerol to 1,3- propanediol by an engineered strain of Escherichia coli. Appl Environ Microbiol 75: 1628-1634.
225. Tkac, J., M. Navratil, E. Sturdik & P. Gemeiner, (2001) Monitoring of dihydroxyacetone production during oxidation of glycerol by immobilized Gluconobacter oxydans cells with an enzyme biosensor. Enzyme Microb Tech 28: 383-388.
226. Tobimatsu, T., H. Kajiura & T. Toraya, (2000) Specificities of reactivating factors for adenosylcobalamin-dependent diol dehydratase and glycerol dehydratase. Archives of Microbiology 174: 81-88.
227. Tobimatsu, T., H. Kajiura, M. Yunoki, M. Azuma & T. Toraya, (1999) Identification and expression of the genes encoding a reactivating factor for adenosylcobalamindependent glycerol dehydratase. J Bacteriol 181: 4110-4113.
228. Tong, I. T. & D. C. Cameron, (1992) Enhancement of 1,3-propanediol production by cofermentation in Escherichia coli expressing Klebsiella pneumoniae dha regulon genes. Appl Biochem Biotechnol 34-35: 149-159.
229. Tong, I. T., H. H. Liao & D. C. Cameron, (1991) 1,3-Propanediol production by Escherichia coli expressing genes from the Klebsiella pneumoniae dha regulon. Appl Environ Microbiol 57: 3541-3546.
230. Toraya, T., S. Honda, S. Kuno & S. Fukui, (1978) Coenzyme B12-

dependent diol dehydratase: regulation of apoenzyme synthesis in Klebsiella pneumoniae (Aerobacter aerogenes) ATCC 8724. J Bacteriol 135: 726-729.

231. Tribe, D. E. & J. Pittard, (1979) Hyperproduction of tryptophan by Escherichia coli: genetic manipulation of the pathways leading to tryptophan formation. Appl Environ Microbiol 38: 181-190.

232. Trinh, C. T. & F. Srienc, (2009) Metabolic engineering of Escherichia coli for efficient conversion of glycerol to ethanol. Appl Environ Microbiol 75: 6696-6705.

233. Turner, K. W. & A. M. Roberton, (1979) Xylose, arabinose, and rhamnose fermentation by Bacteroides ruminicola. Appl Environ Microbiol 38: 7-12.

234. Um, Y., S. A. Jun, C. Moon, C. H. Kang, S. W. Kong & B. I. Sang, (2010) Microbial Fed-batch Production of 1,3-Propanediol Using Raw Glycerol with Suspended and Immobilized Klebsiella pneumoniae. Appl Biochem Biotech 161: 491-501.

235. Van der Werf, M. J., M. V. Guettler, M. K. Jain & J. G. Zeikus, (1997) Environmental and physiological factors affecting the succinate product ratio during carbohydrate fermentation by Actinobacillus sp. 130Z. Arch Microbiol 167: 332-342.

236. Wang, Z. X., J. Zhuge, H. Fang & B. A. Prior, (2001) Glycerol production by microbial fermentation: a review. Biotechnol Adv 19: 201-223.

237. Wang F, Qu H, Zhang D, Tian P, Tan T (2007) Production of 1,3-propanediol from glycerol by recombinant E. coli using incompatible plasmids system. Mol Biotechnol 37 (2):112-119.

238. Wei, D. Z., Y. Zheng, H. Y. Zhang, L. Zhao, L. J. Wei & X. Y. Ma, (2008) One-step production of 2,3-butanediol from starch by secretory over-expression of amylase in Klebsiella pneumoniae. J Chem Technol Biot 83: 1409-1412.

239. Wei, S., Q. Song & D. Wei, (2007) Repeated use of immobilized Gluconobacter oxydans cells for conversion of glycerol to dihydroxyacetone. Prep Biochem Biotechnol 37: 67-76.

240. Wendisch, V. F., (2007) Amino Acid Biosynthesis – Pathways, Regulation and Metabolic Engineering. In: Microbiology Monographs. A. Steinbüchel (ed). Berlin: Springer, pp.

241. Wendisch, V. F., M. Bott & B. J. Eikmanns, (2006) Metabolic engineering of Escherichia coli and Corynebacterium glutamicum for biotechnological production of organic acids and amino acids. Curr Opin Microbiol 9: 268-274.

242. Werkman, C. H. & G. F. Gillen, (1932) Bacteria Producing Trimethylene Glycol. J Bacteriol 23: 167-182.

243. Wethmar, M. & W. D. Deckwer, (1999) Semisynthetic culture medium for growth and dihydroxyacetone production by Gluconobacter oxydans. Biotechnol Tech 13: 283-287.

244. Weusthuis, R. A., I. Lamot, J. van der Oost & J. P. Sanders, (2011) Microbial production of bulk chemicals: development of anaerobic processes. Trends Biotechnol 29: 153-158.

245. Whinfield, J. & J. Dickinson, (1946) Improvements relating to the manufacture of highly polymeric substances. UK Patent No. GB 578079

246. Wicke, C., M. Huners, V. Wray, M. Nimtz, U. Bilitewski & S. Lang, (2000) Production and structure elucidation of glycoglycerolipids from a marine sponge-associated Microbacterium species. J Nat Prod 63: 621-626.

247. Wu, J. Y., K. L. Yeh, W. B. Lu, C. L. Lin & J. S. Chang, (2008) Rhamnolipid production with indigenous Pseudomonas aeruginosa EM1 isolated from oil-contaminated site. Bioresour Technol 99: 1157-1164.

248. Xiaohu Fan, R. B. a. Y. Z., (2010) Glycerol (Byproduct of Biodiesel Production) as a Source for Fuels and Chemicals – Mini Review. The Open Fuels & Energy Science Journal 3: 17-22.

249. Xiu, Z. L., Y. H. Wang & H. Teng, (2011) Effect of aeration strategy on the metabolic flux of Klebsiella pneumoniae producing 1,3-propanediol in continuous cultures at different glycerol concentrations. J Ind Microbiol Biot 38: 705-715.

250. Xu, Y. Z., N. N. Guo, Z. M. Zheng, X. J. Ou, H. J. Liu & D. H. Liu, (2009) Metabolism in 1,3- propanediol fed-batch fermentation by a D-lactate deficient mutant of Klebsiella pneumoniae. Biotechnol Bioeng 104: 965-972.

251. Yang, G., J. Tian & J. Li, (2007) Fermentation of 1,3-propanediol by a lactate deficient mutant of Klebsiella oxytoca under microaerobic conditions. Appl Microbiol Biotechnol 73: 1017-1024.

252. Yu, K. O., S. W. Kim & S. O. Han, (2010) Engineering of glycerol utilization pathway for ethanol production by Saccharomyces cerevisiae. Bioresour Technol 101: 4157-4161.

253. Zeikus, J. G., M. K. Jain & P. Elankovan, (1999) Biotechnology of succinic acid production and markets for derived industrial products. Appl Microbiol Biot 51: 545-552.

254. Zeikus, J. G., J. B. McKinlay & C. Vieille, (2007) Prospects for a bio-based succinate industry. Appl Microbiol Biot 76: 727-740.
255. Zhang, X., K. Jantama, J. C. Moore, L. R. Jarboe, K. T. Shanmugam & L. O. Ingram, (2009) Metabolic evolution of energy-conserving pathways for succinate production in Escherichia coli. Proc Natl Acad Sci U S A 106: 20180-20185.
256. Zhang, X., K. T. Shanmugam & L. O. Ingram, (2010) Fermentation of glycerol to succinate by metabolically engineered strains of Escherichia coli. Appl Environ Microbiol 76: 2397-2401.
257. Zhang, Y., Y. Li, C. Du, M. Liu & Z. Cao, (2006) Inactivation of aldehyde dehydrogenase: a key factor for engineering 1,3-propanediol production by Klebsiella pneumoniae. Metab Eng 8: 578-586.
258. Zheng, Y., X. Chen & Y. Shen, (2008) Commodity chemicals derived from glycerol, an important biorefinery feedstock. Chem Rev 108: 5253-5277.
259. Zhu, M. M., P. D. Lawman & D. C. Cameron, (2002) Improving 1,3-propanediol production from glycerol in a metabolically engineered Escherichia coli by reducing accumulation of sn-glycerol-3-phosphate. Biotechnol Prog 18: 694-699.
260. Zhu, M. M., F. A. Skraly & D. C. Cameron, (2001) Accumulation of methylglyoxal in anaerobically grown Escherichia coli and its detoxification by expression of the Pseudomonas putida glyoxalase I gene. Metab Eng 3: 218-225.
261. Zhuge, B., C. Zhang, H. Y. Fang, J. A. Zhuge & K. Permaul, (2010) Expression of 1,3- propanediol oxidoreductase and its isoenzyme in Klebsiella pneumoniae for bioconversion of glycerol into 1,3-propanediol. Appl Microbiol Biot 87: 2177-2184.

Chapter 2

PHAGE DISPLAY AND SYNTHETIC PEPTIDES AS PROMISING BIOTECHNOLOGICAL TOOLS FOR THE SEROLOGICAL DIAGNOSIS OF LEPROSY

Silvana Maria Alban[1], Juliana Ferreira de Moura[2], Vanete Thomaz-Soccol[1], Samira Bührer Sékula[3], Larissa Magalhães Alvarenga[3], Marcelo Távora Mira[4], Carlos Chávez Olortegui[5], and João Carlos Minozzo[6]

[1]Department of Biotechnology and Bioprocess Engineering, Federal University of Parana, Curitiba, Parana, Brazil

[2]Basic Pathology Department, Federal University of Parana, Curitiba, Parana, Brazil

[3]Immunology Department, Tropical Pathology and Public Health Institute, Federal University of Goias, Goiania, Goias, Brazil

[4]Health Sciences Postgraduate Program, School of Medicine, Pontifical Catholic University of Parana, Curitiba, Parana, Brazil

[5]Department of Biochemistry and Immunology, Federal University of Minas Gerais, Belo Horizonte, Minas Gerais, Brazil

[6]Center for Production and Research of Immunobiological Products, Parana State Department of Health, Piraquara, Parana, Brazil

ABSTRACT

Background

The diagnosis of leprosy is primarily based on clinical manifestations, and there is no widely available laboratory test for the early detection of this disease, which is caused by *Mycobacterium leprae*. In fact, early detection and treatment are the key elements to the successful control of leprosy.

Methodology/Principal Findings

Peptide ligands for antibodies from leprosy patients were selected from phage-displayed peptide libraries. Three peptide sequences expressed by reactive phage clones were chemically synthesized. Serological assays that used synthetic peptides were evaluated using serum samples from leprosy patients, household contacts (HC) of leprosy patients, tuberculosis patients and endemic controls (EC). A pool of three peptides identified 73.9% (17/23) of multibacillary (MB) leprosy patients using an enzyme-linked immunosorbent assay (ELISA). These peptides also showed some seroreactivities to the HC and EC individuals. The peptides were not reactive to rabbit polyclonal antisera against the different environmental mycobacteria. The same peptides that were conjugated to the carrier protein bovine serum albumin (BSA) induced the production of antibodies in the mice. The anti-peptide antibodies that were used in the Western blotting analysis of *M. leprae* crude extracts revealed a single band of approximately 30 kDa in one-dimensional electrophoresis and four 30 kDa isoforms in the two-dimensional gel. The Western blotting data indicated that the three peptides are derived from the same bacterial protein.

Conclusions/Significance

These new antigens may be useful in the diagnosis of MB leprosy patients. Their potentials as diagnostic reagents must be more extensively evaluated in future studies using a large panel of positive and negative sera. Furthermore, other test approaches using peptides should be assessed to increase their sensitivity and specificity in detecting leprosy patients. We have revealed evidence in support of phage-displayed peptides as promising biotechnological tools for the design of leprosy diagnostic serological assays.

INTRODUCTION

Leprosy, which is caused by *Mycobacterium leprae* [1], remains a currently relevant disease. As of 2012, cases were reported in 115 countries with the vast majority being concentrated in India, Brazil, and Indonesia [2]. Despite the reduction in the global prevalence rate from 5.4 million in 1985 to 181,941 at the beginning of 2012, the number of new cases that are detected has remained stable over recent years [2], indicating the continuity of its transmission [3].

The clinical manifestations are determined by the immune response of the patient to *M. leprae*. Individuals with lepromatous leprosy present with high bacterial loads and exacerbated humoral immune responses. In contrast, those with tuberculoid leprosy show few bacilli in their lesions and intense cellular immune responses that can be evaluated by the lepromin test [1], [4].

Most patients present borderline leprosy clinical forms [1]. The diagnosis of this disease is essentially clinical and is occasionally accompanied by bacteriological or histological examinations [5]. With regard to treatments, the World Health Organization (WHO) has proposed clinical classifications, including the numbers of skin lesions and nerves that are involved, for grouping the leprosy patients into MB and paucibacillary (PB) categories [6]. Patients with up to five lesions are classified as PB, and those with more than five lesions are classified as MB. However, classifications that are based solely on the numbers of lesions impair the proper diagnosis of this disease. Many MB populations with few skin lesions are incorrectly classified as PB; therefore, they are inadequately treated and run the risk of relapse [7].

Regarding immunological diagnostic assays, the presence of antibodies against phenolic glycolipid I (PGL-I) has been extensively studied. While anti-PGL-I serology can detect the majority of MB patients, it has limited value in identifying PB patients. In addition, false positive results in the areas of endemicity are relatively high (>10%) [8], [9]. Consequently, it has been recommended that PGL-I-based tests be used in support of the clinical examinations to direct the clinicians towards appropriate treatment and none of these PGL-I-based tests have been widely implemented in field situations [10]. Thus, tools that permit the correct and early diagnosis of *M. leprae* infection in patients who are displaying symptoms are a priority in leprosy research. The search for antigens for immunological diagnoses was initially based on research using total extracts and subcellular fractions of *M. leprae* followed by advances that were achieved using recombinant DNA technology and, more recently, on studies involving comparative genomic analyses and bioinformatics. The main difficulty that is encountered involves obtaining reagents that are more sensitive and specific or that distinguish *M. leprae* exposure from infection. The low specificity of the antigens is a result of cross-reactivity with other mycobacteria, which becomes even more problematic in countries with high incidence rates of tuberculosis and routine *M. bovis*bacillus Calmette-Guerin (BCG) vaccinations [11].

Accordingly, this study proposes the use of the phage-display technique as a tool to identify new reagents that may be effectively used in immunological assays. We have extended our previous observations by evaluating the potentials of peptide mimotopes of *M. leprae*antigens selected by the screening of phage-displayed random peptide libraries as potential serological test reagents for leprosy diagnosis. The mimotopes were tested on leprosy patients, HC, EC, tuberculosis patients and with immune sera that were raised against several mycobacteria in animals. The results indicate that peptide mimotopes may be useful for the diagnosis of leprosy.

METHODS

Ethics Statement

All of the animal care and experimental procedures and human blood sample collections were performed in accordance with the institutional guidelines. All of the individuals provided written informed consent prior to venipuncture. The animal protocols were conducted in compliance with the Brazilian rules of animals used for experimental purposes. These protocols are based on national and international guidelines, such as those that have been disclosed by the following organizations: the International Guiding Principles for Biomedical Research Involving Animals (CIOMS), the American Association for Laboratory Animal Science (AALAS), and the Brazilian College of Animal Experimentation (COBEA).

Both animal procedure and experiment involving human subjects were approved by the Research Ethics Committee of the Federal University of Parana (UFPR), Curitiba, Brazil (protocol number CEP/SD 428.108.07.10).

Human Sera

The leprosy patients and HC were recruited from the Clinical Dermatology Hospital of Parana (Piraquara, Brazil), the Barão Regional Specialties Center (Curitiba, Brazil), the Pro-Hansen Foundation (Curitiba, Brazil), and the Humanitas Philanthropic Society (São Jerônimo da Serra, Brazil). Sera from 10 PB and 23 MB leprosy patients were included, of whom 14 were newly diagnosed and 19 had received up to four months of treatment. In addition, sera were included from 26 HC, 30 patients with pulmonary tuberculosis (TB) who had been treated for up to three months in specialized centers, and 30 healthy EC with no known previous exposure to *M. leprae* or *M. tuberculosis*.

Synthetic Phage Library Derived-Peptides Evaluated by ELISA Assay

Procedures for the biopanning of phage libraries with the anti-*M. leprae* antibodies of MB patients were performed as previously described [12], [13]. The peptide sequences in the reactive phage clones were synthesized using a 9-fluorenylmethoxycarbonyl (Fmoc)-based solid-phase synthesis technique.

The ELISA test was optimized with regard to antigen concentration and serum and conjugate dilutions. Microtiter plates (Corning) were coated with 100 µL of the peptide pool at 1.5 µg/mL in 0.05 M carbonate buffer (pH 9.6) overnight at 4°C. After washing with a solution containing 0.9% NaCl and 0.05% Tween 20, the plates were blocked with Protein-Free Blocking Buffer

(Thermo Fisher Scientific) for 1 h at 37°C. Then, the plates were washed and incubated for 1 h at 37°C with sera in duplicate dilutions of 1:50 in a phosphate-buffered saline (PBS) solution at pH 7.4, which also contained 0.1% BSA. The plates were washed and incubated with anti-human IgG (Fc-specific)-biotin antibodies (Sigma-Aldrich) that were diluted 1:2 500 in PBS with 0.1% BSA for 1 h at 37°C. The plates were then washed and incubated with streptavidin-peroxidase (Sigma-Aldrich) that was diluted 1:2 500 in PBS with 0.1% BSA for 1 h at 37°C and detected using o-phenylenediamine (OPD) dihydrochloride (Sigma-Aldrich). The reactivities of the rabbit anti-mycobacterial sera with the peptides were also analyzed (data not shown). In this analysis, the antibodies that were bound to the peptides were detected by anti-rabbit IgG-peroxidase antibodies that were diluted to 1:4 000.

Immunoassay with Cellulose Membrane-Bound Peptides

The peptides and their alanine analogs were prepared by Spot Synthesis [14]. Peptide synthesis was performed using Fmoc-protection chemistry on cellulose membranes that had been derivatized with polyethylene glycol spacers (Intavis Bioanalytical Instruments AG) and MultiPep RS (Intavis Bioanalytical Instruments AG). Following overnight incubation with a blocking solution (0.05% phosphate buffered saline tween-20 (PBST) and 3% BSA), the membrane was probed with the sera of the MB patients at 1:250 dilutions in PBST for 1 h at room temperature. The reaction was developed with the anti-human IgG (Fc-specific)-peroxidase antibody that was diluted to 1:120 000 in PBST for 1 h in addition to ECL Plus (GE Healthcare) and Hyperfilm ECL (GE Healthcare). The spot intensities were determined using the ImageJ software (version 1.43).

Research of Native Proteins in *M. Leprae* Using Anti-Peptide Antibodies

The synthetic peptides were conjugated to BSA using glutaraldehyde as a cross-linker [15]. Conjugates (20 µg) that were emulsified in incomplete Freund›s adjuvant (Sigma-Aldrich) were administered via subcutaneous (first and second doses) and intraperitoneal (third to fifth doses) injections to eight-week-old Swiss female mice (n=5 per group). The injections were performed on days 1, 14, 28, 42, and 56. Non-immune serum was used as the negative control, and immune blood was collected one week after the final dose. To assess the production of anti-peptide antibodies, microtiter plates (Corning) were coated with 100 µL of the peptide at final concentrations of 1.5 µg/mL in 0.05 M carbonate buffer (pH 9.6) overnight at 4°C. After washing in a solution containing 0.9% NaCl and 0.05% Tween 20, the plate was blocked (2% casein in PBS, pH 7.4) for 1 h at 37°C. The plate was then washed and incubated for 1 h at 37°C with sera that had been diluted in incubation buffer (0.25% casein in 0.05% PBST). After washing, the detection of the reaction was performed using the anti-mouse IgG (γ-specific)-peroxidase antibody (Sigma-Aldrich) at a 1:2 500 dilution and OPD dihydrochloride as a chromogen.

The *M. leprae* whole cells were provided by Dr. J. S. Spencer from Colorado State University, Fort Collins, USA through the National Institute of Allergy and Infectious Diseases/National Institutes of Health under the contract N01-AI-25469 for the production of soluble antigens for use in electrophoresis. The cells were resuspended in 0.9% NaCl, mixed with 100 µg/mL phenylmethanesulfonyl fluoride (PMSF) and 2 mM ethylenediaminetetraacetic acid (EDTA) and sonicated (Sonopuls HD 2200) for four cycles of 15 min each [16]. The protein concentration of the soluble fraction that was obtained by centrifugation at 10 000 g for 20 min at 4°C was determined using the Quant-iT Protein Assay Kit (Invitrogen). *M. leprae* protein (40 µg) were then electrophoresed on a 15% sodium dodecyl sulfate-polyacrylamide gel electrophoresis (SDS-PAGE) and transferred to a PVDF membrane that was blocked with 0.3% PBST, washed with 0.05% PBST and incubated with anti-peptide serum in 0.05% PBST. The membrane was washed, incubated with anti-mouse IgG (γ-specific)-peroxidase antibodies that were diluted to 1:5 000 in 0.05% PBST, and processed using ECL Plus.

For the identification of the proteins in *M. leprae*, 2-dimensional gel electrophoresis (2-DE) followed by Western blotting were performed. Two hundred µg of *M. leprae* protein were purified using a 2D Clean-Up Kit (GE Healthcare) and solubilized in rehydration buffer (8 M urea, 2% CHAPS, 20 mM DTT, 0.5% IPG buffer, and 0.002% bromophenol blue). Samples were in gel rehydrated and run on linear 13 cm IPG strips at pH levels ranging from 4–7 (GE Healthcare) with the Ettan IPGphor II (GE Healthcare) for 21400 Vh. The focused strips were equilibrated in a solution that contained 6 M urea, 50 mM Tris (pH 8.8), 30% glycerol, 2% SDS, 0.002% bromophenol blue and 0.065 M DTT for 15 min, followed by incubation in the same solution that was prepared with 0.14 M iodoacetamide instead of DTT for an additional 15 min. The second-dimensional separation was performed using 15% SDS-PAGE by the SE 600 Ruby (GE Healthcare) electrophoresis unit. Western blotting was performed as previously described.

Statistical Analyses

A cut-off point for optimal sensitivity and specificity for the ELISA tests was determined using the Receiver Operating Characteristic (ROC) curve analysis

[17], [18]. The accuracy of each test was evaluated according to the area under the curve (AUC) and the Youden index J[19]. The statistical analyses were performed using the MedCalc statistical software (version 13.2.0).

The differences in seroreactivities between the groups were analyzed with the two-tailed Mann-Whitney U test for nonparametric distribution using the GraphPad Prism software (version 5.03). The P values were corrected for multiple comparisons. The P value indicating statistical significance was set to <0.05.

RESULTS

Peptide Sequences

Table 1 shows the peptide sequences that were selected by phage display using the affinity purified anti-*M. leprae* MB patients' antibodies as previously reported [12]. The consensus sequence DPAW was observed between peptides 5A and 6A.

Table 1. Amino acid sequences of the peptides obtained by phage display.

Peptide	Amino acid sequence	Length
5A	APDDPAWQNIFNLRR	15
6A	ICPRDPAWSYCN	12
1B	NNIHAHRYWGTDLNA	15

doi:10.1371/journal.pone.0106222.t001

ELISA Using Synthetic Peptides

The peptides 5A, 6A, and 1B were chemically synthesized and evaluated by ELISA against the sera from the leprosy patients. The peptides were capable of detecting 30% (3/10) of the PB and 73.9% (17/23) of the MB patients (Table 2). Compared with the PB leprosy patients, MB patients, in general, showed strong positive responses. Of the 30 serum samples from the tuberculosis patients, 16 (53.3%) were reactive to the peptides (Table 2). Weak positive responses were observed in 12 out of 26 of the HC sera and in 5 out of the 30 EC sera that were tested in this experiment (Table 2 and Fig. 1). These findings indicate the existence of significant differences between the leprosy (PB and MB groups) and EC groups and the TB and EC groups. No significant differences were observed between HC and EC groups. The sensitivity, specificity, AUC and Youden index J values were calculated for ELISA assay (Table 3). The peptides were not reactive against the rabbit antimycobacterial sera using the ELISA (data not shown).

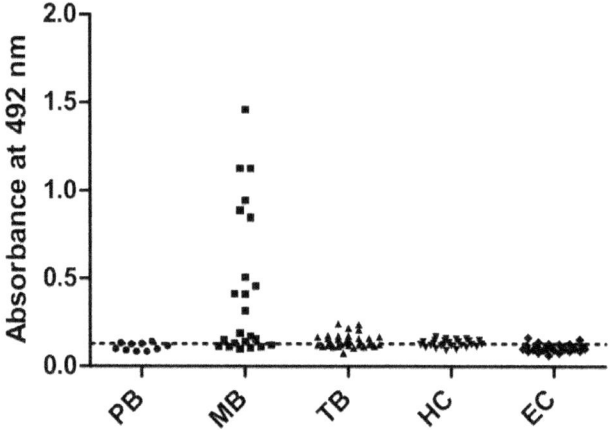

Figure 1. ELISA reactivities of synthetic peptides with sera from leprosy patients and controls. Microtiter plates were coated with 1.5 μg/mL of the peptide pool (5A, 6A, and 1B) and incubated with serum that was diluted 1:50. The detection of the reaction was performed with the anti-human IgG (Fc-specific)-biotin antibody and streptavidin-peroxidase. MB, MB leprosy patients (n=23); PB, PB leprosy patients (n=10); TB, TB patients (n=30); HC, healthy household contacts of MB leprosy patients (n=26); EC, endemic controls (n=30). The dotted line represents the cut-off value (as determined using the ROC curve with serum samples from the EC). Each symbol represents the absorbance obtained with a single serum.

Table 2. ELISA results using peptides in the groups studied.

Group	Number of serum samples tested	Number (%) of positive serum samples by ELISA with peptides
PB	10	3 (30)
MB	23	17 (73.9)
TB	30	16 (53.3)
HC	26	12 (46.2)
EC	30	5 (16.7)

MB: MB leprosy patients. PB: PB leprosy patients. TB: TB patients. HC: healthy household contacts of MB leprosy patients. EC: endemic control.
doi:10.1371/journal.pone.0106222.t002

Table 3. Diagnostic performance of ELISA using peptides for the diagnosis of leprosy.

Serological test	Sensitivity in PB+MB groups % (95% CI)	Specificity in PB+MB groups % (95% CI)	AUC	Youden index J
ELISA-Peptides	60.6 (42.1–77.1)	90.0 (73.5–97.9)	0.779	0.5061

Samples from leprosy patients (n = 33, PB+MB leprosy patients) and endemic control (n = 30) were tested. CI: confidence interval. AUC: area under the ROC curve.
doi:10.1371/journal.pone.0106222.t003

Identification of Key Residues for Antibody Binding in Peptides

For peptide 5A, the proline, aspartate and alanine residues were important for antibody recognition. The substitution of these amino acids by alanine or serine reduced the reactivity of the peptide compared with that of the unmodified peptide. No reactivity was observed for peptide 6A. For peptide 1B, the residues of greatest importance to the antibody interactions were tyrosine, tryptophan, glycine, aspartate, leucine and asparagine (Fig. 2).

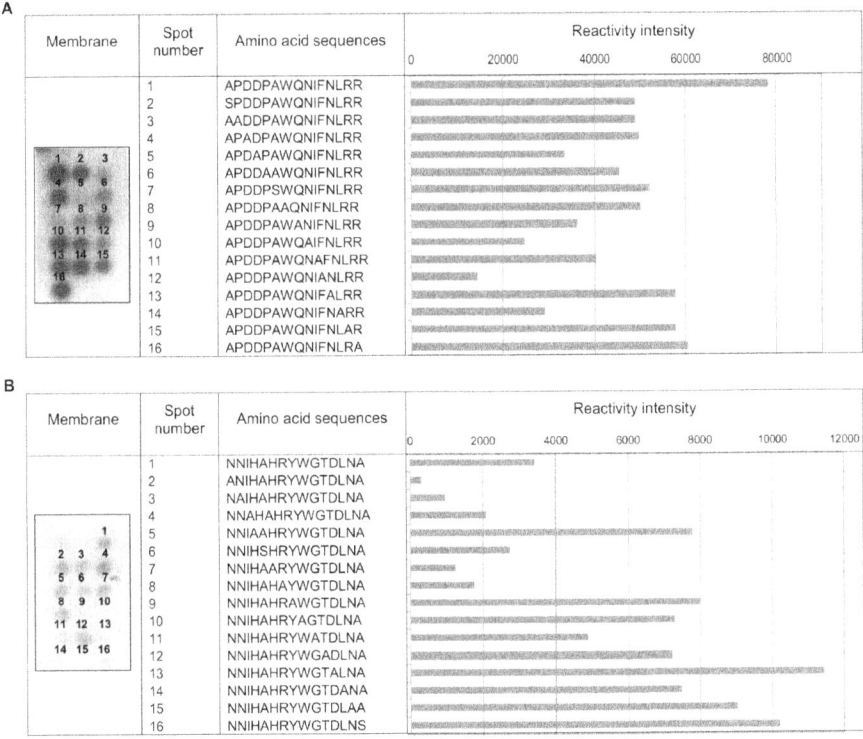

Figure 2. Identification by alanine scanning of key residues in peptides. Reactivities of antibodies from leprosy patients (1:250) with membrane-bound peptides detected using the anti-human (Fc-specific)-peroxidase antibody (1:120000). (A) The reference peptide 5A is in position 1, and the other spots correspond with the alanine analogs of peptide 5A. (B) The reference peptide 1B is in position 1 followed by its alanine analogues. Each bar represents the intensity of the antibody binding to the reference peptide or an analog peptide. When alanine was present in the generation position of the analog, the alanine residue was substituted by serine.

Identification of Natural Antigens in *M. Leprae*

The peptides that were conjugated to BSA induced mouse anti-peptide antibody production and these antibodies reacted with antigens of *M. leprae* after electrophoresis, which was followed by Western blotting (Fig. 3). The Western blotting of the *M. leprae* proteins that was resolved by 2 DE and probed with the anti-peptide antibody revealed four spots of approximately 30 kDa in size (Fig. 4). The three anti-peptide antisera showed the same reactivity profiles with *M. leprae* (data not shown). The estimated isoelectric points (pI) for the four spots were 4.56, 4.64, 4.75 and 4.9.

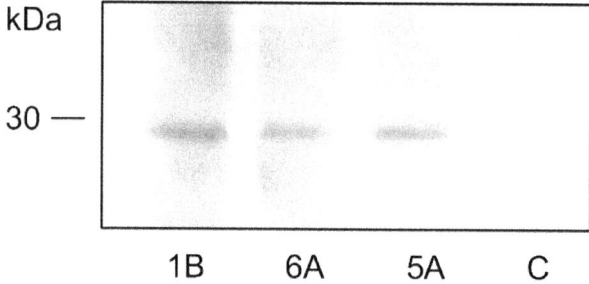

Figure 3. Reactivity of anti-peptide antibodies to *M. leprae* by Western blotting. Forty micrograms of total *M. leprae* extract were separated by 15% SDS-PAGE and, following Western blotting, were reacted with anti-peptide serum 1B (1:12000), anti-peptide serum 6A (1:9000), anti-peptide serum 5A (1:4000), and non-immune serum (C) (1:4000). The detection of the reaction was performed with the anti-mouse IgG (γ-specific)-peroxidase antibody and chemiluminescence.

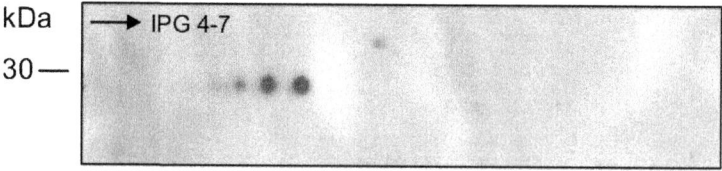

Figure 4. Western blotting analysis of total *M. leprae* extract 2D gel using mouse anti-peptide antibodies. Two hundred micrograms of protein were separated in the first dimension by isoelectric focusing on a 13 cm IPG strip at pH 4-7 and in the second dimension by SDS-PAGE on a 15% gel. Western blotting with anti-peptide serum 1B at 1:4000 is shown. The other two anti-peptide sera showed the same reactivity profiles as that of anti-peptide serum 1B (data not shown). The detection of the reaction was performed with the anti-mouse IgG (γ-specific)-peroxidase antibody and chemiluminescence.

In silico Analyses of Peptides

The three peptide sequences that were identified demonstrated significant alignments with the antigen 85B (fbpB) of *M. leprae* using BLASTp. Antigen 85B is a member of the antigen 85 complex. It has a molecular mass of 31 kDa and a theoretical isoelectric point of 4.9.

DISCUSSION

In this study we evaluated the serological responses to peptides that were selected from random peptide phage-display libraries in an attempt to identify antigens that could be useful for the serological diagnoses of leprosy patients. The MB leprosy patients were easier to identify and the ELISA showed positive results in 73.9% of the samples that were tested. For the PB leprosy patients, 30% of the serum samples were reactive against the peptides as shown by the peptide-based ELISA and weaker responses were observed in this group compared with the MB group.

Protein antigens require binding to human leukocyte antigen (HLA) molecules to be presented by T cells. Because responses to mycobacterial antigens are HLA-restricted, not all individuals respond to the same antigens [20]. Reasons other than HLA restriction exist that explain the seronegativities between the patients that have been diagnosed with MB leprosy by the peptide-based ELISA. Recent research suggests that for tuberculosis, heterogeneous antigen recognition may result from the production of different antigens at different stages of the disease [21]. Another reason is related to differences in the strains of *M. leprae* [8]. However, research has shown that different strains have highly conserved genomes, which indicates that antigenic variations are likely to be negligible [22]. Because of the differences in the immune responses that occur among individuals, it has been suggested that immunological tests should be based on a combination of several different antigens to permit greater coverage in individuals that have been infected with *M. leprae* [3],[23].

The detection of *M. leprae* infection in PB patients through serological assays is difficult because once the immune response develops in these individuals, it is predominantly cellular. Analyses of T cell responses as interferon-gamma (IFN-γ) release assays are required for the diagnosis of PB [9]. Thus, the immunological diagnosis of leprosy may be based on both antibody and T cell assays [24]. The combination of these assays would permit the correct classification of patients and consequently, the initiation of effective treatments that may improve the prognoses for each case.

Seroreactivities against peptides were observed in the tuberculosis and HC and EC individuals. The recognition of peptides by tuberculosis patient sera was expected because *M. leprae* shares antigens with *M. tuberculosis*. Moreover, the diagnosis of leprosy or tuberculosis does not rely on a single test but must be confirmed by a combination of several tests and further associated with a clinical examination.

The capacity of the peptide-based ELISA test to predict disease development in the HC patients was not possible in this work because it requires that they be monitored over a prolonged period of time. Seroreactivity with some endemic control samples suggests that peptides not are capable of discriminate infection of exposure to *M. leprae*. Further investigation needs to be conducted testing the peptides with non-endemic controls to clarify this issue.

The understanding of the molecular basis of peptide recognition by patient antibodies may assist in the design of novel peptides with improvements in specificity and sensitivity for diagnostic purposes. The contribution of each amino acid residue in the peptide sequence to antibody recognition was deduced through alanine scanning. In this assay, for peptide 5A, the residues that were critical for antibody recognition were proline, alanine, and aspartate. Peptide 6A did not indicate reactivity. One potential explanation for these findings is that the conformational flexibility of the peptides on the membranes may be restricted, thus reducing the affinities of the peptides for the antibody [25]. Peptide 1B showed low levels of reactivity. However, sufficient levels were present for the identification of important residues that interacted with the antibody. These results indicate that the residues of the C-terminal region were responsible for binding to the antibody. Because this portion of the sequence is immobilized in the membrane, the accessibility of the antibody to the residues is more difficult and most likely explains the low reactivities of the peptides that were derived from 1B in the membrane assay.

The peptides that were evaluated in this study were shown to be antigenic and immunogenic. This latter property allowed for the characterization of these peptides as mimotopes of *M. leprae*.

The *M. leprae* database analysis suggests that the epitope protein that is mimicked by the peptides is antigen 85B, which is a 31 kDa protein with a pI of 4.9 that possesses properties that are consistent with the information that was experimentally obtained (Fig. 4).

Antigen 85B is a component of the fibronectin-binding protein complex that presents with very interesting antigenic and immunogenic properties because of its capacity to induce strong humoral and cellular responses, including the *in vitro* proliferation of T cells [26]. Its immunological features and predominant role as a *M. leprae* antigen explain why only mimotopes of this antigen have been bioselected. These epitope-mimicking peptides contain amino acid sequences that are part of two sequences (AA 206–230 and 291–325) from three regions of antigen 85B, which have been recognized as important antigenic domains for the diagnosis of patients with lepromatous leprosy [27].

The search for antigens may not only improve the sensitivities and specificities of the diagnostic tests but also contribute to a better understanding of the immune response that is developed against the antigens in affected individuals. The results of these observations may be used to improve antigenic recognition and identify novel peptides. In addition, the peptides are very feasible for widespread use in serodiagnostic tests because they can be easily prepared in sufficient purities and large quantities [27].

This work demonstrates the usefulness of this biotechnological tool in overcoming diagnostic constraints by selecting, synthesizing, and confirming certain peptide sequences *in vitro* for their uses as diagnostic laboratory tests. The peptides that were identified were able to recognize antibodies that are primarily found in MB leprosy patients. MB cases are the principal sources of leprosy infection in the population, and affected individuals are at greater risks of developing disabilities compared with those with PB leprosy [23]. Therefore, the early detection and treatment of MB cases will contribute to a reduction in the transmission of the bacterium. Ideally, a serological test should identify all leprosy patients (MB and PB). In fact, it may be possible to diagnose all of the patients by combining peptides and assays that are based on humoral and cellular immunity. Hence, it is important to identify new peptides using the approach that is described in this work while also evaluating different diagnostic approaches using these peptides to increase the specificity and sensitivity of the detection of MB and particularly PB leprosy patients.

ACKNOWLEDGMENTS

The authors would like to thank the representatives and members of the Clinical Dermatology Hospital of Parana, the Barão Regional Specialties Center, the Humanitas Philanthropic Society, São Sebastião da Lapa Regional Hospital, and the Pro-Hansen Foundation for allowing us access to the premises to collect patient samples. The authors are also grateful to the Center for the Production and Research of Immunobiological Products for their collaboration.

AUTHOR CONTRIBUTIONS

Conceived and designed the experiments: SMA JFM VTS. Performed the experiments: SMA SBS JCM. Analyzed the data: SMA JFM VTS. Contributed reagents/materials/analysis tools: SBS MTM CCO JFM VTS. Contributed to the writing of the manuscript: SMA VTS JFM SBS LMA MTM.

REFERENCES

1. Britton WJ, Lockwood DNJ (2004) Leprosy. Lancet 363: 1209–1219. doi: 10.1016/s0140-6736(04)15952-7
2. World Health Organization (2012) Global leprosy situation, 2012. Available:http://www.who.int/wer/2012/wer8734/en/index.html. Accessed 21 September 2012.
3. Geluk A, van der Ploeg J, Teles ROB, Franken KLMC, Prins C, et al. (2008) Rational combination of peptides derived from different *Mycobacterium leprae* proteins improves sensitivity for immunodiagnosis of *M. leprae* infection. Clin Vaccine Immunol 15: 522–533. doi: 10.1128/cvi.00432-07
4. Alban SM, Sella SR, Miranda RN, Mira MT, Thomaz-Soccol V (2009) PCR-restriction fragment length polymorphism analysis as a tool for *Mycobacterium* species identification in lepromas for lepromin production. Lepr Rev 80: 129–142.
5. Lockwood DNJ (2005) Leprosy. Medicine 33: 26–29. doi: 10.1383/medc.2005.33.7.26
6. World Health Organization (1997) WHO Seventh Expert Committee Report, June 1997. Available: http://www.who.int/lep/resources/Expert.pdf. Accessed 1 July 2010..
7. Bührer-Sékula S, Visschedijk J, Grossi MA, Dhakal KP, Namadi AU, et al. (2007) The ML flow test as a point of care test for leprosy control programmes: potential effects on classification of leprosy patients. Lepr Rev 78: 70–79.
8. Duthie MS, Goto W, Ireton GC, Reece ST, Cardoso LPV, et al. (2007) Use of protein antigens for early serological diagnosis of leprosy. Clin Vaccine Immunol 14: 1400–1408. doi: 10.1128/cvi.00299-07
9. Stefani MM (2008) Challenges in the post genomic era for the development of tests for leprosy diagnosis. Rev Soc Bras Med Trop (Suppl 2): 89–94.
10. Duthie MS, Ireton GC, Kanaujia GV, Goto W, Liang H, et al. (2008) Selection of antigens and development of prototype tests for point-of-care leprosy diagnosis. Clin Vaccine Immunol 15: 1590–1597. doi: 10.1128/cvi.00168-08
11. Geluk A, Spencer JS, Bobosha K, Pessolani MC, Pereira GM, et al. (2009) From genome-based in silico predictions to ex vivo verification

of leprosy diagnosis. Clin Vaccine Immunol 16: 352–359. doi: 10.1128/cvi.00414-08

12. Alban SM, de Moura JF, Minozzo JC, Mira MT, Thomaz Soccol V (2013) Identification of mimotopes of *Mycobacterium leprae* as potential diagnostic reagents. BMC Infect Dis 13: 42. doi: 10.1186/1471-2334-13-42

13. Thomaz Soccol V, Alban SM, de Moura JF (2011) Uso de peptídeos miméticos de *Mycobacterium leprae* para diagnóstico e vacinas. Brazilian Patent PI015110000997.

14. Frank R (1992) Spot-synthesis: an easy technique for the positionally addressable, parallel chemical synthesis on a membrane support. Tetrahedron 48: 9217–9232. doi: 10.1016/s0040-4020(01)85612-x

15. Harlow E, Lane D (1988)Antibodies: A laboratory manual. Cold Spring Harbor Laboratory Press. 726 p.

16. Pessolani MC, Brennan PJ (1992) *Mycobacterium leprae* produces extracellular homologs of the antigen 85 complex. Infect Immun 60: 4452–4459.

17. Metz CE (1978) Basic principles of ROC analysis. Semin Nucl Med 8: 283–289. doi: 10.1016/s0001-2998(78)80014-2

18. Zweig MH, Campbell G (1993) Receiver-operating characteristic (ROC) plots: a fundamental evaluation tool in clinical medicine. Clin Chem 39: 561–577.

19. Youden WJ (1950) Index for rating diagnostic tests. Cancer 3: 32–35. doi: 10.1002/1097-0142(1950)3:1<32::aid-cncr2820030106>3.0.co;2-3

20. Meyer CG, May J, Stark K (1998) Human leukocyte antigens in tuberculosis and leprosy. Trends Microbiol 6: 148–154. doi: 10.1016/s0966-842x(98)01240-2

21. Lyashchenko K, Colangeli R, Houde M, Al Jahdali H, Menzies D, et al. (1998) Heterogeneous antibody responses in tuberculosis. Infect Immun 66: 3936–3940.

22. Monot M, Honoré N, Garnier T, Zidane N, Sherafi D, et al. (2009) Comparative genomic and phylogeographic analysis of *Mycobacterium leprae*. Nat Genet 41: 1282–1289. doi: 10.1038/ng.477

23. Aráoz R, Honoré N, Banu S, Demangel C, Cissoko Y, et al. (2006) Towards an immunodiagnostic test for leprosy. Microbes Infect 8: 2270–2276. doi: 10.1016/j.micinf.2006.04.002

24. Aseffa A, Brennan P, Dockrell H, Gillis T, Hussain R, et al. (2005) Report on the first meeting of the IDEAL (Initiative for Diagnostic and Epidemiological Assays for Leprosy) consortium held at Armauer Hansen Research Institute, ALERT, Addis Ababa, Ethiopia on 24–27 October 2004. Lepr Rev 76: 147–159.

25. Van Regenmortel MH (2001) Antigenicity and immunogenicity of synthetic peptides. Biologicals 29: 209–213. doi: 10.1006/biol.2001.0308

26. Thole JE, Schöningh R, Janson AA, Garbe T, Cornelisse YE, et al. (1992) Molecular and immunological analysis of a fibronectin-binding protein antigen secreted by*Mycobacterium leprae*. Mol Microbiol 6: 153–163. doi: 10.1111/j.1365-2958.1992.tb01996.x

27. Filley E, Thole JE, Rook GA, Nagai S, Waters M, et al. (1994) Identification of an antigenic domain on *Mycobacterium leprae* protein antigen 85B, which is specifically recognized by antibodies from patients with leprosy. J Infect Dis 169: 162–169. doi: 10.1093/infdis/169.1.162

# Chapter 3

DNA REPAIR AND CHEMOTHERAPY

Seiya Sato and Hiroaki Itamochi

Department of Obstetrics and Gynecology, Tottori University School of Medicine, Nishicho, Yonago-City, Tottori, Japan

INTRODUCTION

Cancer chemotherapy is designed to kill cancer cells, with most agents inducing DNA damage. Highly conserved DNA repair machinery that process DNA damage and maintain genomic integrity developed during the evolution of mammalian cells. Interestingly, in established tumors, DNA repair activity is required to counteract oxidative DNA damage that is prevalent within the tumor microenvironment. If the damaged DNA is successfully repaired, the cell will survive.

In order to specifically and effectively kill cancer cells using chemotherapy that induce DNA damage, it is important to take advantage of specific abnormalities in the DNA damage response machinery that are present in cancer cells but not in normal cells. Such properties of cancer cells may be targets for sensitization and lead to the development of biomarkers. Furthermore, inhibition of a DNA damage response pathway may enhance the therapeutic effects of DNA-damaging agents when employed in combination with these agents.

Recently, DNA repair inhibition has emerged as a promising strategy for personalized cancer therapy. Synthetic lethality exploits inter-gene relationships where the loss of function of either one of two related genes is nonlethal, but loss of both causes cell death. Emerging clinical data provide compelling evidence that overexpression of DNA repair factors may have prognostic and predictive significance in patients.

In this chapter, we will provide an overview of major DNA repair pathways and describe recent advances in anticancer therapy with a focus on DNA repair in cancer.

DNA DAMAGE RESPONSE

The cellular DNA damage response (DDR) involves activation of cell cycle checkpoints to induce cell cycle arrest while repair mechanisms, transcriptional modulation, and/or apoptotic pathways are activated. DNA damage induced by anticancer agents triggers recruitment of multiprotein complexes and activates a number of pathways, including ataxia telangiectasia mutated (ATM) and ATM and Rad3-related (ATR) signaling pathways. Cell cycle checkpoint kinases (Chk) of Chk1 and Chk2 are functionally redundant protein kinases that respond to checkpoint signals, initiate ATM and ATR, and play a critical role in determining cellular responses to DNA damage [1, 2]. Chk1 is mainly activated through ATR-mediated phosphorylation. Activated Chk1 phosphorylate Cdc25A, which leads to ubiquitin-and proteasome-dependent protein degradation, and downstream to increased phosphorylation of cyclin-dependent kinase (CDK) 2. In contrast, Chk2 is activated mainly by ATM, and activated Chk2 phosphorylates Cdc25A. Activated Chk1 and Chk2 then phosphorylates diverse downstream effectors, which in turn are involved in cell cycle checkpoints (i.e., G1/S-phase, intra-S-phase, and G2/M-phase checkpoints), the DNA replication checkpoint, and the mitotic spindle checkpoint, as well as DNA repair and apoptosis.

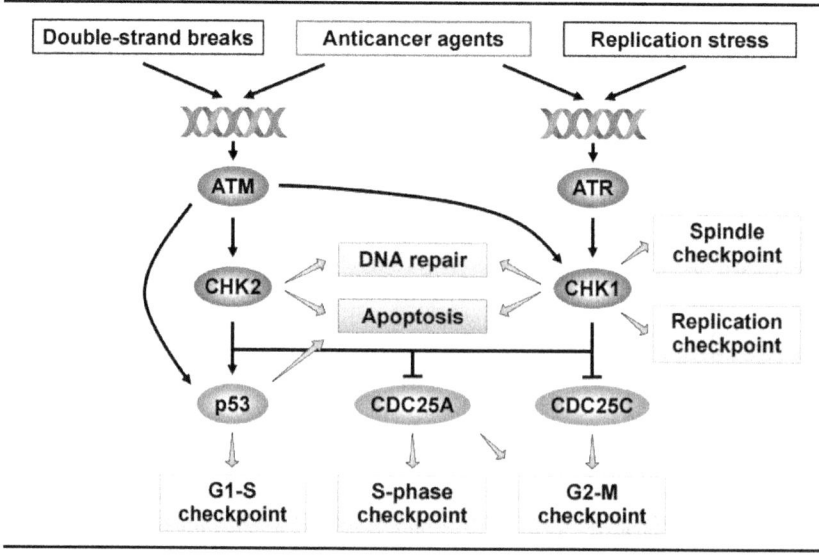

ATM, ataxia telangiectasia–mutated; ATR, ATM and Rad3–related; CHK, checkpoint kinase.

Figure 1. DNA-damage response signaling pathways.

Consequently, through regulating the activity of CDKs, the progression from one cell cycle phase to another is delayed. The resulting cell cycle arrest allows time for repair, thereby preventing genome duplication or cell division in the presence of damaged DNA.

DNA REPAIR PATHWAYS

DNA repair pathways in mammalian cells maintain genomic integrity. Depending on the type of DNA damage, cells invoke specific DNA repair pathways in order to restore genetic information.

Minor changes to DNA such as oxidized or alkylated bases, small base adducts and single-strand breaks (SSBs) are restored by the base excision repair (BER) pathway [3]. Poly(adenosine diphosphate ribose) (PAR) polymerase (PARP) is important in this process. Upon detection of SSBs, PARP covalently transfers PAR chains to itself and to acceptor proteins in the vicinity of the lesion, thereby facilitating the repair of SSBs. More complex, DNA helix-distorting base lesions, such as those induced by UV light, are repaired by nucleotide excision repair (NER) [4]. Another kind of damage disturbing the helical structure of DNA is represented by base mismatches. Mismatch repair factors recognize and process misincorporated nucleotides as well as insertion or deletion loops that arise during recombination or from errors of DNA polymerases [5].

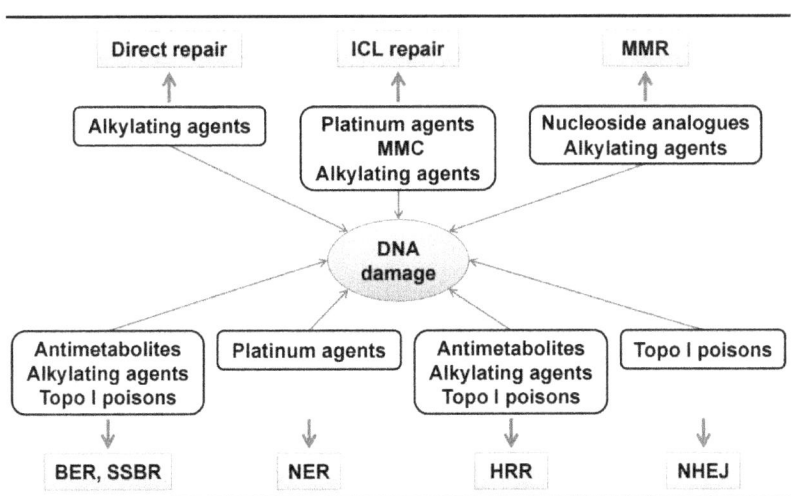

BER, base excision repair; HRR, homologous recombination repair; ICL, inter-strand crosslink; MMR, mismatch repair; MMC, mytomycin C; NER, nucleotide excision repair; NHEJ, non-homologous end joining; SSBR, single-strand break repair; Topo, topoisomerase.

Figure 2. DNA repair pathways and chemotherapeutic agents.

Covalent links between the two strands of the double helix represent a type of DNA damage referred to as interstrand crosslinks (ICLs). ICLs represent the most deleterious lesions produced by chemotherapeutic agents such as mitomycin C (MMC), cisplatin and cyclophosphamide. ICL repair is complex and involves the collaboration of several repair pathways, namely Fanconi anaemia, NER, translesion synthesis (TLS) and homologous recombination (HR) [6].

So far, four mechanistically distinct DNA double-strand break (DSB) repair mechanisms in mammalian cells have been described: non-homologous end joining (NHEJ), alternative NHEJ, single-strand annealing and HR [7]. NHEJ and HR represent the two major DSB repair pathways, with NHEJ operating throughout the cell cycle and HR being the most active during S-phase [8].

CANCER THERAPIES TARGETING DNA REPAIR MECHANISM

Alterations in expression of DNA repair may influence cancer biology and aggressive phenotypes. Clinical evidence supports the hypothesis that overexpression of DNA repair factors may have prognostic and predictive significance in patients [9]. Furthermore, highly proliferative cancer cells are hypersensitive to DNA damage because the S-phase is the most vulnerable period of the cell cycle. Therefore, DDR pathways make an ideal target for therapeutic intervention.

Dysfunction of one DNA repair pathway may be compensated by the function of another compensatory DDR pathway, which may be increased and contribute to resistance to DNA-damaging chemotherapy. So, inhibition of the pathway in combination with DNA damage agents will selectively kill cancer cells. These hypotheses are currently being tested in the laboratory and are being translated into clinical studies.

DIRECT REPAIR

The simplest form of DNA repair is direct reversal of the lesion. Direct reversal of the oxidative lesion O6-methylguanine is carried out by the suicide enzyme methylguanine methyltransferase (MGMT) via an active site Cys145 that acts as a methyl recipient, followed by rapid ubiquitin-induced degradation. MGMT expression is one of several factors governing the response to alkylating chemotherapy agents [10, 11].

MGMT demethylates O6-methylguanine lesions, which are formed as a result of erroneous methylation by S-adenosylmethionine (SAM) and other alkylations at the O6 position of guanine that are induced by dietary nitrosamines or chemotherapy agents such as temozolomide (TMZ), dacarbazine (DTIC) and nitrosoureas [12, 13]. The higher levels of MGMT that are frequently observed in tumor tissue compared with normal tissue suggest that its depletion with pseudo-substrates that resemble O6-methylguanine might be a viable strategy to sensitize tumor cells to O6 alkylating agents. However, these pseudo-substrates have shown only marginal clinical benefit [14, 15].

A more promising approach may be the exploitation of reduced MGMT activity owing to epigenetic silencing in some cancers [16]. MGMT promoter methylation correlated with sensitivity to BCNU in patients with astrocytomas and also correlated with sensitivity to TMZ plus radiotherapy in patients with gliomas [17]. Therefore, MGMT promoter methylation could be useful for stratifying patients for TMZ treatment.

BASE EXCISION REPAIR

BER is responsible for detection and repair of damage caused by a number of mechanisms including alkylation, oxidation by reactive oxygen species (ROS), SSBs and base deamination. BER repairs DNA damage that is therapeutically induced by ionizing radiation, DNA-methylating agents, topoisomerase I poisons such as camptothecin, irinotecan and topotecan [18]. Single-strand break repair (SSBR) and BER are often assumed to be synonymous because they involve the same components and are similar after the initial recognition step. The main components of the pathway are glycosylases, endonucleases, DNA polymerases and DNA ligases, with PARP1 and PARP2 facilitating the process. Damaged bases are first removed by BER glycosylases to form apurinic or apyrimidinic (AP) sites. BER endonucleases then generate an SSB, which along with directly induced SSBs and those generated by topoisomerase (topo) I poisons [19, 20], are the substrates for SSBR. On detecting SSBs, PARP1 rapidly becomes bound and poly(ADP-ribosyl)ated, protecting the nick ends from undesirable recombination and allowing the recruitment of the molecular scaffold protein X-ray repair cross-complementing protein (XRCC) 1 for ongoing repair [21].

The BER pathway is an attractive target for the modulation of chemosensitivity. Early inhibitors of DNA polymerase-β (Pol β), flap endonuclease 1 (FEN1), ligase 1 and ligase 3 enhance sensitivity to ionizing radiation and TMZ. However, the most advanced drugs that target this pathway are AP endonuclease 1 (APE1) inhibitors and PARP-inhibitors (PARP-i, described later). Both APE1 and PARP expression and/or activity are generally higher in tumors [9, 22, 23].

There are two classes of APE1 inhibitor: methoxyamine, which binds the AP site in DNA, and inhibitors of APE1 endonuclease activity. Preclinically, methoxyamine potentiates the cytotoxicity of TMZ [24] and pemetrexed. In a phase I trial of methoxyamine, responses were seen in combination with pemetrexed, and there is an ongoing study with TMZ. Lucanthone, a topo II inhibitor, also inhibits APE1 endonuclease activity and potentiates the cytotoxicity of DNA-methylating agents in breast cancer cells [25]. Novel, more specific, APE1 endonuclease inhibitors increased the persistence of AP sites in vitro and increased the cytotoxicity of alkylating agents [26]. The synthetic lethality relationship between HR and APE1 was confirmed by the observed cytotoxicity following ATM inhibitor exposure in $APE1^{-/-}$ cells [27].

NUCLEOTIDE EXCISION REPAIR

NER recognizes and repairs base lesions associated with distortion of the DNA helical structure, including UV-induced photoproducts not eliminated by direct repair, and an array of bulky adducts induced by various exogenous chemical agents. NER removes helix-distorting adducts on DNA and contributes to the repair of intrastrand and ICLs; the xeroderma pigmentosum (XP) proteins and excision repair cross-complementation group 1 (ERCC1) also have crucial roles in both the NER and ICL repair pathways [28]. Deficiency in NER confers sensitivity to platinum agent therapy, which reflects a reduced capacity to repair ICLs [29, 30]. There are currently no small molecule inhibitors of NER, although cyclosporine and cetuximab might down-regulate XPG and ERCC1–XPF expression, respectively. Recent evidence suggests that the efficacy of PARP-i–topo I poison combinations may be most effective in tumors that lack ERCC1–XPF, which are involved in the NER pathway [31].

MISMATCH REPAIR

Mismatch repair (MMR) recognizes and repairs errors introduced during replication. MMR also recognizes and repairs insertion/deletion loops (IDLs), particularly within microsatellite DNA. Hence, "microsatellite instability" (MSI) is recognized as a hallmark of MMR failure [32, 33]. If MSI manifests within tumor suppressor genes, it can produce frameshift mutations that contribute to carcinogenesis in colorectal, endometrial, ovarian, and gastric cancers [34]. Defective MMR increases mutation rates up to 1,000-fold, results in MSI, and is associated with cancer development [35].

Several DDR genes have microsatellites and could be mutated in MSI-high cancer, potentially conferring sensitivity to some DNA-damaging agents [36, 37]. However, defects in MMR cause tolerance to TMZ, platinum agents and some nucleoside analogues, which leads to drug resistance [38,39].

Some researchers have focused on attempts to reactivate epigenetically silenced MLH1. However, after promising preclinical data that demonstrated chemosensitization [40], clinical trials have shown adverse reactions.

HOMOLOGOUS RECOMBINATION REPAIR

HR repair (HRR) is crucial for the maintenance of genomic stability, and is the predominant mechanism for DSB. HRR pathway for DSB repair is a highly complex process that involves multiple proteins, and occurs during the S and G2 phases of the cell cycle [41]. Many tumor suppressors participate in this pathway, including BRCA1, BRCA2 and ATM. As heterozygosity at a BRCA allele is associated with effective HR, DSB accumulation induced by PARP-inhibition specifically occurs only in tumor cells with acquired BRCA$^{-/-}$ homozygosity [42, 43]. Reasons for "BRCAness" are inactivation of BRCA1 or BRCA2 function caused by aberrant epigenetic or posttranslational modifications, and a wider range of mutations in other genes resulting in defective DSB signaling and HRR. Tumors with HRR defects are highly sensitive to crosslinking agents such as cisplatin, carboplatin and nitrosoureas, and DSBs that are induced by ionizing radiation and topo I poisons.

The high frequency of HRR defects in tumors may underlie the efficacy of cytotoxic therapy and provide a rationale for the use of inhibitors of HRR in the sensitization of tumors with functional HRR to conventional chemotherapy. Recent evidence suggests that PARP-i induces single agent cytotoxicity in cells with reduced expression of ATM, the checkpoint activator that is activated by DSBs [27, 44]. There are few HRR inhibitors, but mirin is an inhibitor of MRE11 endonuclease activity and thus inhibits HRR function [45]. Germline mutations in the HR protein RAD51D confer susceptibility to ovarian cancer and may be a target for PARP-i in a small subset of women [46]. Other prototype RAD51 inhibitors have been identified but the most common way to target HRR is by inhibition of the ATM–ChK2 or ATR–Chk1 pathways. Hyperactive growth factor signaling and oncogene-induced replicative stress increase DNA breakage that activates the ATR–Chk1 pathway, and some examples of synthetic lethality of checkpoint or DNA repair inhibitors in cells harbouring activated oncogenes have been shown. ATR knockdown was synthetically lethal in cells that were transformed with mutant KRAS [47], and inhibition of Chk1 and Chk2 significantly delayed disease progression of transplanted MYC-overexpressing lymphoma cells in vivo [48].

NON-HOMOLOGOUS END JOINING

NHEJ is thought to be the major pathway for DSB repair. Damage recognition in NHEJ is performed by the Ku70/Ku80 heterodimer, which binds to the

DSB ends with high affinity, possibly tethering the broken ends together. Ku binding recruits and activates the DNA-dependent protein kinase catalytic subunit (DNA-PKcs), forming the DNA-PK complex that phosphorylates other repair proteins including XRCC4-like factor (XLF), Werner syndrome helicase, DNA ligase IV and XRCC4.

DNA-PKcs is a member of the PI3K-related protein kinase family of enzymes that also includes ATM, ATR and mammalian target of rapamycin (mTOR). PI3K inhibitors, such as wortmannin and LY294002, also inhibit DNA-PKcs, and in proof-of-concept studies, these drugs hindered DSB rejoining and enhanced the cytotoxicity of DSB-inducing agents [49, 50]. More potent and specific DNA-PKcs inhibitors have been developed [51, 52] that substantially slow DSB repair and increase the cytotoxicity and antitumor activity of IR, radiomimetics and topo II poisons in cells and xenografts [53,54]. However, none of these agents have reached the clinical testing stage.

TRANSLESION SYNTHESIS

If damaged DNA bases or adducts are not repaired, they may stall replication forks, which could contribute to genomic instability [55]. Several DNA polymerases can synthesize DNA past DNA lesions. Such TLS contributes to survival. However, errors can occur because these polymerases have no proofreading function and therefore, TLS should be considered a DNA damage tolerance mechanism rather than a DNA repair mechanism. Defects in TLS polymerases contribute to carcinogenesis but also confer sensitivity to DNA-damaging agents, and inhibitors of these polymerases are starting to emerge [56, 57].

SYNTHETIC LETHAL STRATEGIES

Perhaps the most promising prospect for cancer treatment is the exploitation of dysregulated DDR by the synthetic lethality approach. Synthetic lethality exploits inter-gene relationships where the loss of function of either one of two related genes is nonlethal, but loss of both causes cell death. Loss of some elements of one DNA repair pathway may be compensated by the increased activity of other elements or pathways. The discovery of the synthetic lethality relationship between PARP1 and BRCA suggests that other tumor-specific defects in DSB repair factors may be therapeutically targeted by PARP inhibition.

The best characterized synthetic lethality relationship is between BRCA mutation and PARP1 inhibition [58-60]. BRCA1 and-2 have long been known as tumor suppressors, and their inherited mutation increases susceptibility to breast and ovarian tumors [61]. Both BRCA gene products have a role in the HRR pathway [62]. In BRCA-deficient cells, loss of effective HR leads to DSB persistence and cell death. However, resistance to PARP-i can develop owing to secondary mutations in BRCA1 or BRCA2 that restore their function [63, 64]. In addition, even in BRCA-mutant cells, HRR function and PARP-i resistance can be restored if 53BP1 or DNA-PKcs are also inactivated [65, 66].

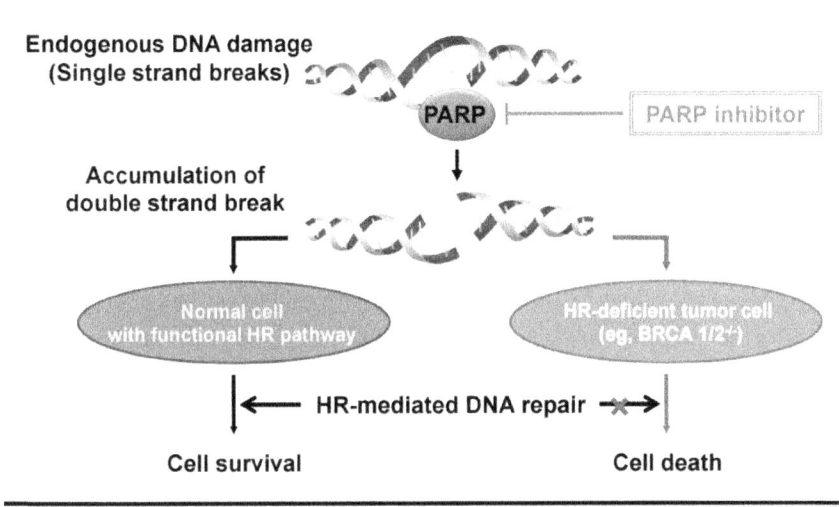

HR, homologous recombination; PARP, poly-adenosine-diphosphate-ribose (PAR) polymerase.

Figure 3. Tumor selective synthetic lethality.

Beyond BRCA1 and BRCA2, their joint interaction partner PALB2 is emerging as a breast cancer susceptibility gene, thus providing another opportunity for PARP-i-based therapies [67]. NVP-BEZ235, a recognized dual PI3K/mTOR inhibitor, was also reported to efficiently block ATM, ATR and DNA-PK activity. Furthermore, NVP-BEZ235 was found to act as a radio-and chemosensitizer in various cancer cell lines [68, 69] and is currently being tested as a single agent in various phase I/II clinical trials [70, 71].

Synthetic lethality of components of the cell cycle checkpoint machinery could be exploited in cancers harbouring activated oncogenes, since oncogene-induced replication stress activates the ATR-Chk1 signaling pathway. Importantly, more than 50% of human tumors are defective in p53 tumor suppressor function and cell cycle checkpoint inhibitors have been demonstrated to sensitize p53-deficient cancer cells to various anticancer agents in clinical use [72]. The two transducer kinases Chk1 and Chk2 are downstream of ATM and ATR, and several inhibitors of transducer kinases have emerged in recent years. Recently, three novel Chk1 inhibitors, GDC-0425, SCH900776 and LY-2606368, have entered phase I clinical trials either as single agents or in combination with gemcitabine, a nucleoside analogue [73]. Another promising drug that interferes with checkpoint activation is the WEE1 tyrosine kinase inhibitor MK-1775 [74]. MK-1775 is already under investigation in a phase II trial combined with carboplatin in order to assess the benefit for patients with p53-mutated epithelial ovarian cancer. Several agents targeting CDC25 phosphatases that represent key molecules in checkpoint regulation have also been developed [75, 76].

SSBR factors other than PARP1 are potential synthetic lethality partners in DSB repair loss, which is supported by the observed cytotoxicity induced by inhibitors of ATM or DNA-PKcs following knockdown of the BER protein XRCC1 [77]. Recent evidence suggests that relationships between BER and non-HR DNA repair pathways may have potential synthetic lethality. The ATR inhibitor NU6027 was also more profoundly cytotoxic to BER-defective cells and in BER-functional cells treated with a PARP-i, reflecting the complementarity of HRR and BER [78].

Phosphatase and tensin homolog (PTEN) is a negative regulator of the anti-apoptotic PI3K/Akt/mTOR pathway. PTEN has recently been implicated in the maintenance of genomic integrity [79-83]. In the nucleus, PTEN promotes chromosome stability and DNA repair. PTEN loss-of-function could be an effective target for treatment strategies. Since PTEN deficiency causes a defect in HR, cells rely on PARP for the repair of DSBs. PTEN deficiency therefore sensitizes cancer cells to PARP inhibition [84-86]. Mendes-Pereira et al. [84] tested for synthetic lethality in HCT116 colorectal tumor cells transfected with a PTEN-mutant cDNA clone. Homozygosity for PTEN mutation was associated with a 20-fold increase in sensitivity to PARP-i in vitro and in vivo. Ectopic expression of RAD51 in a PTEN-deficient cell line overcame PARP-i sensitivity, supporting the proposed link between PTEN mutation and reduced RAD51 expression. Similar results were demonstrated in uterine endometrial carcinoma [85]. In primary PTEN−/− mouse astrocytes, reduced transcription of the RAD51 paralogs was associated with sensitivity to PARP inhibition [86], while PTEN disruption in colorectal cancer cells resulted in reduced MRE11 accumulation at DSBs that is also associated with PARP-i sensitivity [87]. Prostate cancers exhibiting PTEN loss often harbor a genetic rearrangement leading to TMPRSS22-ERG fusion. The TMPRSS22-ERG protein product promotes the formation of DNA DSBs and interacts with PARP, thus sensitizing cells to PARP inhibition [88, 89]. In lung cancer cells, PTEN deficiency potentiated the synergistic effect of olaparib and cisplatin combination treatment [90], while rucaparib sensitized PTEN-deficient prostate cancer cells to ionizing radiation [91]. In melanoma cells, PTEN loss may contribute to BRAF and APE1 inhibition [92, 93]. Retrospective analysis of genetic alterations and PTEN status in tumors taken from patients who are participating in an ongoing clinical trial will provide information for the development of synthetic lethal treatment involving PTEN [94].

Mutations of the von Hippel–Lindau (VHL) tumor suppressor gene occur in the majority of sporadic renal cell carcinomas (RCC). The lack of VHL function in cells results in decreased repair capacity [95]. For example, the suppressor of cytokine signaling 1 (SOCS1) promotes nuclear redistribution and K63 ubiquitylation of VHL in response to DSBs. Loss of VHL function or VHL mutation that compromises K63 ubiquitylation attenuates the DDR, resulting in decreased HRR and persistence of DSBs [96]. Furthermore, loss of VHL function is associated with stabilization of hypoxia-inducible factor α (HIFα). The exposure of cells to hypoxia markedly enhances genetic instability caused by exogenous genotoxins, and HIF activation decreased NER [97]. Recently, synthetic lethal (SL) partner of VHL was identified from a screening of large volumes of cancer genomic data using a small interfering RNA screen. The VHL-deficient cells are significantly more sensitive to the knockdown of the predicted VHL-SL partners [98]. DNA repair pathway abnormalities involving VHL dysfunction might be therapeutic targets.

Many strategies based on the concept of synthetic lethality have so far only been investigated in preclinical settings.

PARP INHIBITOR

A number of potential PARP-i have been identified. In xenograft and in vitro models, PARP-i have been demonstrated to potentiate the action of a wide variety of damaging agents including platinums, the alkylating agents TMZ and cyclophosphamide, the nucleoside analogue gemcitabine, the topo inhibitor irinotecan, and ionizing radiation [90]. Furthermore, preclinical studies also suggested the potential use of PARP-i in sporadic cancers that share phenotypical features with cancers arising from hereditary BRCA mutations, a phenomenon that is referred to as "BRCAness" [91]. Many additional phase I and II trials are currently underway, examining the combination with a variety of agents including carboplatin, 5-fluorouracil and oxaliplatin, cisplatin and paclitaxel, topotecan, gemcitabine, and radiotherapy [92]. For example, rucaparib has been evaluated in phase I and II studies in combination with TMZ for malignant melanoma, demonstrating successful PARP inhibition at the tissue level and probable anticancer activity, but significant myelosuppression caused dose-limiting toxicity [93].

An initial phase I study of olaparib in a cohort enriched for BRCA1/2 mutation carriers demonstrated evidence of in vivo anti-PARP activity and evidence of response in 40% of BRCA carriers [60]. Phase II trials of olaparib for breast or ovarian cancer associated with BRCA1/2 mutations were favorable, suggesting antitumor efficacy [94, 95]. Good responses were also seen in patients with BRCA-associated breast and ovarian cancers, and even in unselected patients with high-grade serous ovarian cancer [96, 97]. However, olaparib did not progress to a phase III trial for hereditary BRCA mutation-associated breast cancer due to economic concerns [98].

Iniparib has been evaluated in a phase II study of metastatic triple-negative breast cancer treatment in combination with gemcitabine and carboplatin. A significantly improved median overall survival was demonstrated compared with gemcitabine and carboplatin, without increased toxicity. However, a phase III trial failed to meet co-primary endpoints of overall and progression-free survival improvement, and after further disappointing results in a phase III non-small cell lung cancer trial, iniparib has been suspended from further development [99].

A good safety profile was also observed with veliparib in combination with TMZ. This was associated with early positive results in metastatic colorectal and BRCA-deficient breast cancers, although the combination was associated with poor response and no progression-free or overall survival improvement in advanced melanoma. Likewise, phase II investigation of rucaparib in BRCA1/2-mutated breast or ovarian cancer demonstrated PARP activity inhibition and evidence of a tumor response. The oral PARP1/2 inhibitor niraparib has also been evaluated at phase I and was shown to possess an acceptable safety profile and probable antitumor activity. Other PARP-i including orally bioavailable agents are currently being tested in clinical trials [100].

Clinical trials of PARP-i have generally been disappointing owing to toxicity, which may be due to use of a dose of PARP-i that was established as safe when used as a single agent. In general, preclinical data indicated that the MTD of single agent PARP-i was much higher than MTD of PARP-i when combined with another cytotoxic agent such as TMZ [101, 102]. This is because almost total inhibition of PARP-i is needed to render endogenous DNA damage cytotoxic, but this level of inhibition is not necessary to render the additional burden of deliberately introduced DNA damage cytotoxic, both in the tumor and in proliferating normal tissues. In addition, secondary BRCA2 mutations have been identified, which restore the full-length protein, thereby re-establishing BRCA2 functions and conferring PARP-i resistance [103]. A major challenge of using PARP-i is the acquired resistance of initially PARP-i-sensitive cancer cells due, for example, to the loss of p53-binding protein-1 (53BP1) or to overexpression of multidrug-resistance efflux transporters [104, 105]. The data described above suggest that the clinical utility of PARP-i in combination with chemotherapy may be limited in tumors in view of its narrow therapeutic index.

PREDICTIVE BIOMARKERS

Relevant biomarker assays should predict the functionality of DNA repair pathways, rather than just providing information about mutations or expression levels of proteins involved in the DNA repair pathway. Furthermore, such detailed molecular profiling of cancer versus normal tissue from a given patient is critical to maximize the potential of personalized cancer drugs in terms of both therapeutic success and cost-effectiveness.

A general marker of DNA damage is the phosphorylation of histone H2AX by ATM, ATR and DNA-PK. γH2AX foci, formed at sites of DSBs, or increased levels of γH2AX, may be measured by immunofluorescence microscopy, flow cytometry or immunoblotting and used to detect DNA damage [106]. The increase and/or persistence of γH2AX can be used to demonstrate the inhibition of PARP, DNA-PK, ATR and Chk1. To directly measure the effect of a molecularly targeted agent, immunological methods may be used to detect the product. For example, activation of DNA-PK and ATM in response to DNA damage can be determined by measuring their autophosphorylation with phospho-specific antibodies, and PARP activity may be measured by immunodetection of the ADP-ribose polymer product, to guide PARP-i clinical trials [60, 107-109]. In multiple clinical trials, PARP activity in peripheral mononuclear blood cells has been used as a marker of effective inhibition [110,111].

An alternative approach is to assess HRR function in fresh viable tumor material by measuring the number of RAD51 foci following ex vivo DNA damage induction [112-114]. In the ovarian cancer study, this was further analyzed in BRCA2-mutated pancreatic cancer cell clones to predict RAD51 foci formation as a marker of HR, and to examine for sensitivity to PARP inhibition.

When inactivation of a single gene has been identified as a crucial determinant of sensitivity, it may then be used to select patients for the appropriate therapy. For example, low levels of the NER endonuclease ERCC1 correlate with cisplatin sensitivity in several cancers [30, 115, 116]. Several studies report methods to identify tumors with non-germline HRR defects: gene expression profiling, methylation-specific arrays, immunohistochemistry analysis of tissue microarrays and copy number aberrations by array comparative genomic hybridization. [117-121]

Immunohistochemistry analysis of formalin-fixed, paraffin-embedded samples may be a useful tool for identifying DDR defects in order to stratify patients. To measure the effect of an agent that directly causes DNA DSBs in all phases of the cell cycle, patient-derived lymphocytes can be used [122]. Owing to the invasive procedures that are needed to obtain tumor material, except in the case of hematological malignancies, circulating tumor cells offer the best hope of routinely obtaining suitable material [123].

CONCLUSION

DNA repair mechanisms play an essential role in promoting genomic stability. On the other hand, impaired DNA repair capacity in cancer cells may result in a favorable response to chemotherapy. Many conventional therapeutic regimens that effectively kill cancer cells are based on DNA damage. However, most chemotherapeutic regimens cause severe side effects that limit their therapeutic potential. Inhibition of DNA repair is a new paradigm in cancer therapy, and there is heightened interest in the therapeutic potential of these inhibitors that selectively target tumors with minimal host toxicity.

The synthetic lethal approaches targeting the individual genetic profile of the tumors are under clinical development. The molecular characterization of tumors and reliable biomarkers are needed for effective personalized therapy. Further research is necessary in order to determine the most appropriate treatment for patients.

REFERENCES

1. Derheimer FA. Multiple roles of ATM in monitoring and maintaining DNA integrity. FEBS letters 2010; 584(17) 3675-3681.
2. Sancar A. Molecular mechanisms of mammalian DNA repair and the DNA damage checkpoints. Annual review of biochemistry 2004; 73 39-85.
3. Barnes DE. Repair and genetic consequences of endogenous DNA base damage in mammalian cells. Annual review of genetics 2004; 38 445-476.
4. Hoeijmakers JH. DNA damage, aging, and cancer. The New England journal of medicine 2009; 361(15) 1475-1485.
5. Jiricny J. The multifaceted mismatch-repair system. Nature reviews Molecular cell biology 2006; 7(5) 335-346.

6. Scharer OD. DNA interstrand crosslinks: natural and drug-induced DNA adducts that induce unique cellular responses. Chembiochem : a European journal of chemical biology 2005; 6(1) 27-32.
7. Ciccia A. The DNA damage response: making it safe to play with knives. Molecular cell 2010; 40(2) 179-204.
8. Chapman JR. Playing the end game: DNA double-strand break repair pathway choice. Molecular cell 2012; 47(4) 497-510.
9. Abbotts R. Human AP endonuclease 1 (APE1): from mechanistic insights to druggable target in cancer. Cancer treatment reviews 2010; 36(5) 425-435.
10. Eker AP. DNA repair in mammalian cells: Direct DNA damage reversal: elegant solutions for nasty problems. Cellular and molecular life sciences : CMLS 2009; 66(6) 968-980.
11. Lord CJ. Biology-driven cancer drug development: back to the future. BMC biology 2010; 8 38.
12. Tubbs JL. DNA binding, nucleotide flipping, and the helix-turn-helix motif in base repair by O6-alkylguanine-DNA alkyltransferase and its implications for cancer chemotherapy. DNA repair 2007; 6(8) 1100-1115.
13. Tricker AR. Carcinogenic N-nitrosamines in the diet: occurrence, formation, mechanisms and carcinogenic potential. Mutation research 1991; 259(3-4) 277-289.
14. Ranson M. Lomeguatrib, a potent inhibitor of O6-alkylguanine-DNA-alkyltransferase: phase I safety, pharmacodynamic, and pharmacokinetic trial and evaluation in combination with temozolomide in patients with advanced solid tumors. Clinical cancer research : an official journal of the American Association for Cancer Research 2006; 12(5) 1577-1584.
15. Watson AJ. Tumor O(6)-methylguanine-DNA methyltransferase inactivation by oral lomeguatrib. Clinical cancer research : an official journal of the American Association for Cancer Research 2010; 16(2) 743-749.
16. Esteller M. Inactivation of the DNA repair gene O6-methylguanine-DNA methyltransferase by promoter hypermethylation is a common event in primary human neoplasia. Cancer research 1999; 59(4) 793-797.
17. Hegi ME. MGMT gene silencing and benefit from temozolomide in glioblastoma. The New England journal of medicine 2005; 352(10) 997-1003.

18. Plo I. Association of XRCC1 and tyrosyl DNA phosphodiesterase (Tdp1) for the repair of topoisomerase I-mediated DNA lesions. DNA repair 2003; 2(10) 1087-1100.
19. Wang JC. Cellular roles of DNA topoisomerases: a molecular perspective. Nature reviews Molecular cell biology 2002; 3(6) 430-440.
20. Pommier Y. Repair of and checkpoint response to topoisomerase I-mediated DNA damage. Mutation research 2003; 532(1-2) 173-203.
21. El-Khamisy SF. A requirement for PARP-1 for the assembly or stability of XRCC1 nuclear foci at sites of oxidative DNA damage. Nucleic acids research 2003; 31(19) 5526-5533.
22. Hirai K. Aberration of poly(adenosine diphosphate-ribose) metabolism in human colon adenomatous polyps and cancers. Cancer research 1983; 43(7) 3441-3446.
23. Zaremba T. Poly(ADP-ribose) polymerase-1 polymorphisms, expression and activity in selected human tumour cell lines. British journal of cancer 2009; 101(2) 256-262.
24. Taverna P. Methoxyamine potentiates DNA single strand breaks and double strand breaks induced by temozolomide in colon cancer cells. Mutation research 2001; 485(4) 269-281.
25. Luo M. Inhibition of the human apurinic/apyrimidinic endonuclease (APE1) repair activity and sensitization of breast cancer cells to DNA alkylating agents with lucanthone. Anticancer research 2004; 24(4) 2127-2134.
26. Mohammed MZ. Development and evaluation of human AP endonuclease inhibitors in melanoma and glioma cell lines. British journal of cancer 2011; 104(4) 653-663.
27. Sultana R. Synthetic lethal targeting of DNA double-strand break repair deficient cells by human apurinic/apyrimidinic endonuclease inhibitors. International journal of cancer Journal international du cancer 2012; 131(10) 2433-2444.
28. Naegeli H. The xeroderma pigmentosum pathway: decision tree analysis of DNA quality. DNA repair 2011; 10(7) 673-683.
29. Koberle B. DNA repair capacity and cisplatin sensitivity of human testis tumour cells. International journal of cancer Journal international du cancer 1997; 70(5) 551-555.
30. Usanova S. Cisplatin sensitivity of testis tumour cells is due to deficiency in interstrand-crosslink repair and low ERCC1-XPF expression. Molecular cancer 2010; 9 248.

31. Zhang YW. Poly(ADP-ribose) polymerase and XPF-ERCC1 participate in distinct pathways for the repair of topoisomerase I-induced DNA damage in mammalian cells. Nucleic acids research 2011; 39(9) 3607-3620.
32. Iorns E. Integrated functional, gene expression and genomic analysis for the identification of cancer targets. Plos One 2009; 4(4) e5120.
33. Sourisseau T. Aurora-A expressing tumour cells are deficient for homology-directed DNA double strand-break repair and sensitive to PARP inhibition. EMBO molecular medicine 2010; 2(4) 130-142.
34. Li GM. Mechanisms and functions of DNA mismatch repair. Cell research 2008; 18(1) 85-98.
35. Umar A. Revised Bethesda Guidelines for hereditary nonpolyposis colorectal cancer (Lynch syndrome) and microsatellite instability. Journal of the National Cancer Institute 2004; 96(4) 261-268.
36. Wu X. Causal link between microsatellite instability and hMRE11 dysfunction in human cancers. Molecular cancer research : MCR 2011; 9(11) 1443-1448.
37. Ham MF. Impairment of double-strand breaks repair and aberrant splicing of ATM and MRE11 in leukemia-lymphoma cell lines with microsatellite instability. Cancer science 2006; 97(3) 226-234.
38. Karran P. DNA damage tolerance, mismatch repair and genome instability. BioEssays : news and reviews in molecular, cellular and developmental biology 1994; 16(11) 833-839.
39. Fordham SE. DNA mismatch repair status affects cellular response to Ara-C and other anti-leukemic nucleoside analogs. Leukemia 2011; 25(6) 1046-1049.
40. Plumb JA. Reversal of drug resistance in human tumor xenografts by 2'-deoxy-5-azacytidine-induced demethylation of the hMLH1 gene promoter. Cancer research 2000; 60(21) 6039-6044.
41. Shrivastav M. Regulation of DNA double-strand break repair pathway choice. Cell research 2008; 18(1) 134-147.
42. Farmer H. Targeting the DNA repair defect in BRCA mutant cells as a therapeutic strategy. Nature 2005; 434(7035) 917-921.
43. Bryant HE. Specific killing of BRCA2-deficient tumours with inhibitors of poly(ADP-ribose) polymerase. Nature 2005; 434(7035) 913-917.
44. Williamson CT. Enhanced cytotoxicity of PARP inhibition in mantle cell lymphoma harbouring mutations in both ATM and p53. EMBO molecular medicine 2012; 4(6) 515-527.

45. Dupre A. A forward chemical genetic screen reveals an inhibitor of the Mre11-Rad50-Nbs1 complex. Nature chemical biology 2008; 4(2) 119-125.

46. Loveday C. Germline mutations in RAD51D confer susceptibility to ovarian cancer. Nature genetics 2011; 43(9) 879-882.

47. Gilad O. Combining ATR suppression with oncogenic Ras synergistically increases genomic instability, causing synthetic lethality or tumorigenesis in a dosage-dependent manner. Cancer research 2010; 70(23) 9693-9702.

48. Ferrao PT. Efficacy of CHK inhibitors as single agents in MYC-driven lymphoma cells. Oncogene 2012; 31(13) 1661-1672.

49. Rosenzweig KE. Radiosensitization of human tumor cells by the phosphatidylinositol3-kinase inhibitors wortmannin and LY294002 correlates with inhibition of DNA-dependent protein kinase and prolonged G2-M delay. Clinical cancer research : an official journal of the American Association for Cancer Research 1997; 3(7) 1149-1156.

50. Boulton S. Mechanisms of enhancement of cytotoxicity in etoposide and ionising radiation-treated cells by the protein kinase inhibitor wortmannin. Eur J Cancer 2000; 36(4) 535-541.

51. Hardcastle IR. Discovery of potent chromen-4-one inhibitors of the DNA-dependent protein kinase (DNA-PK) using a small-molecule library approach. Journal of medicinal chemistry 2005; 48(24) 7829-7846.

52. Shinohara ET. DNA-dependent protein kinase is a molecular target for the development of noncytotoxic radiation-sensitizing drugs. Cancer research 2005; 65(12) 4987-4992.

53. Zhao Y. Preclinical evaluation of a potent novel DNA-dependent protein kinase inhibitor NU7441. Cancer research 2006; 66(10) 5354-5362.

54. Munck JM. Chemosensitization of cancer cells by KU-0060648, a dual inhibitor of DNA-PK and PI-3K. Molecular cancer therapeutics 2012; 11(8) 1789-1798.

55. Lange SS. DNA polymerases and cancer. Nature reviews Cancer 2011; 11(2) 96-110.

56. Mizushina Y. 3-O-methylfunicone, a selective inhibitor of mammalian Y-family DNA polymerases from an Australian sea salt fungal strain. Marine drugs 2009; 7(4) 624-639.

57. Dorjsuren D. A real-time fluorescence method for enzymatic characterization of specialized human DNA polymerases. Nucleic acids research 2009; 37(19) e128.

58. Tuma RS. Combining carefully selected drug, patient genetics may lead

to total tumor death. Journal of the National Cancer Institute 2007; 99(20) 1505-1506, 1509.
59. Lord CJ. Targeted therapy for cancer using PARP inhibitors. Current opinion in pharmacology 2008; 8(4) 363-369.
60. Fong PC. Inhibition of poly(ADP-ribose) polymerase in tumors from BRCA mutation carriers. The New England journal of medicine 2009; 361(2) 123-134.
61. Miki Y. A strong candidate for the breast and ovarian cancer susceptibility gene BRCA1. Science 1994; 266(5182) 66-71.
62. Venkitaraman AR. Cancer susceptibility and the functions of BRCA1 and BRCA2. Cell 2002; 108(2) 171-182.
63. Sakai W. Secondary mutations as a mechanism of cisplatin resistance in BRCA2-mutated cancers. Nature 2008; 451(7182) 1116-1120.
64. Swisher EM. Secondary BRCA1 mutations in BRCA1-mutated ovarian carcinomas with platinum resistance. Cancer research 2008; 68(8) 2581-2586.
65. Bouwman P. 53BP1 loss rescues BRCA1 deficiency and is associated with triple-negative and BRCA-mutated breast cancers. Nature structural & molecular biology 2010; 17(6) 688-695.
66. Patel AG. Nonhomologous end joining drives poly(ADP-ribose) polymerase (PARP) inhibitor lethality in homologous recombination-deficient cells. Proceedings of the National Academy of Sciences of the United States of America 2011; 108(8) 3406-3411.
67. Poumpouridou N. Hereditary breast cancer: beyond BRCA genetic analysis; PALB2 emerges. Clinical chemistry and laboratory medicine : CCLM / FESCC 2012; 50(3) 423-434.
68. Kudoh A. Dual inhibition of phosphatidylinositol 3'-kinase and mammalian target of rapamycin using NVP-BEZ235 as a novel therapeutic approach for mucinous adenocarcinoma of the ovary. International journal of gynecological cancer : official journal of the International Gynecological Cancer Society 2014; 24(3) 444-453.
69. Oishi T. The PI3K/mTOR dual inhibitor NVP-BEZ235 reduces the growth of ovarian clear cell carcinoma. Oncology reports 2014; 32(2) 553-558.
70. Yang F. Dual Phosphoinositide 3-Kinase/Mammalian Target of Rapamycin Inhibitor NVP-BEZ235 Has a Therapeutic Potential and Sensitizes Cisplatin in Nasopharyngeal Carcinoma. Plos One 2013; 8(3).
71. Mukherjee B. The Dual PI3K/mTOR Inhibitor NVP-BEZ235 Is a Potent

Inhibitor of ATM- and DNA-PKCs-Mediated DNA Damage Responses. Neoplasia 2012; 14(1) 34-U53.

72. Ma CX. Targeting Chk1 in p53-deficient triple-negative breast cancer is therapeutically beneficial in human-in-mouse tumor models. The Journal of clinical investigation 2012; 122(4) 1541-1552.

73. Guzi TJ. Targeting the Replication Checkpoint Using SCH 900776, a Potent and Functionally Selective CHK1 Inhibitor Identified via High Content Screening. Molecular cancer therapeutics 2011; 10(4) 591-602.

74. Hamer PCD. WEE1 Kinase Targeting Combined with DNA-Damaging Cancer Therapy Catalyzes Mitotic Catastrophe. Clinical Cancer Research 2011; 17(13) 4200-4207.

75. Lavecchia A. CDC25 Phosphatase Inhibitors: An Update. Mini-Rev Med Chem 2012; 12(1) 62-73.

76. Brezak MC. IRC-083864, a novel his quinone inhibitor of CDC25 phosphatases active against human cancer cells. International Journal of Cancer 2009; 124(6) 1449-1456.

77. Sultana R. Targeting XRCC1 deficiency in breast cancer for personalized therapy. Cancer research 2013; 73(5) 1621-1634.

78. Peasland A. Identification and evaluation of a potent novel ATR inhibitor, NU6027, in breast and ovarian cancer cell lines. British journal of cancer 2011; 105(3) 372-381.

79. Gupta A. Cell cycle checkpoint defects contribute to genomic instability in PTEN deficient cells independent of DNA DSB repair. Cell Cycle 2009; 8(14) 2198-2210.

80. Puc J. Lack of PTEN sequesters CHK1 and initiates genetic instability. Cancer cell 2005; 7(2) 193-204.

81. Shen WH. Essential role for nuclear PTEN in maintaining chromosomal integrity. Cell 2007; 128(1) 157-170.

82. Yin Y. PTEN: a new guardian of the genome. Oncogene 2008; 27(41) 5443-5453.

83. Planchon SM. The nuclear affairs of PTEN. Journal of cell science 2008; 121(Pt 3) 249-253.

84. Mendes-Pereira AM. Synthetic lethal targeting of PTEN mutant cells with PARP inhibitors. EMBO molecular medicine 2009; 1(6-7) 315-322.

85. Dedes KJ. PTEN deficiency in endometrioid endometrial adenocarcinomas predicts sensitivity to PARP inhibitors. Science translational medicine 2010; 2(53) 53ra75.

86. McEllin B. PTEN loss compromises homologous recombination repair in

87. Fraser M. PTEN deletion in prostate cancer cells does not associate with loss of RAD51 function: implications for radiotherapy and chemotherapy. Clinical cancer research : an official journal of the American Association for Cancer Research 2012; 18(4) 1015-1027.
88. King JC. Cooperativity of TMPRSS2-ERG with PI3-kinase pathway activation in prostate oncogenesis. Nature genetics 2009; 41(5) 524-526.
89. Brenner JC. Mechanistic rationale for inhibition of poly(ADP-ribose) polymerase in ETS gene fusion-positive prostate cancer. Cancer cell 2011; 19(5) 664-678.
90. Minami D. Synergistic effect of olaparib with combination of cisplatin on PTEN-deficient lung cancer cells. Molecular cancer research : MCR 2013; 11(2) 140-148.
91. Chatterjee P. PARP inhibition sensitizes to low dose-rate radiation TMPRSS2-ERG fusion gene-expressing and PTEN-deficient prostate cancer cells. Plos One 2013; 8(4) e60408.
92. Paraiso KH. PTEN loss confers BRAF inhibitor resistance to melanoma cells through the suppression of BIM expression. Cancer research 2011; 71(7) 2750-2760.
93. Abbotts R. Targeting human apurinic/apyrimidinic endonuclease 1 (APE1) in phosphatase and tensin homolog (PTEN) deficient melanoma cells for personalized therapy. Oncotarget 2014; 5(10) 3273-3286.
94. Dillon LM. Therapeutic targeting of cancers with loss of PTEN function. Current drug targets 2014; 15(1) 65-79.
95. Schults MA. Loss of VHL in RCC Reduces Repair and Alters Cellular Response to Benzo[a]pyrene. Frontiers in oncology 2013; 3 270.
96. Metcalf JL. K63-ubiquitylation of VHL by SOCS1 mediates DNA double-strand break repair. Oncogene 2014; 33(8) 1055-1065.
97. Schults MA. Diminished carcinogen detoxification is a novel mechanism for hypoxia-inducible factor 1-mediated genetic instability. The Journal of biological chemistry 2010; 285(19) 14558-14564.
98. Jerby-Arnon L. Predicting cancer-specific vulnerability via data-driven detection of synthetic lethality. Cell 2014; 158(5) 1199-1209.
99. Donawho CK. ABT-888, an orally active poly(ADP-ribose) polymerase inhibitor that potentiates DNA-damaging agents in preclinical tumor models. Clinical cancer research : an official journal of the American

Association for Cancer Research 2007; 13(9) 2728-2737.
100. Turner N. Hallmarks of 'BRCAness' in sporadic cancers. Nature reviews Cancer 2004; 4(10) 814-819.
101. Davar D. Role of PARP inhibitors in cancer biology and therapy. Current medicinal chemistry 2012; 19(23) 3907-3921.
102. Plummer R. A phase II study of the potent PARP inhibitor, Rucaparib (PF-01367338, AG014699), with temozolomide in patients with metastatic melanoma demonstrating evidence of chemopotentiation. Cancer chemotherapy and pharmacology 2013; 71(5) 1191-1199.
103. Tutt A. Oral poly(ADP-ribose) polymerase inhibitor olaparib in patients with BRCA1 or BRCA2 mutations and advanced breast cancer: a proof-of-concept trial. Lancet 2010; 376(9737) 235-244.
104. Audeh MW. Oral poly(ADP-ribose) polymerase inhibitor olaparib in patients with BRCA1 or BRCA2 mutations and recurrent ovarian cancer: a proof-of-concept trial. Lancet 2010; 376(9737) 245-251.
105. Fong PC. Poly(ADP)-ribose polymerase inhibition: frequent durable responses in BRCA carrier ovarian cancer correlating with platinum-free interval. Journal of clinical oncology : official journal of the American Society of Clinical Oncology 2010; 28(15) 2512-2519.
106. Gelmon KA. Olaparib in patients with recurrent high-grade serous or poorly differentiated ovarian carcinoma or triple-negative breast cancer: a phase 2, multicentre, open-label, non-randomised study. The Lancet Oncology 2011; 12(9) 852-861.
107. Guha M. PARP inhibitors stumble in breast cancer. Nature biotechnology 2011; 29(5) 373-374.
108. O'Shaughnessy J. Iniparib plus chemotherapy in metastatic triple-negative breast cancer. The New England journal of medicine 2011; 364(3) 205-214.
109. Glendenning J. PARP inhibitors--current status and the walk towards early breast cancer. Breast 2011; 20 Suppl 3 S12-19.
110. Drew Y. Therapeutic potential of poly(ADP-ribose) polymerase inhibitor AG014699 in human cancers with mutated or methylated BRCA1 or BRCA2. Journal of the National Cancer Institute 2011; 103(4) 334-346.
111. Thomas HD. Preclinical selection of a novel poly(ADP-ribose) polymerase inhibitor for clinical trial. Molecular cancer therapeutics 2007; 6(3) 945-956.
112. Sandhu SK. Poly (ADP-ribose) polymerase (PARP) inhibitors for the treatment of advanced germline BRCA2 mutant prostate cancer. Annals

of oncology : official journal of the European Society for Medical Oncology / ESMO 2013; 24(5) 1416-1418.
113. Oplustilova L. Evaluation of candidate biomarkers to predict cancer cell sensitivity or resistance to PARP-1 inhibitor treatment. Cell Cycle 2012; 11(20) 3837-3850.
114. Jaspers JE. Loss of 53BP1 causes PARP inhibitor resistance in Brca1-mutated mouse mammary tumors. Cancer discovery 2013; 3(1) 68-81.
115. Bonner WM. GammaH2AX and cancer. Nature reviews Cancer 2008; 8(12) 957-967.
116. Plummer R. Phase I study of the poly(ADP-ribose) polymerase inhibitor, AG014699, in combination with temozolomide in patients with advanced solid tumors. Clinical cancer research : an official journal of the American Association for Cancer Research 2008; 14(23) 7917-7923.
117. Kummar S. Phase I study of PARP inhibitor ABT-888 in combination with topotecan in adults with refractory solid tumors and lymphomas. Cancer research 2011; 71(17) 5626-5634.
118. Kummar S. A phase I study of veliparib in combination with metronomic cyclophosphamide in adults with refractory solid tumors and lymphomas. Clinical cancer research : an official journal of the American Association for Cancer Research 2012; 18(6) 1726-1734.
119. Menear KA. 4-[3-(4-cyclopropanecarbonylpiperazine-1-carbonyl)-4-fluorobenzyl]-2H-phthalazin- 1-one: a novel bioavailable inhibitor of poly(ADP-ribose) polymerase-1. Journal of medicinal chemistry 2008; 51(20) 6581-6591.
120. Lubbers LS. PISA, a novel pharmacodynamic assay for assessing poly(ADP-ribose) polymerase (PARP) activity in situ. Journal of pharmacological and toxicological methods 2010; 61(3) 319-328.
121. Willers H. Utility of DNA repair protein foci for the detection of putative BRCA1 pathway defects in breast cancer biopsies. Molecular cancer research : MCR 2009; 7(8) 1304-1309.
122. Gaymes TJ. Inhibitors of poly ADP-ribose polymerase (PARP) induce apoptosis of myeloid leukemic cells: potential for therapy of myeloid leukemia and myelodysplastic syndromes. Haematologica 2009; 94(5) 638-646.
123. Mukhopadhyay A. Development of a functional assay for homologous recombination status in primary cultures of epithelial ovarian tumor and correlation with sensitivity to poly(ADP-ribose) polymerase inhibitors. Clinical cancer research : an official journal of the American Association for Cancer Research 2010; 16(8) 2344-2351.

124. Olaussen KA. DNA repair by ERCC1 in non-small-cell lung cancer and cisplatin-based adjuvant chemotherapy. The New England journal of medicine 2006; 355(10) 983-991.

125. Sato S. Combination chemotherapy of oxaliplatin and 5-fluorouracil may be an effective regimen for mucinous adenocarcinoma of the ovary: a potential treatment strategy. Cancer science 2009; 100(3) 546-551.

126. Jazaeri AA. Gene expression profiles of BRCA1-linked, BRCA2-linked, and sporadic ovarian cancers. Journal of the National Cancer Institute 2002; 94(13) 990-1000.

127. Konstantinopoulos PA. Gene expression profile of BRCAness that correlates with responsiveness to chemotherapy and with outcome in patients with epithelial ovarian cancer. Journal of clinical oncology : official journal of the American Society of Clinical Oncology 2010; 28(22) 3555-3561.

128. Ahluwalia A. DNA methylation in ovarian cancer. II. Expression of DNA methyltransferases in ovarian cancer cell lines and normal ovarian epithelial cells. Gynecologic oncology 2001; 82(2) 299-304.

129. Vollebergh MA. Genomic instability in breast and ovarian cancers: translation into clinical predictive biomarkers. Cellular and molecular life sciences : CMLS 2012; 69(2) 223-245.

130. Rodriguez AA. DNA repair signature is associated with anthracycline response in triple negative breast cancer patients. Breast cancer research and treatment 2010; 123(1) 189-196.

131. Tanaka T. Induction of ATM activation, histone H2AX phosphorylation and apoptosis by etoposide: relation to cell cycle phase. Cell Cycle 2007; 6(3) 371-376.

132. Pantel K. Circulating tumour cells in cancer patients: challenges and perspectives. Trends in molecular medicine 2010; 16(9) 398-406.

Chapter 4

THERAPEUTIC STRATEGIES BASED ON POLYMERIC MICROPARTICLES

C. Vilos[1,2] and L. A. Velasquez[1,2]

[1]Center for Integrative Medicine and Innovative Science (CIMIS), Facultad de Medicina, Universidad Andrés Bello, Santiago, Echaurren 183, 8370071 Santiago, Chile

[2]Center for the Development of Nanoscience and Nanotechnology (CEDENNA), Avenida Ecuador 3493, 9170124 Santiago, Chile

ABSTRACT

The development of the field of materials science, the ability to perform multidisciplinary scientific work, and the need for novel administration technologies that maximize therapeutic effects and minimize adverse reactions to readily available drugs have led to the development of delivery systems based on microencapsulation, which has taken one step closer to the target of personalized medicine. Drug delivery systems based on polymeric microparticles are generating a strong impact on preclinical and clinical drug development and have reached a broad development in different fields supporting a critical role in the near future of medical practice. This paper presents the foundations of polymeric microparticles based on their formulation, mechanisms of drug release and some of their innovative therapeutic strategies to board multiple diseases.

INTRODUCTION

The discovery and development of new drugs for the treatment of diseases is a lengthy and costly process [1]. The drug development typically requires about 14 years, and studies demonstrated that by the year 2013 the cost to reach phase III of clinical trials will be around $ 1.9 billion [2]. Moreover, the number of drug approvals is minimal, reaching less than 32 new molecular entities per year last decade (NME) [3]. The long time required to develop a new drug application and its high costs illustrate the need to develop new therapeutic strategies, which improve the effectiveness of available drugs. Figure 1 shows a scheme of the different stages of drug development required by the Food and Drug Administration (FDA) from discovery of an NME until its marketing.

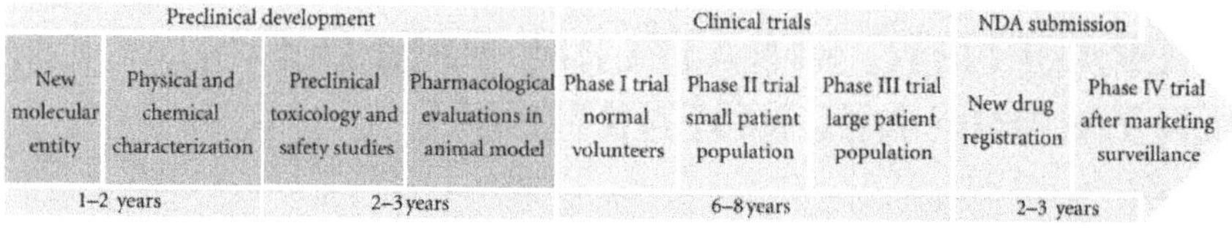

Preclinical development				Clinical trials			NDA submission	
New molecular entity	Physical and chemical characterization	Preclinical toxicology and safety studies	Pharmacological evaluations in animal model	Phase I trial normal volunteers	Phase II trial small patient population	Phase III trial large patient population	New drug registration	Phase IV trial after marketing surveillance
1–2 years		2–3 years			6–8 years		2–3 years	

Figure 1: Schematic description of the stages required by the Food and Drug Administration (FDA) to reach the commercialization of a new drug application (NDA).

The conventional administration of drugs (i.e., tablets, capsules, and injections), and the limited solubility of the drugs often require high doses in order to reach enough concentrations of drug at its site of action to achieve an appropriate therapeutic effect [4]. In other cases, the application of some therapeutic protocols requires the administration of repeated doses to maintain an adequate concentration of drug in the bloodstream and provide therapeutic action for long periods of time [5]. The high blood concentrations of drugs and the administration of multiple doses can generate significative fluctuations of the drug in the bloodstream, which can reach the toxicological parameters, and generate adverse reactions for the patients. All this drawbacks have lead to develop new therapeutic strategies more effective and with fewer side effects for patients.

The advancement of materials science and pharmaceutical technology has allowed the creation of several strategies for drug delivery such as osmotic pumps [6, 7], liposomes [8, 9], hydrogels [10–12], and polymeric microparticles [13, 14]. The main goals of those drug delivery devices are the generation of a sustained release of drug over time, a reduced number of doses required to the treatment of diseases, and the protection of the drugs from inactivation before reaching the target tissue.

The polymeric microparticles (p-MPs) as a drug delivery strategy have advantages over other systems since they do not require surgical procedures for their application or removal from the body like the osmotic pumps. Furthermore, the p-MPs have exhibited a better stability in the biological environment than liposomes, and their highly reproducible formulation methods provide support to encapsulate hydrophilic and hydrophobic drugs, which gives them a wide range of therapeutic applications.

On the other hand, the release of drugs from p-MPs shows several benefits compared with the conventional drug administration methods, which include their ability to modulate the rate of drugs release for a long time periods and their capacity to reduce the drug toxicity.

The extensive benefits of administration of encapsulated drugs into p-MPs serves as the foundation for many future medical endeavors. This paper provides an overview of the basics of polymeric microparticles based on their formulation, their mechanisms of drug delivery, and their applications in the treatment of diseases.

POLYMERS

The use of biodegradable and biocompatible polymers has generated significant advances in modern medicine because it has impacted different fields of biomedicine, which include tissue engineering and diagnostic and therapeutic strategies [15, 16].

The p-MPs, as drug delivery systems, have been developed using different natural and synthetic polymers [17]. The natural polymers include chitosan [18], alginate [19], dextran [20], gelatin [21], and albumin [22], and the synthetic polymers comprise to poly(lactide-co-glycolide) (PLGA) [23], (3-hydroxybutyrate-co-3-hydroxyvalerate) (PHBV) [24], poly(sebacic anhydride) [25], poly(ε-caprolactone), among others [26].

During the last years, the advances in materials sciences have generated different polymers tailored for drug-conjugated, which include smart response that supported the development of novel drug delivery systems [27]. Recently, the use of thermoresponsive (i.e., NIPAAm and CMCTS-g-PDEA) [28, 29] and pH-responsive (i.e., Eudragit L100, Eudragit S and AQOAT AS-MG) [30, 31] polymers in the formulation of p-MPs was described, which promises improved approaches to the delivery of drugs.

MICROENCAPSULATION METHODS

Understanding the physicochemical properties of drugs is essential before determining the appropriate method for the synthesis of the p-MPs because the wide range of pharmaceutical agents such as peptide, proteins, nucleic acids, antibiotics, and chemotherapeutics, have distinctive solubility and stability at different conditions (i.e., temperature, pH, and organic solvents) [32, 33]. On the other hand, the fundamental properties of the polymers for the development of p-MPs involve their solubility and stability, their biodegradability and biocompatibility [34], and their physical (i.e., crystallinity and glass transition temperature) and mechanical properties (i.e., strength, elongation, and Young's modulus) [35].

The microemulsion methods provided a highly reproducible platform to formulate p-MPs with a uniform size and predictable inner structure, which can be determined by the use of single- or double-emulsion process. The single-emulsion method consists in an oil/water (O/W) or water/oil (W/O) emulsion that generates solid spherical shape microparticles, with a polymeric inner core, which is favorable to encapsulate hydrophobic drugs [36]. On the other hand, the proteins and other hydrophilic drugs are usually encapsulated using the water/oil/water (W/O/W) double-emulsion method, because it generates core-shell microparticles characterized by hydrophilic pockets [37]. Figure 2, presents a scheme of the morphology of p-MPs, formulated by the single- and double-emulsion-evaporation method. Studies about the conditions of preparation of p-MPs have shown that high concentrations of polymers generate an increase of the particles size and a decrease-loading yield. This phenomenon may be attributed to the increment in the viscosity of the polymeric phase that emulsified to drug [38]. In addition, other studies have described that the intensity with which it generates the emulsion affects its internal conformation of microparticles. Mao et al. (2007) showed by transmission electron microscopy that a high intensity of emulsion reduced significantly the internal porosity of p-MPs [39].

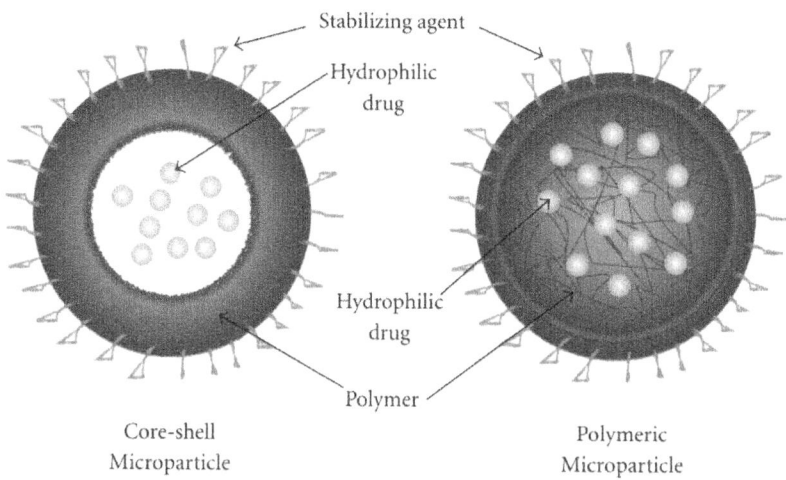

Figure 2: Scheme of the morphology of polymeric microparticles prepared by the single- and double-emulsion method and their internal distribution of drugs with different physicochemical properties.

Despite the high loading efficiency that supports the conventional emulsion methods, recently, innovative procedures based on double-emulsion method such as the solid/oil/water (S/O/W), the solid/oil/oil (S/O/O), and the water/oil/oil (W/O/O) methods have been described, which allows to maintain their complete structural and functional integrity of proteins after the microencapsulation process [40].

Another method to synthesize polymeric micro- and nanoparticles is through microfluidic technology [41–43]. This technique generates droplets or particles in a device (T-junction) supplied with the polymers and drugs dissolved in immiscible solutions, followed by the solidification of the droplets by means of polymerization or solvent evaporation [44]. The main advantage of microfluidics is to obtain large volumes of particles, which have a highly uniform and predictable size, which determines their potential use in the synthesis of multiple polymeric colloids loaded with drugs and pharmaceutical application [45].

Spray-drying is a method widely used in the pharmaceutical and biotechnology industry for the synthesis of p-MPs because it allows to produce large quantities of particles with spherical and amorphous morphology and it can display roughness or porosity in their surface [46]. In the last years, spray-freeze-drying methods were able to formulate p-MPs loaded with poor water-soluble drugs and temperature-sensitive molecules. In addition, these methods produce microparticles with controlled size and porosity, making them particularly attractive to load a wide range of drugs with biomedical interest [47, 48].

Figure 3 illustrates images of p-MPs prepared in our laboratory from PLGA and PHBV and characterized using a confocal laser scanning microscopy, a transmission electron microscopy, and a scanning electron microscopy.

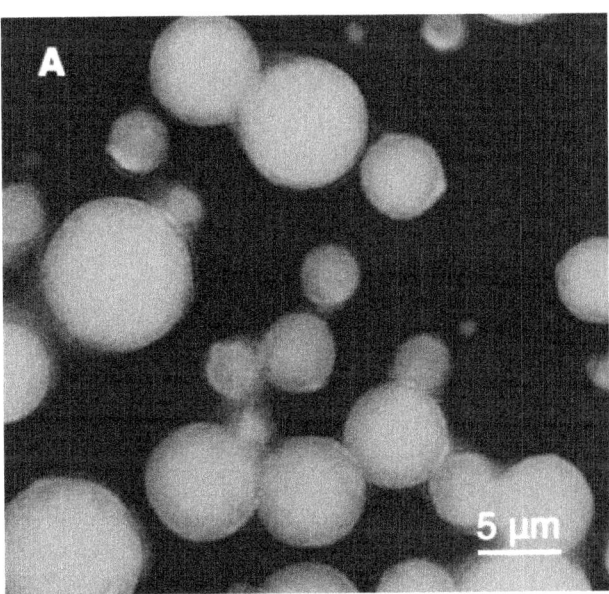

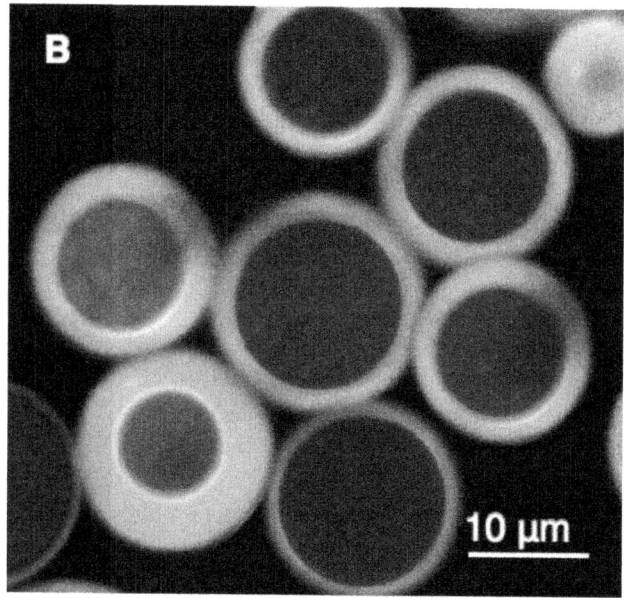

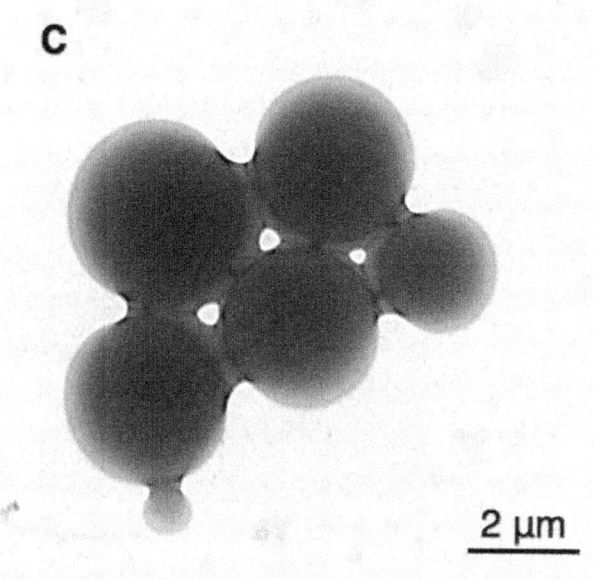

Figure 3: Polymeric microparticles formulated by single- (a) and double- (b, c, and d) emulsion method. Images obtained through confocal laser scanning microscopy of (a) FITC-loaded poly(lactide-co-glycolide) (PLGA) microparticles (MPs) (green), (b) NBD-cholesterol (green), and Texas-Red (red) loaded PLGA microparticles. (c) Transmission electron microscopy (TEM) of ceftiofur-loaded poly(3-hydroxybutyrate-co-3-hydroxyvalerate) (PHBV) microparticles; (d) scanning electron microscopy (SEM) of florfenicol-loaded PHBV microparticles.

MECHANISMS OF DRUG RELEASE

The release of drugs from p-MPs arises as a consequence of the degradation and/or erosion of the polymeric device [49]. Therefore, the knowledge about the chemical nature of polymers is essential to understand the mechanism of release. In the cases when degradation of polymeric matrix occurs, the drug diffuses through the channels generated by the breaking of the polymer chains without loss of volume in the particle. In contrast, when the polymeric carrier undergoes erosion, together with the polymer mass loss the drug is released. In this case, there is a decrease in volume of polymeric matrix according to the drug release [50–52].

Studies have demonstrated that the rate of degradation of polyesters such as PLGA or PHBV is inversely proportional to the molecular weight of the polymers. Furthermore, the degradation time of PLGA (copolymer) depends on the ratio of its monomers, poly(lactic acid) and poly(glycolic acid), such that polymers containing a higher concentration of poly (lactic acid) exhibited a slower degradation [49]. Others studies have showed that high temperatures and low pH condition increase the degradation of polymers with a subsequent increment of the release rate of drug encapsulated into polymeric microparticles [53, 54].

THERAPEUTIC STRATEGIES BASED ON POLYMERIC MICROPARTICLES

The p-MPs formulations have unique properties in terms of particle size, shape, inner structure, porosity, drug loading, encapsulation efficiency, and profile of release [55, 56]. Therefore, the selection of an appropriate route of administration of p-MPs (i.e., intramuscular, intraperitoneal, intra-articular, and intrapulmonary) is a critical element to achieve an expected pharmacological action.

Oncologic Disease

Cancer is one of the most significant causes of death worldwide, and the gliomas are the leading brain tumors of the nervous system in adults. It has been described that gliomas have an exceptional ability to infiltrate to healthy tissue, which makes them extremely difficult to be treated [57]. Chemotherapy is one of the most widely used strategies to treat cancer. However, its low specificity and high toxicity generate negative effects for patients that may cause serious complications, affecting in some cases other healthy physiological systems [58–60]. Therefore, the administration of chemotherapeutic agents loaded in polymeric microparticles provides a secure platform to achieve a sustainedrelease in the cancerous tissue, decreasing the use of high doses of drugs and their potential harmful effects [61, 62].

Recently, Y. H. Zhang et al. (2010) described a study using orthotopic implantation of C6 glial cells in a rat brain to evaluate the activity of polymeric microparticles loaded with temozolomide (tm-MPs) injected into the tumor area. The results showed a better survival to the group that received tm-MPs (46 days) than the control group treated orally with nonencapsulated temozolomide (27 days). Moreover, through magnetic resonance imaging (MRI), they found that the group treated with tm-MPs showed the greatest reduction of the tumor size and decrease of the proliferative activity of cells. Furthermore, the cells also presented an increased rate of apoptosis, suggesting that the encapsulation of temozolomide in p-MPs enhanced its chemotherapeutic effect [63]. Other in vitro studies, using similar strategies for the localized release of paclitaxel and cisplatin from polymeric microparticles, also exhibited greater efficacy than the administration of nonencapsulated drug [64,65].

In the last few decades, the use of intraperitoneal chemotherapy has showed high efficacy in the treatment of peritoneal and ovarian cancer, which has allowed enhancing the survival of many patients [66–68]. However, the use of intraperitoneal therapy also has presented some limitations that increase the risk of infection due to the use of catheters for the administration of drugs [69]. Other drawbacks have been associated with the use of chemotherapeutic agents that present hematologic and hepatic toxicity such as cisplatin, melphalan, and etoposide [70–73] and the slow absorption of less toxic drugs, such as paclitaxel, mitoxantrone, and doxorubicin, which do not have a deep tumor penetration [74–77]. Studies have shown that intraperitoneal treatment of ovarian cancer in mice model with paclitaxel-loaded p-MPs has overcome the limitations of free paclitaxel therapy. The administration of paclitaxel-loaded polymeric microparticles exhibited biphasic release kinetics, characterized by a rapid initial release that was sufficient to prevent tumor proliferation and a second phase of sustained release that allowed for the gradual eradication of the tumor [78]. Furthermore, intraperitoneal chemotherapy based on microparticles has reduced the removal of the drug from the peritoneal cavity, leading to slow systemic absorption and maintaining the therapeutic concentrations for longer periods of time (10 to 45 times) in the intraperitoneal region, which generated a significant increase of survival groups treated with p-MPs [79].

Cardiac Disease

Cardiac dysfunction followed by acute myocardial infarction is one of the leading causes of death worldwide [80, 81]. The excessive inflammatory response after the ischemic heart disease generates a chronic elevation of inflammatory cytokines and reactive oxygen species, which may lead to cardiac dysfunction [82–84]. Recently, the release of anti-inflammatory drugs from polymeric microparticles administrated via intracardiac injection has shown promising results to treat the myocardial infarction and other inflammatory diseases, due to blocking the activation of macrophages and thereby reducing the apoptosis or necrosis of cardiomyocytes [85,86].

Recent therapeutic approaches to prevent the development of cardiac failure after myocardial infarct include the direct administration of proangiogenic growth factors [87] and stem cell therapy [88, 89]. However, despite the promising results obtained in animal models and clinical trials [90, 91], some studies have shown limited effectiveness with the administration of growth factors because the native and recombinant proteins exhibited a short half-life and instability [92, 93]. In order to improve those drawbacks, Formiga et al. (2010) have described the synthesis of PLGA microparticles loaded with the cytokine $VEGF_{165}$, a proangiogenic growth factor, and evaluated their vasculogenic effect in a rat model of myocardial infarction. The results obtained showed an excellent angiogenic and arteriogenic effect induced by the sustained release of the cytokine $VEGF_{165}$ from the polymeric microparticles [94].

Immunological Response

Studies under preclinical drug development based on p-MPs have been focusing on the development of strategies that reduce organ rejection and prevent autoimmune diseases. Wu and Horuzsko (2009) proposed a method for improving immune tolerance by dendritic cell receptor stimulation with ILTs (immunoglobulin-like transcripts). Dual coating the surface of p-MPs with the HLA-G1-peptide, an ILTS receptor ligand, and a monoclonal antibody against the CD11c marker improved the modulation of dendritic cells. This system could provide a method to regulate specific immune responses that occur during transplantation, autoimmunity, and allergy [95].

New approaches in the vaccine field include polymeric microparticles loaded with antigens against bacterial pathogens such as Vibrio cholerae [96], Pseudomonas aeruginosa [97], and Bordetella pertussis [98], providing a potent and long-time immune response.

On the other hand, the gene delivery from p-MPs provides a highly attractive strategy because it can generate the in situ expression of target antigens and preserve the native structure of proteins [99]. In addition, the p-MPs can codeliver DNA and adjuvants generating an improved immune response [100, 101]. The current strategies have used polymers with cationic charge such polyethyleneimine to increase the loading and encapsulation efficiency of DNA inside particles [102]. Despite great advances in the development of DNA vaccines and their potential against several diseases, the biggest challenge is to establish the safety of using DNA vaccines in human medicine [103].

Diabetes

In the last decade, there was a notable increase of diabetes around the world [104]. The islet transplantation to patients with severe diabetes has improved their quality of life [105, 106]. However, these transplanted cells are highly susceptible to oxidative stress, which may decrease their proliferative capacity and lead to cellular death [107, 108]. The antioxidant effect of vitamin D3-loaded polymeric microparticles was evaluated in cultured islets isolated from adult rat. The results exhibited a significantly increased insulin production compared to the untreated control groups [109].

Other studies have described novel strategies for the oral and parenteral administration of insulin-loaded PLGA and poly(N-vinylcaprolactam-co-methacrylic acid) microparticles [110]. The particles were synthesized using flow focusing, double-emulsion-solvent evaporation method, and the free radical polymerization procedure [111, 112].

Recently, Technosphere/Insulin, an inhalable formulation under development by MannKind Corporation (Valencia, CA), have initiated the Phase III in both Europe and the US. The Technosphere technology allows to administer insulin via pulmonary and offers several competitive advantages over other pulmonary drug delivery systems. Recent studies have been conducted to analyze the lung deposition and clearance after administration. Their findings showed a uniform distribution throughout the lungs and absorption of insulin into the systemic circulation. Based on the results of clinical trials and on published reports, Technosphere is better than other inhaled insulin platforms [113].

PROSPECTS

Multidisciplinary work in the 21st century of physicians, biomaterials and chemical engineers, and researchers in biotechnology has allowed creating new frontiers to the landscape of pharmaceuticals.

The incorporation of polymeric microparticles as carriers of drugs in medical practice improves the disadvantages generated by elevated plasma levels short-term and adverse reactions caused by the traditional pharmaceutical formulation. It also creates novel strategies for localized and sustained release sites with low vascular permeability. Moreover, the wide range of biomaterials with different physicochemical properties allow the creation of smart systems for drug delivery, which promote an optimal response and long-term efficacy in the treatments of different diseases.

The development of polymeric microparticles, as drug delivery systems, has set the foundation for the emerging and significant role of nanomedicine based on polymeric nanoparticles as carriers of drugs [114–116]. We are optimistic about the marketing in the near future of innovative technology based on polymeric microparticles because it may generate a new era in modern medicine.

ACKNOWLEDGMENTS

Support by FONDECYT Grant 1090589, by BASAL Grant FB0807, and by CONICYT under "Proyecto Tesis en la Industria TPI06" is gratefully acknowledged.

REFERENCES

1. J. M. Reichert, "Trends in development and approval times for new therapeutics in the United States,"Nature Reviews Drug Discovery, vol. 2, no. 9, pp. 695–702, 2003.
2. J. A. DiMasi, R. W. Hansen, and H. G. Grabowski, "The price of innovation: new estimates of drug development costs," Journal of Health Economics, vol. 22, no. 2, pp. 151–185, 2003.
3. A. Mullard, "2011 FDA drug approvals," Nature Reviews Drug Discovery, vol. 11, pp. 91–94, 2012.
4. R. Ottenbrite, "Controlled release technology," in Encyclopedia of Polymer Science and Engineering, J. I. Kroschwitz, Ed., Wiley, New York, NY, USA, 1990.
5. P. A. Sales-Junior, F. Guzman, M. I. Vargas, et al., "Use of biodegradable PLGA microspheres as a slow release delivery system for the Boophilus microplus synthetic vaccine SBm7462," Veterinary Immunology and Immunopathology, vol. 107, no. 3-4, pp. 281–290, 2005.
6. J. Urquhart, "Controlled drug delivery: therapeutic and pharmacological aspects," Journal of Internal Medicine, vol. 248, no. 5, pp. 357–376, 2000.
7. R. K. Verma, S. Arora, and S. Garg, "Osmotic pumps in drug delivery," Critical Reviews in Therapeutic Drug Carrier Systems, vol. 21, no. 6, pp. 477–520, 2004.
8. W. T. Al-Jamal and K. Kostarelos, "Liposomes: from a clinically established drug delivery system to a nanoparticle platform for theranostic nanomedicine," Accounts of Chemical Research, vol. 44, no. 10, pp. 1094–1104, 2011.
9. A. Jesorka and O. Orwar, "Liposomes: technologies and analytical applications," Annual Review of Analytical Chemistry, vol. 1, no. 1, pp. 801–832, 2008.
10. A. S. Hoffman, "Hydrogels for biomedical applications," Advanced Drug Delivery Reviews, vol. 54, no. 1, pp. 3–12, 2002.
11. J. Cabral and S. C. Moratti, "Hydrogels for biomedical applications," Future Medicinal Chemistry, vol. 3, pp. 1877–1888, 2011.
12. N. A. Peppas, Y. Huang, M. Torres-Lugo, J. H. Ward, and J. Zhang, "Physicochemical foundations and structural design of hydrogels in medicine and biology," Annual Review of Biomedical Engineering, vol. 2, no. 2000, pp. 9–29, 2000.

13. W. Jiang, R. K. Gupta, M. C. Deshpande, and S. P. Schwendeman, "Biodegradable poly(lactic-co-glycolic acid) microparticles for injectable delivery of vaccine antigens," Advanced Drug Delivery Reviews, vol. 57, no. 3, pp. 391–410, 2005.
14. E. Mathiowitz, J. S. Jacob, Y. S. Jong et al., "Biologically erodable microspheres as potential oral drug delivery systems," Nature, vol. 386, no. 6623, pp. 410–414, 1997.
15. B. D. Ulery, L. S. Nair, and C. T. Laurencin, "Biomedical applications of biodegradable polymers," Journal of Polymer Science B, vol. 49, no. 12, pp. 832–864, 2011.
16. A. D. Bendrea, L. Cianga, and I. Cianga, "Review paper: progress in the field of conducting polymers for tissue engineering applications," Journal of Biomaterials Applications, vol. 26, no. 1, pp. 3–84, 2011.
17. O. Pillai and R. Panchagnula, "Polymers in drug delivery," Current Opinion in Chemical Biology, vol. 5, no. 4, pp. 447–451, 2001.
18. G. M. Keegan, J. D. Smart, M. J. Ingram, L. M. Barnes, G. R. Burnett, and G. D. Rees, "Chitosan microparticles for the controlled delivery of fluoride," Journal of Dentistry, vol. 40, no. 3, pp. 229–240, 2012.
19. K. Moebus, J. Siepmann, and R. Bodmeier, "Novel preparation techniques for alginate-poloxamer microparticles controlling protein release on mucosal surfaces," European Journal of Pharmaceutical Sciences, vol. 45, no. 3, pp. 358–366, 2012.
20. S. A. Meenach, Y. J. Kim, K. J. Kauffman, N. Kanthamneni, E. M. Bachelder, and K. M. Ainslie, "Synthesis, optimization, and characterization of camptothecin-loaded acetalated dextran porous microparticles for pulmonary delivery," Molecular Pharmacology, vol. 9, no. 2, pp. 290–298, 2012.
21. Z. S. Patel, H. Ueda, M. Yamamoto, Y. Tabata, and A. G. Mikos, "In vitro and in vivo release of vascular endothelial growth factor from gelatin microparticles and biodegradable composite scaffolds,"Pharmaceutical Research, vol. 25, no. 10, pp. 2370–2378, 2008.
22. K. N. Lee, Y. Ye, J. H. Carr, K. Karem, and M. J. D'Souza, "Formulation, pharmacokinetics and biodistribution of Ofloxacin-loaded albumin microparticles and nanoparticles," Journal of Microencapsulation, vol. 28, no. 5, pp. 363–369, 2011.
23. E. M. Fernandez, J. Chang, J. Fontaine, et al., "Activation of invariant Natural Killer T lymphocytes in response to the alpha-galactosylceramide analogue KRN7000 encapsulated in PLGA-based nanoparticles and

microparticles," International Journal of Pharmaceutics, vol. 423, no. 1, pp. 45–54, 2012.

24. W. Chen and Y. W. Tong, "PHBV microspheres as neural tissue engineering scaffold support neuronal cell growth and axon-dendrite polarization," Acta Biomaterialia, vol. 8, no. 2, pp. 540–548, 2012.

25. N. B. Shelke and T. M. Aminabhavi, "Synthesis and characterization of novel poly(sebacic anhydride-co-Pluronic F68/F127) biopolymeric microspheres for the controlled release of nifedipine," International Journal of Pharmaceutics, vol. 345, no. 1-2, pp. 51–58, 2007.

26. E. R. Balmayor, G. A. Feichtinger, H. S. Azevedo, M. Van Griensven, and R. L. Reis, "Starch-poly-ε-caprolactone microparticles reduce the needed amount of BMP-2," Clinical Orthopaedics and Related Research, vol. 467, no. 12, pp. 3138–3148, 2009.

27. W. B. Liechty, D. R. Kryscio, B. V. Slaughter, and N. A. Peppas, "Polymers for drug delivery systems,"Annual Review of Chemical and Biomolecular Engineering, vol. 1, pp. 149–173, 2010.

28. M. Curcio, U. Gianfranco Spizzirri, F. Iemma et al., "Grafted thermoresponsive gelatin microspheres as delivery systems in triggered drug release," European Journal of Pharmaceutics and Biopharmaceutics, vol. 76, no. 1, pp. 48–55, 2010.

29. L. Ma, M. Liu, and X. Shi, "pH- and temperature-sensitive self-assembly microcapsules/microparticles: synthesis, characterization, in vitro cytotoxicity, and drug release properties," Journal of Biomedical Materials Research B, vol. 100, no. 2, pp. 305–313, 2012.

30. K. Rizi, R. J. Green, O. Khutoryanskaya, M. Donaldson, and A. C. Williams, "Mechanisms of burst release from pH-responsive polymeric microparticles," Journal of Pharmacy and Pharmacology, vol. 63, no. 9, pp. 1141–1155, 2011.

31. M. A. Alhnan, E. Kidia, and A. W. Basit, "Spray-drying enteric polymers from aqueous solutions: a novel, economic, and environmentally friendly approach to produce pH-responsive microparticles," European Journal of Pharmaceutics and Biopharmaceutics, 2011.

32. M. N. Aamir and M. Ahmad, "Production and stability evaluation of modified-release microparticles for the delivery of drug combinations," AAPS PharmSciTech, vol. 11, no. 1, pp. 351–355, 2010.

33. A. Wieber, T. Selzer, and J. Kreuter, "Characterisation and stability studies of a hydrophilic decapeptide in different adjuvant drug delivery systems: a comparative study of PLGA nanoparticles versus chitosan-

dextran sulphate microparticles versus DOTAP-liposomes," International Journal of Pharmaceutics, vol. 421, no. 1, pp. 151–159, 2011.

34. G. Winzenburg, C. Schmidt, S. Fuchs, and T. Kissel, "Biodegradable polymers and their potential use in parenteral veterinary drug delivery systems," Advanced Drug Delivery Reviews, vol. 56, no. 10, pp. 1453–1466, 2004.

35. I. Engelberg and J. Kohn, "Physico-mechanical properties of degradable polymers used in medical applications: a comparative study," Biomaterials, vol. 12, no. 3, pp. 292–304, 1991.

36. C. Yang, D. Plackett, D. Needham, and H. M. Burt, "PLGA and PHBV microsphere formulations and solid-state characterization: possible implications for local delivery of fusidic acid for the treatment and prevention of orthopaedic infections," Pharmaceutical Research, vol. 26, no. 7, pp. 1644–1656, 2009.

37. X. Jia, D. Chen, X. Jiao, and S. Zhai, "Environmentally-friendly preparation of water-dispersible magnetite nanoparticles," Chemical Communications, no. 8, pp. 968–970, 2009.

38. H. Zhao, J. Gagnon, and U. O. Hafeli, "Process and formulation variables in the preparation of injectable and biodegradable magnetic microspheres," BioMagnetic Research and Technology, vol. 5, p. 2, 2007.

39. S. Mao, J. Xu, C. Cai, O. Germershaus, A. Schaper, and T. Kissel, "Effect of WOW process parameters on morphology and burst release of FITC-dextran loaded PLGA microspheres," International Journal of Pharmaceutics, vol. 334, no. 1-2, pp. 137–148, 2007.

40. D. Yegian and V. Budd, "Novobiocin: activity in vitro and in experimental tuberculosis," American Review of Tuberculosis, vol. 76, no. 2, pp. 272–278, 1957.

41. Z. T. Cygan, J. T. Cabral, K. L. Beers, and E. J. Amis, "Microfluidic platform for the generation of organic-phase microreactors," Langmuir, vol. 21, no. 8, pp. 3629–3634, 2005.

42. R. Karnik, F. Gu, P. Basto et al., "Microfluidic platform for controlled synthesis of polymeric nanoparticles," Nano Letters, vol. 8, no. 9, pp. 2906–2912, 2008.

43. P. M. Valencia, P. A. Basto, L. Zhang et al., "Single-step assembly of homogenous lipid-polymeric and lipid-quantum dot nanoparticles enabled by microfluidic rapid mixing," ACS Nano, vol. 4, no. 3, pp. 1671–1679, 2010.

44. G. F. Christopher, N. N. Noharuddin, J. A. Taylor, and S. L. Anna, "Experimental observations of the squeezing-to-dripping transition in T-shaped microfluidic junctions," Physical Review E, vol. 78, no. 3, Article ID 036317, 2008.

45. Q. Xu, M. Hashimoto, T. T. Dang et al., "Preparation of monodisperse biodegradable polymer microparticles using a microfluidic flow-focusing device for controlled drug delivery," Small, vol. 5, no. 13, pp. 1575–1581, 2009.

46. R. Vehring, "Pharmaceutical particle engineering via spray drying," Pharmaceutical Research, vol. 25, no. 5, pp. 999–1022, 2008.

47. T. Niwa, H. Shimabara, M. Kondo, and K. Danjo, "Design of porous microparticles with single-micron size by novel spray freeze-drying technique using four-fluid nozzle," International Journal of Pharmaceutics, vol. 382, no. 1-2, pp. 88–97, 2009.

48. S. M. D'Addio, J. G. Chan, P. C. Kwok, R. K. Prud'homme, and H. K. Chan, "Constant size, variable density aerosol particles by ultrasonic spray freeze drying," International Journal of Pharmaceutics, vol. 427, no. 2, pp. 185–191, 2012.

49. A. Göpferich, "Mechanisms of polymer degradation and erosion," Biomaterials, vol. 17, no. 2, pp. 103–114, 1996.

50. A. Göpferich and J. Tessmar, "Polyanhydride degradation and erosion," Advanced Drug Delivery Reviews, vol. 54, no. 7, pp. 911–931, 2002.

51. F. V. Burkersroda, L. Schedl, and A. Göpferich, "Why degradable polymers undergo surface erosion or bulk erosion," Biomaterials, vol. 23, no. 21, pp. 4221–4231, 2002.

52. X. Xu and P. I. Lee, "Programmable drug delivery from an erodible assocation polymer system,"Pharmaceutical Research, vol. 10, no. 8, pp. 1144–1152, 1993.

53. N. Faisant, J. Siepmann, and J. P. Benoit, "PLGA-based microparticles: elucidation of mechanisms and a new, simple mathematical model quantifying drug release," European Journal of Pharmaceutical Sciences, vol. 15, no. 4, pp. 355–366, 2002.

54. B. S. Zolnik and D. J. Burgess, "Effect of acidic pH on PLGA microsphere degradation and release,"Journal of Controlled Release, vol. 122, no. 3, pp. 338–344, 2007.

55. H. T. Wang, H. Palmer, R. J. Linhardt, D. R. Flanagan, and E. Schmitt, "Degradation of poly(ester) microspheres," Biomaterials, vol. 11, no. 9, pp. 679–685, 1990.

56. R. van Dijkhuizen-Radersma, S. C. Hesseling, P. E. Kaim, K. De Groot, and J. M. Bezemer, "Biocompatibility and degradation of poly(ether-ester) microspheres: in vitro and in vivo evaluation,"Biomaterials, vol. 23, no. 24, pp. 4719–4729, 2002.

57. T. Demuth and M. E. Berens, "Molecular mechanisms of glioma cell migration and invasion," Journal of Neuro-Oncology, vol. 70, no. 2, pp. 217–228, 2004.

58. C. H. Chang, J. Horton, D. Schoenfeld, et al., "Comparison of postoperative radiotherapy and combined postoperative radiotherapy and chemotherapy in the multidisciplinary management of malignant gliomas. A joint radiation therapy oncology group and Eastern cooperative oncology group study,"Cancer, vol. 52, no. 6, pp. 997–1007, 1983.

59. V. R. Recinos, B. M. Tyler, K. Bekelis et al., "Combination of intracranial temozolomide with intracranial carmustine improves survival when compared with either treatment alone in a rodent glioma model,"Neurosurgery, vol. 66, no. 3, pp. 530–537, 2010.

60. T. Walbert, M. R. Gilbert, M. D. Groves et al., "Combination of 6-thioguanine, capecitabine, and celecoxib with temozolomide or lomustine for recurrent high-grade glioma," Journal of Neuro-Oncology, vol. 102, no. 2, pp. 273–280, 2011.

61. P. Menei and J. P. Benoit, "Implantable drug-releasing biodegradable microspheres for local treatment of brain glioma," Acta Neurochirurgica, supplement 88, pp. 51–55, 2003.

62. A. J. Sawyer, J. M. Piepmeier, and W. M. Saltzman, "New methods for direct delivery of chemotherapy for treating brain tumors," Yale Journal of Biology and Medicine, vol. 79, no. 3-4, pp. 141–152, 2006.

63. M. Henze, W. Pietsch, V. Burwitz, et al., "Confirmation of a recent optical nova candidate in M 31 and H-alpha identification of seven M 31 novae," The Astronomer's Telegram #1602, 2008.

64. J. Xie, S. T. Ruo, and C. H. Wang, "Biodegradable microparticles and fiber fabrics for sustained delivery of cisplatin to treat C6 glioma in vitro," Journal of Biomedical Materials Research A, vol. 85, no. 4, pp. 897–908, 2008.

65. J. Xie, J. C. M. Marijnissen, and C. H. Wang, "Microparticles developed by electrohydrodynamic atomization for the local delivery of anticancer drug to treat C6 glioma in vitro," Biomaterials, vol. 27, no. 17, pp. 3321–3332, 2006.

66. A. G. Zeimet, D. Reimer, A. C. Radl et al., "Pros and cons of intraperitoneal chemotherapy in the treatment of epithelial ovarian cancer," Anticancer

Research, vol. 29, no. 7, pp. 2803–2808, 2009.

67. D. K. Armstrong and M. F. Brady, "Intraperitoneal therapy for ovarian cancer: a treatment ready for prime time," Journal of Clinical Oncology, vol. 24, no. 28, pp. 4531–4533, 2006.

68. Z. Lu, J. Wang, M. G. Wientjes, and J. L. S. Au, "Intraperitoneal therapy for peritoneal cancer," Future Oncology, vol. 6, no. 10, pp. 1625–1641, 2010.

69. C. W. E. Redman, F. G. Lawton, D. M. Luesley, E. J. Buxton, and G. Blackledge, "Problems of peritoneal access in intraperitoneal treatment and monitoring of ovarian cancer," British Journal of Obstetrics and Gynaecology, vol. 96, no. 1, pp. 97–101, 1989.

70. M. Markman and J. L. Walker, "Intraperitoneal chemotherapy of ovarian cancer: a review, with a focus on practical aspects of treatment," Journal of Clinical Oncology, vol. 24, no. 6, pp. 988–994, 2006.

71. D. S. Alberts, E. A. Surwit, Y. M. Pen et al., "Phase I clinical and pharmacokinetic study of mitoxantrone given to patients by intraperitoneal administration," Cancer Research, vol. 48, no. 20, pp. 5874–5877, 1988.

72. R. Demicheli, G. Bonciarelli, A. Jirillo, et al., "Pharmacologic data and technical feasibility of intraperitoneal doxorubicin administration," Tumori, vol. 71, no. 1, pp. 63–68, 1985.

73. W. R. Robinson, N. Davis, and A. S. Rogers, "Paclitaxel maintenance chemotherapy following intraperitoneal chemotherapy for ovarian cancer," International Journal of Gynecological Cancer, vol. 18, no. 5, pp. 891–895, 2008.

74. S. B. Howell, C. E. Pfeifle, and R. A. Olshen, "Intraperitoneal chemotherapy with Melphalan," Annals of Internal Medicine, vol. 101, no. 1, pp. 14–18, 1984.

75. R. J. Morgan Jr, J. H. Doroshow, T. Synold et al., "Phase I trial of intraperitoneal docetaxel in the treatment of advanced malignancies primarily confined to the peritoneal cavity: dose-limiting toxicity and pharmacokinetics," Clinical Cancer Research, vol. 9, no. 16, pp. 5896–5901, 2003.

76. P. J. O'Dwyer, F. P. LaCreta, J. P. Daugherty et al., "Phase I pharmacokinetic study of intraperitoneal etoposide," Cancer Research, vol. 51, no. 8, pp. 2041–2046, 1991.

77. E. F. McClay, R. Goel, P. Andrews et al., "A phase I and pharmacokinetic study of intraperitoneal carboplatin and etoposide," British Journal of Cancer, vol. 68, no. 4, pp. 783–788, 1993.

78. Z. Lu, M. Tsai, D. Lu, J. Wang, M. G. Wientjes, and J. L. S. Au, "Tumor-penetrating microparticles for intraperitoneal therapy of ovarian cancer," Journal of Pharmacology and Experimental Therapeutics, vol. 327, no. 3, pp. 673–682, 2008.

79. M. Tsai, Z. Lu, J. Wang, T. K. Yeh, M. G. Wientjes, and J. L. S. Au, "Effects of carrier on disposition and antitumor activity of intraperitoneal paclitaxel," Pharmaceutical Research, vol. 24, no. 9, pp. 1691–1701, 2007.

80. P. Gaudron, C. Eilles, G. Ertl, and K. Kochsiek, "Adaptation to cardiac dysfunction after myocardial infarction," Circulation, vol. 87, no. 5, pp. IV83–IV89, 1993.

81. H. Zhang, X. Chen, E. Gao et al., "Increasing cardiac contractility after myocardial infarction exacerbates cardiac injury and pump dysfunction," Circulation Research, vol. 107, no. 6, pp. 800–809, 2010.

82. P. Anversa, "Myocyte death in the pathological heart," Circulation Research, vol. 86, no. 2, pp. 121–124, 2000.

83. P. Anversa, A. Leri, and J. Kajstura, "Cardiac Regeneration," Journal of the American College of Cardiology, vol. 47, no. 9, pp. 1769–1776, 2006.

84. R. Bolli, M. O. Jeroudi, B. S. Patel et al., "Direct evidence that oxygen-derived free radicals contribute to postischemic myocardial dysfunction in the intact dog," Proceedings of the National Academy of Sciences of the United States of America, vol. 86, no. 12, pp. 4695–4699, 1989.

85. J. C. Sy, G. Seshadri, S. C. Yang et al., "Sustained release of a p38 inhibitor from non-inflammatory microspheres inhibits cardiac dysfunction," Nature Materials, vol. 7, no. 11, pp. 863–869, 2008.

86. A. Lamprecht, H. Rodero Torres, U. Schäfer, and C. M. Lehr, "Biodegradable microparticles as a two-drug controlled release formulation: a potential treatment of inflammatory bowel disease," Journal of Controlled Release, vol. 69, no. 3, pp. 445–454, 2000.

87. N. Maulik and M. Thirunavukkarasu, "Growth factor/s and cell therapy in myocardial regeneration," Journal of Molecular and Cellular Cardiology, vol. 44, no. 2, pp. 219–227, 2008.

88. R. Passier, L. W. Van Laake, and C. L. Mummery, "Stem-cell-based therapy and lessons from the heart," Nature, vol. 453, no. 7193, pp. 322–329, 2008.

89. V. F. M. Segers and R. T. Lee, "Stem-cell therapy for cardiac disease," Nature, vol. 451, no. 7181, pp. 937–942, 2008.

90. T. T. Rissanen, J. E. Markkanen, K. Arve et al., "Fibroblast growth factor 4 induces vascular permeability, angiogenesis and arteriogenesis in a rabbit hindlimb ischemia model," The FASEB Journal, vol. 17, no. 1, pp. 100–102, 2003.

91. M. Hedman, J. Hartikainen, M. Syvanne, et al., "Safety and feasibility of catheter-based local intracoronary vascular endothelial growth factor gene transfer in the prevention of postangioplasty and in-stent restenosis and in the treatment of chronic myocardial ischemia: phase II results of the Kuopio Angiogenesis Trial (KAT)," Circulation, vol. 107, no. 21, pp. 2677–2683, 2003.

92. M. Simons, B. H. Annex, R. J. Laham et al., "Pharmacological treatment of coronary artery disease with recombinant fibroblast growth factor-2: double-blind, randomized, controlled clinical trial," Circulation, vol. 105, no. 7, pp. 788–793, 2002.

93. T. D. Henry, B. H. Annex, G. R. McKendall et al., "The VIVA trial: vascular endothelial growth factor in ischemia for vascular angiogenesis," Circulation, vol. 107, no. 10, pp. 1359–1365, 2003.

94. F. R. Formiga, B. Pelacho, E. Garbayo et al., "Sustained release of VEGF through PLGA microparticles improves vasculogenesis and tissue remodeling in an acute myocardial ischemia-reperfusion model,"Journal of Controlled Release, vol. 147, no. 1, pp. 30–37, 2010.

95. J. Wu and A. Horuzsko, "Expression and function of immunoglobulin-like transcripts on tolerogenic dendritic cells," Human Immunology, vol. 70, no. 5, pp. 353–356, 2009.

96. G. Ano, A. Esquisabel, M. Pastor, et al., "A new oral vaccine candidate based on the microencapsulation by spray-drying of inactivated Vibrio cholerae," Vaccine, vol. 29, no. 34, pp. 5758–5764, 2011.

97. S. Taranejoo, M. Janmaleki, M. Rafienia, M. Kamali, and M. Mansouri, "Chitosan microparticles loaded with exotoxin A subunit antigen for intranasal vaccination against Pseudomonas aeruginosa: an in vitro study," Carbohydrate Polymers, vol. 83, no. 4, pp. 1854–1861, 2011.

98. S. Garlapati, N. F. Eng, T. G. Kiros et al., "Immunization with PCEP microparticles containing pertussis toxoid, CpG ODN and a synthetic innate defense regulator peptide induces protective immunity against pertussis," Vaccine, 2011.

99. D. T. O'Hagan, M. Singh, and J. B. Ulmer, "Microparticles for the delivery of DNA vaccines,"Immunological Reviews, vol. 199, pp. 191–200, 2004.

100. R. K. Evans, D. M. Zhu, D. R. Casimiro et al., "Characterization and biological evaluation of a microparticle adjuvant formulation for plasmid DNA vaccines," Journal of Pharmaceutical Sciences, vol. 93, no. 7, pp. 1924–1939, 2004.

101. A. Caputo, K. Sparnacci, B. Ensoli, and L. Tondelli, "Functional polymeric nano/microparticles for surface adsorption and delivery of protein and DNA vaccines," Current Drug Delivery, vol. 5, no. 4, pp. 230–242, 2008.

102. S. P. Kasturi, K. Sachaphibulkij, and K. Roy, "Covalent conjugation of polyethyleneimine on biodegradable microparticles for delivery of plasmid DNA vaccines," Biomaterials, vol. 26, no. 32, pp. 6375–6385, 2005.

103. D. N. Nguyen, J. J. Green, J. M. Chan, R. Langer, and D. G. Anderson, "Polymeric materials for gene delivery and DNA vaccination," Advanced Materials, vol. 21, no. 8, pp. 847–867, 2009.

104. I. G. de Quevedo, L. Siminerio, R. L'Heveder, and K. M. Narayan, "Challenges in real-life diabetes translation research: early lessons from BRIDGES projects," Diabetes Research and Clinical Practice, vol. 95, no. 3, pp. 317–325, 2012.

105. S. Matsumoto, H. Noguchi, Y. Yonekawa et al., "Pancreatic islet transplantation for treating diabetes,"Expert Opinion on Biological Therapy, vol. 6, no. 1, pp. 23–37, 2006.

106. N. Onaca, G. B. Klintmalm, and M. F. Levy, "Pancreatic islet cell transplantation: a treatment strategy for type I diabetes mellitus," Nutrition in Clinical Practice, vol. 19, no. 2, pp. 154–164, 2004.

107. R. P. Robertson and J. S. Harmon, "Pancreatic islet β-cell and oxidative stress: the importance of glutathione peroxidase," FEBS Letters, vol. 581, no. 19, pp. 3743–3748, 2007.

108. H. Kaneto, Y. Kajimoto, Y. Fujitani et al., "Oxidative stress induces p21 expression in pancreatic islet cells: possible implication in beta-cell dysfunction," Diabetologia, vol. 42, no. 9, pp. 1093–1097, 1999.

109. G. Luca, G. Basta, R. Calafiore et al., "Multifunctional microcapsules for pancreatic islet cell entrapment: design, preparation and in vitro characterization," Biomaterials, vol. 24, no. 18, pp. 3101–3114, 2003.

110. J. Emami, H. Hamishehkar, A. R. Najafabadi et al., "A novel approach to prepare insulin-loaded poly (lactic-co-glycolic acid) microcapsules and the protein stability study," Journal of Pharmaceutical Sciences, vol. 98, no. 5, pp. 1712–1731, 2009.

111. R. C. Mundargi, V. Rangaswamy, and T. M. Aminabhavi, "Poly(N-vinylcaprolactam-co-methacrylic acid) hydrogel microparticles for oral insulin delivery," Journal of Microencapsulation, vol. 28, no. 5, pp. 384–394, 2011.

112. M. J. Cozar-Bernal, M. A. Holgado, J. L. Arias, et al., "Insulin-loaded PLGA microparticles: flow focusing versus double emulsion/solvent evaporation," Journal of Microencapsulation, vol. 28, no. 5, pp. 430–441, 2011.

113. S. S. Iyer, W. H. Barr, and H. T. Karnes, "A ‹biorelevant› approach to accelerated in vitro drug release testing of a biodegradable, naltrexone implant," International Journal of Pharmaceutics, vol. 340, no. 1-2, pp. 119–125, 2007.

114. J. M. Chan, P. M. Valencia, L. Zhang, R. Langer, and O. C. Farokhzad, "Polymeric nanoparticles for drug delivery," Methods in Molecular Biology, vol. 624, pp. 163–175, 2010.

115. J. Shi, Z. Xiao, N. Kamaly, and O. C. Farokhzad, "Self-assembled targeted nanoparticles: evolution of technologies and bench to bedside translation," Accounts of Chemical Research, vol. 44, no. 10, pp. 1123–1134, 2011.

116. J. Shi, Z. Xiao, A. R. Votruba, C. Vilos, and O. C. Farokhzad, "Differentially charged hollow core/shell lipid-polymer-lipid hybrid nanoparticles for small interfering rna delivery," Angewandte Chemie, vol. 50, no. 31, pp. 7027–7031, 2011.

Chapter 5

MAMMALIAN MODELS OF DUCHENNE MUSCULAR DYSTROPHY: PATHOLOGICAL CHARACTERISTICS AND THERAPEUTIC APPLICATIONS

Akinori Nakamura[1] and Shin'ichi Takeda[2]

[1]Department of Medicine (Neurology and Rheumatology), School of Medicine Shinshu University, 3-1-1 Ahahi, Matsumoto 390-8621, Japan

[2]Department of Molecular Therapy, National Institute of Neuroscience, National Center of Neurology and Psychiatry, 4-1-1 Ogawa-higashi, Kodaira, Tokyo 187-8502, Japan

ABSTRACT

Duchenne muscular dystrophy (DMD) is a devastating X-linked muscle disorder characterized by muscle wasting which is caused by mutations in the DMD gene. The DMD gene encodes the sarcolemmal protein dystrophin, and loss of dystrophin causes muscle degeneration and necrosis. Thus far, therapies for this disorder are unavailable. However, various therapeutic trials based on gene therapy, exon skipping, cell therapy, read through therapy, or pharmaceutical agents have been conducted extensively. In the development of therapy as well as elucidation of pathogenesis in DMD, appropriate animal models are needed. Various animal models of DMD have been identified, and mammalian (murine, canine, and feline) models are indispensable for the examination of the mechanisms of pathogenesis and the development of therapies. Here, we review the pathological features of DMD and therapeutic applications, especially of exon skipping using antisense oligonucleotides and gene therapies using viral vectors in murine and canine models of DMD.

INTRODUCTION

Duchenne muscular dystrophy (DMD) is a lethal X-linked muscle disease characterized by progressive skeletal muscle atrophy and weakness [1]. Mutations in the causative gene DMD result in loss of the cytoskeletal protein dystrophin. This is accompanied by a defect of dystrophin-glycoprotein complex

(DGC) in the sarcolemma and leads to progressive muscle degeneration [2, 3]. The dystrophic-deficient skeletal muscle exhibits muscle fiber necrosis with invasion of inflammatory cells followed by muscle regeneration. The muscle is progressively replaced by fibrous or fatty tissue. Recent advances in molecular biology have identified murine, canine, and feline DMD animal models [4]. Efforts to develop various therapeutic approaches such as gene therapy, cell therapy, or pharmaceutical agents have been conducted using DMD animal models, although no radical and permanent therapy is available. In this review, we describe the pathological characteristics and availability of therapeutic applications with a focus on gene therapy in mammalian models of DMD.

DUCHENNE MUSCULAR DYSTROPHY

Pathogenesis of DMD

Muscular dystrophies are inherited, and progressive muscle disorders are characterized by muscle fiber degeneration and necrosis. Duchenne muscular dystrophy (DMD) is the most severe and common X-linked disorder (1 in 3,500 male births) [1]. The onset of this disorder is recognized by observation of walking difficulties experienced by children between 2 and 5 years of age. The skeletal muscle degeneration is progressive, resulting in patients being wheelchair bound by the age of 13. DMD patients tend to die by the age of about 30 as a result of respiratory or cardiac failure [1].

DMD is caused by a mutation in the DMD gene, which is one of the largest in the human genome (approximately 2.5 million base pairs, encoding 79 exons). Full-length dystrophin mRNA is approximately 14 kb and is mainly expressed in skeletal, cardiac, and smooth muscles, as well as in the brain [2]. The dystrophin protein encoded by the DMD gene is rod shaped and consists of 4 domains: the N-terminal actin-binding domain, a rod-shaped domain composed of 24 spectrin-like rod repeats and 4 hinges, a cysteine-rich domain that interacts with dystroglycan and sarcoglycan complexes, and the C-terminal domain that interacts with the syntrophin complex and dystrobrevin [3]. Dystrophin is localized at the sarcolemma and forms a dystrophin-glycoprotein complex (DGC) with dystroglycan, sarcoglycan, and syntrophin/dystrobrevin complexes. These associations link the cytoskeletal protein actin to the basal lamina of muscle fibers [3]. DGC is thought to act as a membrane stabilizer during muscle contraction or a transducer of signals from the extracellular matrix to the muscle cytoplasm via intracellular signaling molecules. Loss of dystrophin leads to conditions under which the membrane becomes leaky as a result of mechanical or hypo-osmotic stress. Consequently, Ca^{2+} permeability is increased and various Ca^{2+} dependent proteases such as calpain are activated under conditions of dystrophin deficiency. It has also been proposed that alteration of the expression or function of the plasma membrane proteins associated with dystrophin such as neuronal nitric oxide synthase (nNOS) and various ion channels are involved in the molecular mechanisms of muscle degeneration [5].

The DMD gene mutations include missense, nonsense, deletion, insertion, or duplication mutations. When a mutation of DMD disrupts the reading frame of amino acids (an out-of-frame mutation), the dystrophin defect results in the severe DMD phenotype. On the other hand, a mutations which maintains the reading frame (an in-frame mutation) tends to produce a truncated but functional dystrophin, which leads to the more benign phenotype known as Becker muscular dystrophy (BMD). About 90% of cases of the phenotypes of DMD and BMD can be explained by this frame-shift theory [6]. There are two major hot spots for mutations around exons 3–7 and exons 45–55 in the DMD gene [6].

Investigations of Gene Therapy for Treatment of DMD

For the treatment of DMD, various experimental approaches such as gene therapy using viral vectors, exon-skipping therapy, stem cell transplantation, read-through therapy, and pharmacological agents have been extensively developed, but none of these have met with success in the clinic. Among these therapeutic strategies, exon skipping using antisense oligonucleotides (AOs) has been considered to be one of the most promising therapies for the restoration of dystrophin expression at the sarcolemma in dystrophin-deficient muscle. Exon skipping as treatment for DMD is developed based on the frame-shift theory. AOs are chemically-synthesized single-strand nucleic acids about 25 bases in length which are designed to recognize a specific sequence of the mRNA splicing pattern or of the binding protein. These agents can artificially change the translation of the nucleic acid. Among the various types of AOs, the 2′-O-methyl phosphorothioate AO (2OMeAO) and the phosphorodiamidate morpholino oligomers (PMO: morpholino) have high efficacy, low toxicity, and high stability [7, 8]. Therefore, exon skipping strategies mediated by 2OMeAO and PMO have been extensively performed and clinical studies have been conducted.

Gene therapy using viral vectors has been extensively investigated. Adeno-associated virus (AAV) vectors are the most appropriate tools for viral vector gene therapy because they are nonpathogenic due to a replication defect and have low immunity with an effective ability to infect nondividing cells. This strategy, however, is limited with respect to the size of the inserted exogenous gene. The upper limit of 4.9 kb prevents the full-length DMD cDNA (14 kb) from being successfully incorporated into the vector. Our group and other researchers have designed a shorter but functional microdystrophin, which is driven by the muscle-specific creatine kinase (MCK) promoter in combination with AAV vectors of various serotypes. The efficacy of this system has been examined [9, 10], and it has been found that the immune reaction is one of the critical issues in canine or primates. Other serotypes of AAV with a weaker immune reaction, such as AAV8, have recently been developed. These rAAVs are capable of providing powerful systemic delivery of the DMD gene to muscles throughout the body, including the heart.

MAMMALIAN MODELS OF DMD

Animal models are needed for elucidation of the pathogenesis and assessment of efficacy and toxicity during the development of therapies. In DMD, various animal models have been identified and utilized. In this review, we introduce and discuss not only the pathological characteristics of mammalian (murine, feline, or canine) models of DMD but also recent advances in therapeutic applications using these models.

MURINE MODEL OF DMD

Various mouse models with mutations in the mouse DMD gene have been identified: X-linked muscular dystrophy mouse (mdx) [11] and 4 additional strains of mdx mouse—mdx2cv, mdx3cv, mdx4cv, or mdx5cv mouse [12]. Moreover, the DMD exon 52 knockout (mdx52) mouse [13] and the mdx mouse with additional ablation of the dystrophin homologue utrophin (mdx/utr$^{-/-}$) [14] was produced. Among these models, the mdxmouse is the most commonly used model. The mouse models are indispensable for elucidation of the pathogenic mechanism and for development of therapeutic approaches, since they can be easily and reliably reproduced. In this section, we introduce the pathological features of mdx and mdx52 mice and the availability of therapeutic approaches which have benefited from the use of these mouse models.

Mdx Mice

Pathological Characteristics

X-linked muscular dystrophy mouse (mdx) was found to spontaneously occur in a C57BL/10 strain [15], and a nonsense mutation was identified in exon 23 of the DMD gene. This mutation causes a lack of dystrophin at the sarcolemma [2]. A previous study using mdx mice revealed that muscle necrosis with infiltration of neutrophils or macrophages is recognized at approximately 2 weeks of age and that a massive muscle degeneration/necrosis occurs at approximately 1 month of age. Muscles continue to go through cycles of necrosis and regeneration throughout the lifespan of the mdx mouse; it is just that is slows and is milder after 12 weeks. Necrotic fibers can be found at any age and appear with increasing frequency after 18 months of age and the pathology progresses. The muscle pathology is most pronounced between 2 and 8 weeks of age. This period is characterized by the presence of extensive necrosis, regenerated centrally nucleated fibers, and high levels of serum creatine kinase (CK), a biochemical marker of muscle necrosis [11, 15]. The skeletal and cardiac muscle deterioration of the mdx mouse is relatively much milder than that of human cases of DMD [16, 17]. As a result, the fibrosis and infiltration of inflammatory cells in the skeletal muscle at later stages tend to be much less than that observed in DMD patients.

Therapeutic Applications

Recent advances in therapeutic applications include exon skipping, gene therapy, and cell therapy. In the development of exon skipping therapy, the mdx mouse is available for exon 23 skipping to convert an out-of-frame mutation into an in-frame mutation. Systemic administration of 2'-O-methyl phosphorothioate antisense oligonucleotides (2OMeAO) with a nonion polymer F127 to mdx mice revealed that dystrophin is expressed in the whole-body skeletal muscle with the exception of heart muscle. However, the 2OMeAO did not reach a therapeutic level [18]. Meanwhile, systemic induction of dystrophin expression by phosphorodiamidate morpholino oligomers (PMO: morpholino) in the mdx mouse reached a treatable level in whole-body skeletal muscle with the exception of the heart [19, 20]. The peptide-linked PMO (PPMO) induced high expression of dystrophin not only in whole-body skeletal muscles but also in the heart in mdx mice [21]. A unique exon-skipping method was proposed in which the mutated exon 23 on the mRNA of mdx mice is removed by a single administration of an AAV vector expressing antisense sequences linked to a modified U7 small nuclear RNA [22].

In the development of gene therapy, we and other groups have designed a shorter but functional microdystrophin to incorporate into the recombinant AAV2 (rAAV2). The microdystrophin, which has a large deletion in the central rod domain, was constructed on the basis of the sequence of the mutated DMD gene in a BMD patient with a nearly normal life expectancy. The expression of this construct was driven by the muscle-specific creatine kinase (MCK) promoter, and local and systemic delivery of rAAV2 was found to restore the muscle function and life span of mdx mice [8, 9]. The other serotypes—rAAV6 [23], rAAV8 [24], and rAAV9 [25]—are also expected to be useful for systemic delivery of the DMD gene.

Although certain issues such as body size, genetic background, and pathological features should be addressed, the mdx mouse has been indispensable as a DMD model for development of therapeutic approaches.

Mdx52 Mouse

Pathological Characteristics

Katsuki and colleagues successfully generated a new DMD mouse model known as mdx52. In this model, exon 52 of the murine DMD gene was deleted using a homologous recombination technique [13]. Like the mdxmouse, mdx52 lacks dystrophin and presents dystrophic changes with muscle hypertrophy (Figure 1). In particular, the retina-specific dystrophin isoform Dp260 is absent and abnormalities were observed in electroretinographic analyses [26].

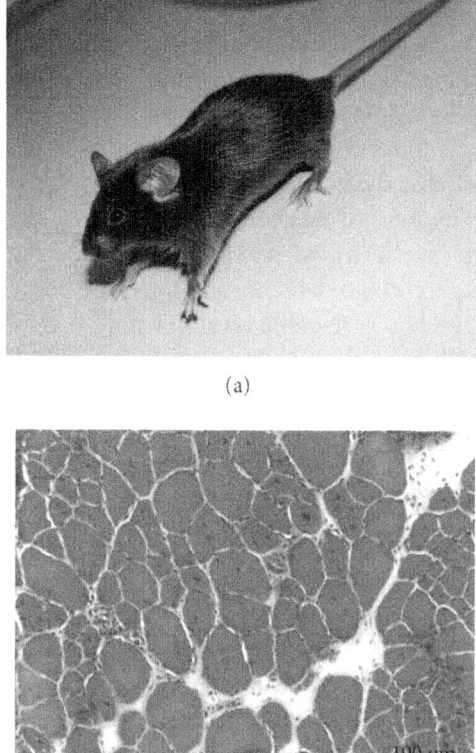

Figure 1: (a) A male mdx52 mouse at 6 weeks of age. The generalized appearance is almost normal. (b) The pathology (H and E) of the tibialis anterior muscle shows mul-

ticentrally nucleated fibers with an increased extent of invasion of inflammatory cells and the interstitial space. Hypertrophic muscle fibers were also observed.

Therapeutic Applications

The targeting of exon 51 for exon skipping is theoretically applicable to the highest percentage (13%) of DMD patients with an out-of frame deletion mutation [27–29]. An in-frame dystrophin mutation performed by an exon 51 skipping technique via local intramuscular injection of 2OMeAOs was found to be successful in some patients with DMD [30, 31]. Recently, we conducted exon 51 skipping using PMO in mdx52 mice to convert an out-of-frame mutation into an in-frame mutation and examined muscle function. The results showed that dystrophin expression was restored in whole-body skeletal muscles with amelioration of the dystrophic pathology and improved muscle function [32]. Thus, mdx52 has proven to be useful for development of the exon skipping technique and was used to obtain proof of concept for ongoing clinical trials.

CANINE MODEL OF DMD

Several different canine models of DMD have been reported thus far [4]. Mutations of the canine DMD gene were identified in Golden Retriever [33], Rottweiler [34], German Short-Hair Pointer [35], and Cavalier King Charles Spaniels [36]. Golden Retriever muscular dystrophy (GRMD) has been the most extensively examined and characterized. We recently described Beagle-based canine X-linked muscular dystrophy ($CXMD_J$) and conducted many studies on the elucidation of pathogenesis and the development of therapeutic approaches using this model. Exon-skipping therapy has been investigated using Cavalier King Charles Spaniels with muscular dystrophy (CKCS-MD). In this section, we describe the pathological characteristics and availability of gene therapies for GRMD, $CXMD_J$, and CKCS-MD.

Golden Retriever Muscular Dystrophy (GRMD)

Pathological Characteristics

GRMD is characterized by progressive skeletal muscle weakness and atrophy as well as cardiac involvement. These characteristics are similar to those of DMD. GRMD is caused by a point mutation at the intron 6 splice acceptor site of the canine DMD gene. This causes skipping of exon 7 and a premature stop codon in exon 8 and results in a lack of dystrophin [33]. The level of serum CK is dramatically increased early at 1-2 days of age and extensive diaphragm damage is observed [37]. The clinical manifestations of GRMD are progressive with the gradual loss of muscle mass and contractures that often lead to skeletal deformities [37]. Extensive muscle degeneration and necrosis of generalized muscles are identified from birth onwards [38]. A distinct feature is enlargement of the base of the tongue, pharyngeal muscle, and esophagus, resulting in dysphagia, drooling, and regurgitation. Severe fibrosis in muscle and joint contracture develops by 6 months of age [33], and respiratory failure or cardiomyopathy are frequently observed at younger ages [39]. GRMD is further characterized by progressive cardiomyopathy as indicated by deeper and narrower Q waves in leads II, III, and aVF relative to cardiac involvement in human DMD patients [38, 40, 41].

The skeletal and cardiac characteristics of GRMD are more similar to those of DMD than of mdx. The genetic background and body size of the Golden Retriever is closer to human than the mouse. Therefore, GRMD has been considered to be a useful animal model for human DMD in recent years.

Therapeutic Applications

Preliminary gene therapy experiments on GRMD performed using adenovectors [42], and high expression of a dystrophin minigene was achieved in GRMD using replication-deficient adenoviral vectors [43]. Very recently, systemic delivery of the AAV9 human mini-dystrophin vector induced widespread muscle expression of the transgene in neonatal dystrophic dogs [44].

In exon-skipping therapy, prolonged maintenance of functional dystrophin in GRMD muscle has additionally been achieved through chimeric RNA/DNA oligonucleotide therapy [45]. GRMD muscle cells were observed in order to compare the effectiveness of 2OMe AO, PMO, or peptide-linked PMO (PPMO) [46]. The PPMO was found to be capable of inducing high and sustained levels of exon skipping and induced the highest level of dystrophin expression with no apparent adverse effects upon the cells.

Canine X-Linked Muscular Dystrophy (CXMD$_j$)

Pathological Characteristics

As described in the previous section, GRMD is well characterized and attractive for research on DMD. However, Golden Retrievers are too large to be treated or raised easily and animal trials employing these dogs have substantial animal welfare implications and high costs associated with both maintenance and treatment. To address these issues, we developed a strain of medium-sized dystrophic Beagles by artificial insemination of two females with thawed spermatozoa obtained from a Golden Retriever with GRMD followed by interbreed crossing of the carrier female dogs with Beagle sires [47]. We designated the dystrophic dogs as canine X-linked muscular dystrophy in Japan ($CXMD_J$). The colony is maintained at the General Animal Research Facility of the National Center of Neurology and Psychiatry in Tokyo [47]. The serum CK level of dogs with $CXMD_J$ at birth is very high, and the mortality rate during the neonatal period is about 32.3%. This is significantly higher than that of normal littermates (13.3%). Diaphragm muscle involvement occurs shortly after birth and is more severe than that of limb muscles. At an age of about 2-3 months, atrophy and weakness of limb muscles appear, followed by development of macroglossia, dysphasia, gait disturbance, and joint contracture from 4 months of age. These symptoms rapidly progress until about 10 months of age, after which the progression of the disease is retarded [48]. The clinical manifestations of third generation $CXMD_J$ are similar but milder than those observed in GRMD (Figure 2).

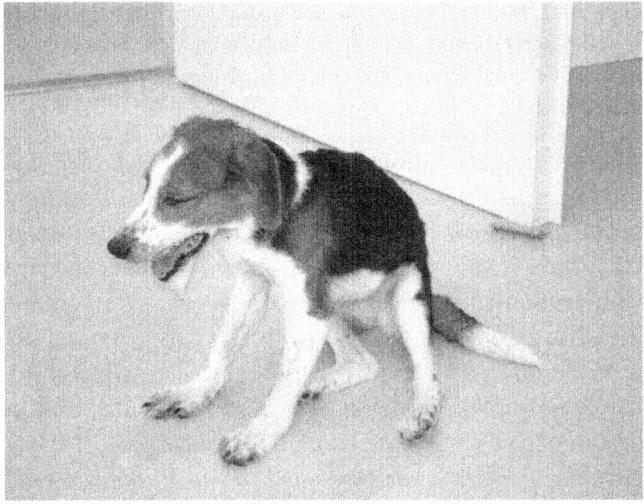

(a)

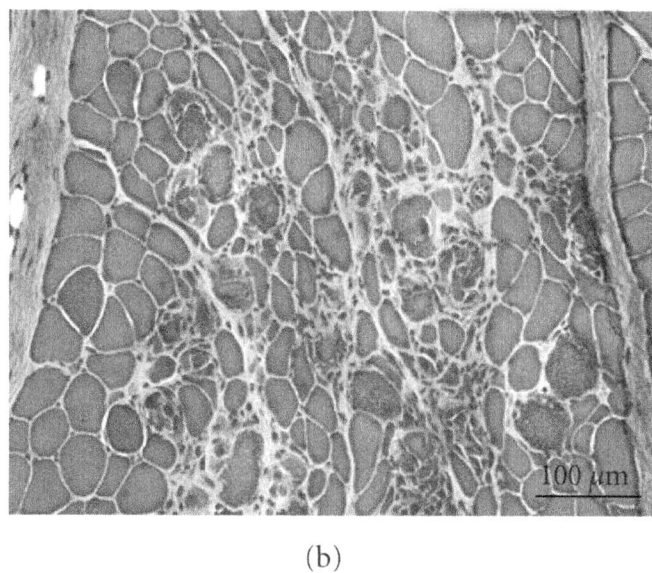

(b)

Figure 2: (a) A male canine with X-linked muscular dystrophy ($CXMD_J$) at 6 months of age. The atrophy of muscles throughout the body (including temporal muscle) is observed. Kyphosis, abnormal sitting posture, and contracture of hindlimb joints are seen. (b) The pathology (H and E) of tibialis cranialis muscle indicates muscle necrosis with infiltration of inflammatory cells, centrally nucleated fibers, and increased interstitial connective tissues.

In the cardiac manifestation, GRMD as well as DMD have particular characteristics such as distinct deep electrocardiogram Q-waves and fibrosis of the left ventricular wall [38, 40, 41]. We regularly examined the cardiac performance of the $CXMD_J$ model by electrocardiogram, echocardiogram, and pathology and found that the deep Q-waves on echocardiograms precede the development of overt fibrosis in histopathology [49]. So far, the pathogenic mechanism of the distinct deep Q-waves has been considered to be ascribed to fibrosis in the posterobasal region of the left ventricular wall in DMD [41]. When we investigated the cardiac conduction system of $CXMD_J$, a remarkable extent of vacuolar degeneration was observed in Purkinje fibers despite the absence of detectable fibrotic lesions in the ventricular myocardium [50]. The degenerated Purkinje fibers were coincident with overexpression of Dp71 (a C-terminal truncated isoform of dystrophin) at the sarcolemma and translocation of µ-calpain (a calcium-dependent protease) to the periphery near the sarcolemma or to the vacuoles. Utrophin, a homologue of dystrophin was upregulated in dystrophin deficiency. Utrophin was found to be highly upregulated in the Purkinje fibers in the early stage, but the expression was dislocated when vacuolar degeneration was recognized at 4 months of age. Based on these findings, we hypothesized that the selective degeneration of Purkinje fibers could be associated with distinct deep Q-waves on electrocardiograms and the fatal arrhythmia which is observed in dystrophinopathy [50]. Thus, our previous investigations of the clinical phenotype in CXMD$_J$ revealed not only that the phenotypes are nearly identical to those of GRMD but also that the model will play a useful role in elucidation of the pathogenic mechanism and in development of therapeutic approaches.

Therapeutic Applications

Recently, in conjunction with our collaborators at the Children's National Medical Center in the USA, we developed a multiexon-skipping technique for targeting exons 6 and 8 to convert an out-of-frame mutation into an in-frame mutation using PMOs. Systemic multiexon-skipping treatment in CXMD$_J$ was found to restore dystrophin in the whole-body skeletal muscle (with the exception of heart). Furthermore, skeletal muscle function was notably improved without any adverse effects [51]. This represents the first report on successful in vivo multiexon-skipping. Furthermore, we have succeeded in producing multiexon skipping by administration of the PMOs used in CXMD$_J$ to MyoD-transduced fibroblasts from a DMD patient with the exon 7 deletion of the DMD gene [52]. This suggests that multiexon skipping is feasible for treatment of DMD patients and that this approach could be applicable to up to 83% of all DMD patients [27].

To examine therapeutic effects and safety in a larger animal model, we have examined the efficiency of rAAV2 infection of canine myotubes and expression of the lacZ gene into normal canine muscle. In contrast to the results of mdx investigations, the rAAV2-mediated gene transfer was found to elicit a severe immune response against the AAV particle, and its gene product in the host [53]. We, therefore, have examined the transgene expression and host immune response to two different rAAV (rAAV2 and rAAV8) in normal canine and in the CXMD$_J$ model after intramuscular injection and systemic administration by limb perfusion to bypass the immune activation of dendritic cells in the injected muscle. In contrast to the results with rAAV2 transduction, intramuscular transduction of rAAV8-lacZ in normal dogs and systemic administration of rAAV8-microdystrophin in CXMD$_J$ were found to improve the expression of both transgenes in the skeletal muscle [54]. The CXMD$_J$ model led us to propose that rAAV8-mediated transduction strategy could have a therapeutic advantage in DMD gene therapy.

Thus, the clinical phenotype of CXMD$_J$ has been well characterized as an appropriate animal model and the similarity of pathology of DMD is regarded as the most appropriate model for DMD in clinical trials. Other therapeutic approaches will be evaluated in the dog model with a view to establishing feasible protocols.

Cavalier King Charles Spaniels with Muscular Dystrophy (CKCS-MD)

Pathological Characteristics

Very recently, it was reported that another dystrophic dog, the Cavalier King Charles Spaniel with dystrophin-deficient muscular dystrophy (CKCS-MD) has a severe phenotype. This canine model has a missense mutation in the 5' donor splice site of exon 50 resulting in deletion of exon 50 in mRNA transcripts and a predicted premature truncation of the translated protein [36].

Therapeutic Applications

The therapeutic strategy of exon skipping in the GRMD or $CXMD_J$ models provides a significant advantage in clinical trials, but also has the unfavorable characteristic that the disease-causing mutation does not include the region of the DMD gene that is most commonly mutated in human DMD patients. About 60% of DMD patients harbor deletions in exons 45–55 of the DMD gene [55, 56]. As described previously, the exon 51 skipping technique, which is being employed in clinical trials [30, 31], is applicable to DMD patients which have deletions of exon 50, exon 52, exons 48–50, or exons 49-50, among others, and is feasible for more patients than other exon skipping strategies [27, 57, 58]. AO-mediated skipping of exon 51 in cultured myoblasts of the CKCS-MD dog restored the reading frame and protein expression [17]. This observation suggests that the use of this canine model would be valuable in the preclinical trials of therapy based upon exon 51 skipping.

FELINE MODEL OF DMD

Hypertrophy Feline Muscular Dystrophy (HFMD)

Feline muscular dystrophy with dystrophin deficiency has been identified [59]. This feline model has a unique phenotypic expression of hypertrophy of the tongue, neck, and shoulder muscles, lingual calcification, excessive salivation, megaesophagus, gait disturbance manifesting as bunny hopping, dilated cardiomyopathy, hepatosplenomegaly and kidney failure [60, 61]. These dystrophic cats have been described as having hypertrophic feline muscular dystrophy (HFMD) because of the distinct hypertrophic feature. The pathology exhibits muscle degeneration and centrally nucleated regeneration and accumulation of calcium deposits within muscle fibers without development of fibrosis. The HFMD cat has a large deletion of muscle and Purkinje promoters resulting in a lack of dystrophin in the skeletal and cardiac muscle [62]. However, the dystrophic cats have not been widely used as a DMD model due to the limited pathological similarity to DMD [4]; there have been no reports on therapeutic approaches using this feline model.

CONCLUSIONS

Murine models will continue to provide important findings for the basic study of pathogenesis and development of therapies, but the clinical phenotype is mild and this is a weak point. On the other hand, canine models have severe skeletal and cardiac defects resembling DMD and the body size and genetic background of canines are more similar to human beings than the murine models. We are, therefore, convinced that canine models will be more useful for future contributions to preclinical study of newly developed therapies.

ACKNOWLEDGMENT

These studies were supported by Health Sciences Research Grants for Research on Psychiatric and Neurological Diseases and Mental Health (nos. H12-kokoro-025, H15-kokoro-021, and H18-kokoro-019), the Human Genome and Gene Therapy (nos. H13-genome-001, and H16-genome-003), and the Health and Labor Sciences Research Grants for Translation Research (no. H19-translational research-003) from the Ministry of Health, Labor and Welfare of Japan, and Grants-in-Aid for Scientific Research from the Ministry of Education, Science, Sports and Culture of Japan (no. 21300157 to A. Nakamura).

REFERENCES

1. H. Moser, "Duchenne muscular dystrophy: pathogenetic aspects and genetic prevention," Human Genetics, vol. 66, no. 1, pp. 17–40, 1984.
2. E. P. Hoffman, R. H. Brown Jr., and L. M. Kunkel, "Dystrophin: the protein product of the Duchenne muscular dystrophy locus," Cell, vol. 51, no. 6, pp. 919–928, 1987.
3. K. P. Campbell, "Three muscular dystrophies: loss of cytoskeleton-extracellular matrix linkage," Cell, vol. 80, no. 5, pp. 675–679, 1995.

4. G. D. Shelton and E. Engvall, "Canine and feline models of human inherited muscle diseases,"Neuromuscular Disorders, vol. 15, no. 2, pp. 127–138, 2005.
5. E. W. Yeung, N. P. Whitehead, T. M. Suchyna, P. A. Gottlieb, F. Sachs, and D. G. Allen, "Effects of stretch-activated channel blockers on [Ca2+]i and muscle damage in the mdx mouse," Journal of Physiology, vol. 562, no. 2, pp. 367–380, 2005.
6. M. Koening, A. H. Beggs, M. Moyer et al., "The molecular basis for Duchenne versus Becker muscular dystrophy: correlation of severity with type of deletion," American Journal of Human Genetics, vol. 45, no. 4, pp. 498–506, 1989.
7. C. Wilson and A. D. Keefe, "Building oligonucleotide therapeutics using non-natural chemistries,"Current Opinion in Chemical Biology, vol. 10, no. 6, pp. 607–614, 2006.
8. S. Karkare and D. Bhatnagar, "Promising nucleic acid analogs and mimics: characteristic features and applications of PNA, LNA, and morpholino," Applied Microbiology and Biotechnology, vol. 71, no. 5, pp. 575–586, 2006.
9. M. Yoshimura, M. Sakamoto, M. Ikemoto et al., "AAV vector-mediated microdystrophin expression in a relatively small percentage of mdx myofibers improved the mdx phenotype," Molecular Therapy, vol. 10, no. 5, pp. 821–828, 2004.
10. P. Gregorevic, J. M. Allen, E. Minami et al., "rAAV6-microdystrophin preserves muscle function and extends lifespan in severely dystrophic mice," Nature Medicine, vol. 12, no. 7, pp. 787–789, 2006.
11. G. R. Coulton, N. A. Curtin, J. E. Morgan, and T. A. Partridge, "The mdx mouse skeletal muscle myopathy: II. Contractile properties," Neuropathology and Applied Neurobiology, vol. 14, no. 4, pp. 299–314, 1988.
12. W. B. Im, S. F. Phelps, E. H. Copen, E. G. Adams, J. L. Slightom, and J. S. Chamberlain, "Differential expression of dystrophin isoforms in strains of mdx mice with different mutations," Human Molecular Genetics, vol. 5, no. 8, pp. 1149–1153, 1996.
13. E. Araki, K. Nakamura, K. Nakao et al., "Targeted disruption of exon 52 in the mouse dystrophin gene induced muscle degeneration similar to that observed in duchenne muscular dystrophy," Biochemical and Biophysical Research Communications, vol. 238, no. 2, pp. 492–497, 1997.

14. A. E. Deconinck, J. A. Rafael, J. A. Skinner et al., "Utrophin-dystrophin-deficient mice as a model for Duchenne muscular dystrophy," Cell, vol. 90, no. 4, pp. 717–727, 1997.

15. G. Bulfield, W. G. Siller, P. A. L. Wight, and K. J. Moore, "X chromosome-linked muscular dystrophy (mdx) in the mouse," Proceedings of the National Academy of Sciences of the United States of America, vol. 81, no. 4 I, pp. 1189–1192, 1984.

16. J. Dangain and G. Vrbova, "Muscle development in mdx mutant mice," Muscle and Nerve, vol. 7, no. 9, pp. 700–704, 1984.

17. Y. Tanabe, K. Esaki, and T. Nomura, "Skeletal muscle pathology in X chromosome-linked muscular dystrophy (mdx) mouse," Acta Neuropathologica, vol. 69, no. 1-2, pp. 91–95, 1986.

18. Q. L. Lu, C. J. Mann, F. Lou et al., "Functional amounts of dystrophin produced by skipping the mutated exon in the mdx dystrophic mouse," Nature Medicine, vol. 9, no. 8, pp. 1009–1014, 2003.

19. D. J. Wells, "Therapeutic restoration of dystrophin expression in Duchenne muscular dystrophy," Journal of Muscle Research and Cell Motility, vol. 27, no. 5-7, pp. 387–398, 2006.

20. S. Fletcher, K. Honeyman, A. M. Fall, P. L. Harding, R. D. Johnsen, and S. D. Wilton, "Dystrophin expression in the mdx mouse after localised and systemic administration of a morpholino antisense oligunucleotide," Journal of Gene Medicine, vol. 8, no. 2, pp. 207–216, 2006.

21. N. Jearawiriyapaisarn, H. M. Moulton, B. Buckley et al., "Sustained dystrophin expression induced by peptide-conjugated morpholino oligomers in the muscles of mdx mice," Molecular Therapy, vol. 16, no. 9, pp. 1624–1629, 2008.

22. A. Goyenvalle, A. Vulin, F. Fougerousse et al., "Rescue of dystrophic muscle through U7 snRNA-mediated exon skipping," Science, vol. 306, no. 5702, pp. 1796–1799, 2004.

23. P. Gregorevic, M. J. Blankinship, J. M. Allen et al., "Systemic delivery of genes to striated muscles using adeno-associated viral vectors," Nature Medicine, vol. 10, no. 8, pp. 828–834, 2004.

24. Z. Wang, T. Zhu, C. Qiao et al., "Adeno-associated virus serotype 8 efficiently delivers genes to muscle and heart," Nature Biotechnology, vol. 23, no. 3, pp. 321–328, 2005.

25. L. T. Bish, K. Morine, M. M. Sleeper et al., "Adeno-associated virus (AAV) serotype 9 provides global cardiac gene transfer superior to

AAV1, AAV6, AAV7, and AAV8 in the mouse and rat," Human Gene Therapy, vol. 19, no. 12, pp. 1359–1368, 2008.

26. S. Kameya, E. Araki, M. Katsuki et al., "Dp260 disrupted mice revealed prolonged implicit time of the b-wave in ERG and loss of accumulation of β-dystroglycan in the outer plexiform layer of the retina,"Human Molecular Genetics, vol. 6, no. 13, pp. 2195–2203, 1997.

27. A. Aartsma-Rus, I. Fokkema, J. Verschuuren et al., "Theoretic applicability of antisense-mediated exon skipping for Duchenne muscular dystrophy mutations," Human Mutation, vol. 30, no. 3, pp. 293–299, 2009.

28. M. Kinali, V. Arechavala-Gomeza, L. Feng et al., "Local restoration of dystrophin expression with the morpholino oligomer AVI-4658 in Duchenne muscular dystrophy: a single-blind, placebo-controlled, dose-escalation, proof-of-concept study," The Lancet Neurology, vol. 8, no. 10, pp. 918–928, 2009.

29. A. T. J. M. Helderman-van den Enden, C. S. M. Straathof, A. Aartsma-Rus et al., "Becker muscular dystrophy patients with deletions around exon 51; a promising outlook for exon skipping therapy in Duchenne patients," Neuromuscular Disorders, vol. 20, no. 4, pp. 251–254, 2010.

30. J. C. van Deutekom, A. A. Janson, I. B. Ginjaar et al., "Local dystrophin restoration with antisense oligonucleotide PRO051," The New England Journal of Medicine, vol. 357, no. 26, pp. 2677–2686, 2007.

31. F. Muntoni, K. Bushby, and G. Van Ommen, "128th ENMC International Workshop on ‹Preclinical optimization and phase I/II clinical trials using antisense oligonucleotides in Duchenne muscular dystrophy› 22-24 October 2004, Naarden, The Netherlands," Neuromuscular Disorders, vol. 15, no. 6, pp. 450–457, 2005.

32. Y. Aoki, A. Nakamura, T. Yokota et al., "In-frame dystrophin following exon 51-skipping improves muscle pathology and function in the exon 52-deficient mdx mouse," Molecular Therapy, vol. 18, no. 11, pp. 1995–2005, 2010.

33. N. J. H. Sharp, J. N. Kornegay, S. D. van Camp et al., "An error in dystrophin mRNA processing in golden retriever muscular dystrophy, an animal homologue of Duchenne muscular dystrophy," Genomics, vol. 13, no. 1, pp. 115–121, 1992.

34. N. Winand, D. Pradham, and B. Cooper, "Molecular characterization of severe Duchenne-type muscular dystrophy in a family of Rottwiler dogs," in Molecular Mechanism of Neuromuscular Disease, Muscular Dystrophy Association, Tucson, Ariz, USA, 1994.

35. S. J. Schatzberg, N. J. Olby, M. Breen et al., "Molecular analysis of a spontaneous dystrophin ‹knockout› dog," Neuromuscular Disorders, vol. 9, no. 5, pp. 289–295, 1999.

36. G. L. Walmsley, V. Arechavala-Gomeza, M. Fernandez-Fuente et al., "A duchenne muscular dystrophy gene hot spot mutation in dystrophin-deficient cavalier king charles spaniels is amenable to exon 51 skipping," PloS one, vol. 5, no. 1, Article ID e8647, 2010.

37. B. J. Cooper, N. J. Winand, H. Stedman et al., "The homologue of the Duchenne locus is defective in X-linked muscular dystrophy of dogs," Nature, vol. 334, no. 6178, pp. 154–156, 1988.

38. B. A. Valentine, B. J. Cooper, A. De Lahunta, R. O›Quinn, and J. T. Blue, "Canine X-linked muscular dystrophy. An animal model of Duchenne muscular dystrophy: clinical studies," Journal of the Neurological Sciences, vol. 88, no. 1-3, pp. 69–81, 1988.

39. F. Nguyen, Y. Cherel, L. Guigand, I. Goubault-Leroux, and M. Wyers, "Muscle lesions associated with dystrophin deficiency in neonatal golden retriever puppies," Journal of Comparative Pathology, vol. 126, no. 2-3, pp. 100–108, 2002.

40. N. S. Moise, B. A. Valentine, C. A. Brown et al., "Duchenne›s cardiomyopathy in a canine model: electrocardiographic and echocardiographic studies," Journal of the American College of Cardiology, vol. 17, no. 3, pp. 812–820, 1991.

41. J. K. Perloff, W. C. Roberts, A. C. de Leon Jr., and D. O›Doherty, "The distinctive electrocardiogram of Duchenne›s progressive muscular dystrophy. An electrocardiographic-pathologic correlative study," The American Journal of Medicine, vol. 42, no. 2, pp. 179–188, 1967.

42. J. M. Howell, S. Fletcher, B. A. Kakulas, M. O›Hara, H. Lochmuller, and G. Karpati, "Use of the dog model for Duchenne muscular dystrophy in gene therapy trials," Neuromuscular Disorders, vol. 7, no. 5, pp. 325–328, 1997.

43. J. M. Howell, H. Lochmüller, A. O›Hara et al., "High-level dystrophin expression after adenovirus-mediated dystrophin minigene transfer to skeletal muscle of dystrophic dogs: prolongation of expression with immunosuppression," Human Gene Therapy, vol. 9, no. 5, pp. 629–634, 1998.

44. J. N. Kornegay, J. Li, J. R. Bogan et al., "Widespread muscle expression of an AAV9 human mini-dystrophin vector after intravenous injection in neonatal dystrophin-deficient dogs," Molecular Therapy, vol. 19, no. 8, pp. 1501–1508, 2010.

45. R. J. Bartlett, S. Stockinger, M. M. Denis et al., "In vivo targeted repair of a point mutation in the canine dystrophin gene by a chimeric RNA/DNA oligonucleotide," Nature Biotechnology, vol. 18, no. 6, pp. 615–622, 2000.

46. G. McClorey, H. M. Moulton, P. L. Iversen, S. Fletcher, and S. D. Wilton, "Antisense oligonucleotide-induced exon skipping restores dystrophin expression in vitro in a canine model of DMD," Gene Therapy, vol. 13, no. 19, pp. 1373–1381, 2006.

47. Y. Shimatsu, K. Katagiri, T. Furuta et al., "Canine X-linked muscular dystrophy in Japan (CXMD),"Experimental Animals, vol. 52, no. 2, pp. 93–97, 2003.

48. Y. Shimatsu, M. Yoshimura, K. Yuasa et al., "Major clinical and histopathological characteristics of canine X-linked muscular dystrophy in Japan, CXMD," Acta Myologica, vol. 24, no. 2, pp. 145–154, 2005.

49. N. Yugeta, N. Urasawa, Y. Fujii et al., "Cardiac involvement in Beagle-based canine X-linked muscular dystrophy in Japan (CXMD): electrocardiographic, echocardiographic, and morphologic studies," BMC Cardiovascular Disorders, vol. 6, article 47, 2006.

50. N. Urasawa, M. R. Wada, N. Machida et al., "Selective vacuolar degeneration in dystrophin-deficient canine Purkinje fibers despite preservation of dystrophin-associated proteins with overexpression of Dp71," Circulation, vol. 117, no. 19, pp. 2437–2448, 2008.

51. T. Yokota, Q. L. Lu, T. Partridge et al., "Efficacy of systemic morpholino exon-skipping in duchenne dystrophy dogs," Annals of Neurology, vol. 65, no. 6, pp. 667–676, 2009.

52. T. Saito, A. Nakamura, Y. Aoki et al., "Antisense PMO found in dystrophic dog model was effective in cells from exon 7-deleted DMD patient," PLoS One, vol. 5, no. 8, Article ID e12239, 2010.

53. K. Yuasa, M. Yoshimura, N. Urasawa et al., "Injection of a recombinant AAV serotype 2 into canine skeletal muscles evokes strong immune responses against transgene products," Gene Therapy, vol. 14, no. 17, pp. 1249–1260, 2007.

54. S. Ohshima, J. H. Shin, K. Yuasa et al., "Transduction efficiency and immune response associated with the administration of AAV8 vector into dog skeletal muscle," Molecular Therapy, vol. 17, no. 1, pp. 73–80, 2009.

55. J. Alter, F. Lou, A. Rabinowitz et al., "Systemic delivery of morpholino oligonucleotide restores dystrophin expression bodywide and improves dystrophic pathology," Nature Medicine, vol. 12, no. 2, pp. 175–177, 2006.

56. A. Aartsma-Rus, A. A. M. Janson, W. E. Kaman et al., "Antisense-induced multiexon skipping for Duchenne muscular dystrophy makes more sense," American Journal of Human Genetics, vol. 74, no. 1, pp. 83–92, 2004.
57. V. Arechavala-Gomeza, I. R. Graham, L. J. Popplewell et al., "Comparative analysis of antisense oligonucleotide sequences for targeted skipping of exon 51 during dystrophin pre-mRNA splicing in human muscle," Human Gene Therapy, vol. 18, no. 9, pp. 798–810, 2007.
58. A. Nakamura and S. Takeda, "Exon-skipping therapy for Duchenne muscular dystrophy," Neuropathology, vol. 29, no. 4, pp. 494–501, 2009.
59. J. H. Vos, J. S. van der Linde-Sipman, and S. A. Goedegebuure, "Dystrophy-like myopathy in the cat," Journal of Comparative Pathology, vol. 96, no. 3, pp. 335–341, 1986.
60. J. L. Carpenter, E. P. Hoffmann, F. C. A. Romanul et al., "Feline muscular dystrophy with dystrophin deficiency," The American Journal of Pathology, vol. 135, no. 5, pp. 909–919, 1989.
61. F. P. Gaschen, E. P. Hoffman, J. R. M. Gorospe et al., "Dystrophin deficiency causes lethal muscle hypertrophy in cats," Journal of the Neurological Sciences, vol. 110, no. 1-2, pp. 149–159, 1992.
62. N. J. Winand, M. Edwards, D. Pradhan, C. A. Berian, and B. J. Cooper, "Deletion of the dystrophin muscle promoter in feline muscular dystrophy," Neuromuscular Disorders, vol. 4, no. 5-6, pp. 433–445, 1994.

Chapter 6

ADVANCED NANOMATERIALS IN MULTIMODAL IMAGING: DESIGN, FUNCTIONALIZATION, AND BIOMEDICAL APPLICATIONS

Zhe Liu, Fabian Kiessling, and Jessica Gätjens

Department of Experimental Molecular Imaging (ExMI), Helmholtz Institute for Biomedical Engineering, Medical Faculty, RWTH Aachen University, Pauwelsstraße 20, 52074 Aachen, Germany

ABSTRACT

The biomedical applications of nanoparticles in molecular imaging, drug delivery, and therapy give rise to the term "nanomedicine" and have led to ever-growing developments in the past decades. New generation of imaging probes (or contrast agents) and state of the art of various strategies for efficient multimodal molecular imaging have drawn much attention and led to successful preclinical uses. In this context, we intend to elucidate the fundamentals and review recent advances as well as to provide an outlook perspective in these fields.

INTRODUCTION

Nanotechnology, abbreviated as "nanotech", has witnessed rapid and broad developments over the past decades [1, 2]. Nowadays, nanoscience is no longer confined to synthesis, characterization and simple derivation of nanomaterials, but has found its way to high-end applications and engineering in industrial sectors, such as electronics, communication, energy, advanced materials, space technology and biomedicine. The National Nanotechnology Initiative (NNI) defines nanosized particles as roughly 1–100 nm in dimensions, but this range might be extended up to 1000 nm [3]. They display novel optical, electronic and structural properties different from individual molecules or bulk particles. When applied in medical science, they give rise to the term of "nanomedicine" which has found more and more applications in preclinical, translational, and clinical research [2, 4, 5]. This will not only stimulate the progress of nanomaterial development, but also implicate unprecedented opportunities in the coming future for individualized diagnostic strategies and treatment methodologies in humans.

Among the vast applications of nanomedicine, molecular imaging is one of the most attractive and fastest growing research fields. As defined by the Society of Nuclear Medicine recently, molecular imaging is "visualization, characterization, and quantification of biological processes at the molecular and cellular levels in humans and other organisms" [6]. It is an emerging interdisciplinary research field that combines chemistry, biology, pharmacology, and medicine to detect biomedical or physiological processes in vitro and in vivo, and therefore allows for early detection of physiological changes, anatomical diagnosis of transformations, and provides valuable clinical information for treatment strategies for various diseases, such as cancer, inflammation, stroke, atherosclerosis, Alzheimer's disease, and many others [2, 7–9]. In addition, customizable nanoparticles have also been employed as efficient carriers for targeted drug delivery and therapeutic agents as well as for gene transportation.

NANOPARTICLES

Particles with nanoscale dimensions are expected to display special physical and biological behavior and unique interactions with biomolecules. Due to the large surface area of nanoparticles and inherent functionalities, structural modifications are readily achieved to alter their pharmacokinetics, prolong their vascular circulation life-time, improve their extravasation capacity, ensure an enhanced biodistribution in vivo, and lead to a sustained and controllable delivering efficacy for drug cargoes [10, 11]. Furthermore, when specific targeting ligands are conjugated to nanoparticles, targeted binding capability to diseased regions can be realized. Nanocarriers will penetrate through microvessels with enhanced permeability and then be taken up by cells, thus offering highly-selective payload accumulation at target sites.

More and more nanoparticle-based structures are currently under investigation and some well-studied nanoparticles include quantum dots, dendrimers, nanotubes, liposomes, micelles, gold nanoparticles, and nano/microbubbles. Different types of nanoparticles have different biomedical purposes and are being used for medical molecular imaging, controlled drug delivery, and targeted therapy. The characteristics and biomedical applications of some representative nanoparticles have been listed in Table 1.

Table 1: Characteristics of several representative nanoparticles and their biomedical applications.

Type of nanoparticle	Synthetic protocol	Size range	Possible surface modifications	Imaging modality applicable	Possible therapeutic strategies	Reference
Quantum dot	colloidal synthesis, self-assembly, viral assembly	several to tens of nm	lipids, polymer, targeting ligands or biomolecules	optical	PDT*	[12]
Dendrimer	organic chemistry techniques	several nm varies from different "generation"	charge, polymer, targeting ligands or biomolecules	MRI, optical	drug and gene delivery	[13]
Liposome	emulsion, polymerization	tens to hundreds of nm	charge, polymer, targeting ligands or biomolecules, viral protein coating,	MRI, optical, radionuclide imaging	drug, gene and protein delivery, PDT,	[14]
Gold nanoparticle	biological reduction, colloidal synthesis, vapor precipitation	several to 100 nm	lipids, polymeric shell, targeting ligands or biomolecules	CT, optical	drug delivery, PTT*	[15]
Carbon Nanotube	arc discharge, laser ablation, vapor precipitation	tens of nm	polymeric shell, targeting ligands or biomolecules	MRI, optical, radionuclide imaging	drug delivery, PTT	[16]
Microbubble	emulsion, layer-by-layer fabrication, polymerization	tens to 1000 nm	polymeric shell, targeting ligands or biomolecules	US	drug, gene and protein delivery	[17]
Iron Oxide	coprecipitation, decomposition, microemulsion, sol-gel, thermal	several to tens of nm	charge, dextran, lipids, polymer, targeting ligands or biomolecules	MRI	siRNA delivery, PTT	[18]
Micelle	microemulsion	tens of nm	charge, polymer, targeting ligands or biomolecules	MRI, optical, radionuclide imaging	drug and gene delivery, PDT	[19]
Adenovirus	replication in host nucleus	tens of to 100 nm	charge, polymer, targeting ligands or biomolecules	MRI, optical	gene delivery	[20]
Silica nanoparticle	chemical polymerization, microemulsion, sol-gel	tens of to 100 nm	charge, polymer, targeting ligands or biomolecules	MRI, optical	drug and gene delivery	[21]

*Abbreviations: PDT= Photodynamic therapy; PTT= Photothermal therapy.

It should be noted that the knowledge on the effect of nanoparticles on human health is very limited. Thus, nanotoxicology, which is the science examining the effects of artificial nanostructures on living organisms is of great importance. For example, nanoparticles can penetrate into the cells and selectively accumulate. They can be transported across the epithelial and endothelial cells via transcytosis. Further, they might travel along the dendrites, axons and the blood and lymphatic vessels provoking oxidative stress and inflammation [22]. However, appropriate chemical modifications on the surface can be readily implemented to render them qualified for medical uses even though some "naked" nanoparticles are highly cytotoxic and cannot be applied in humans directly.

MOLECULAR IMAGING

During the past decades, many traditional medical imaging techniques have been established for routine experimental and clinical uses. These imaging modalities, such as optical imaging (OI), computed tomography (CT), magnetic resonance imaging (MRI), ultrasound (US), and radionuclide imaging (PET/SPECT) have been widely applicable to and made superb performance in experimental small animal imaging, preclinical imaging and clinical human body imaging, diagnosis and treatments [23–26] (Figure 1).

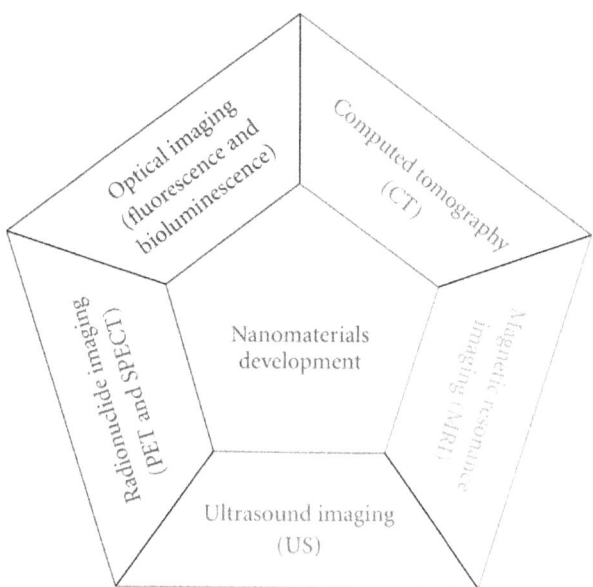

Figure 1: Illustration for nanomaterial applications in various imaging modalities.

Molecular imaging differs from traditional imaging in that probes known as biomarkers are used to help image particular targets or pathways. It is required that biomarkers highly specifically interact with their surroundings, and in turn alter the image according to molecular changes occurring within the area of interest. Molecular imaging agents are endogenous molecules or exogenous probes used to visualize, characterize, and quantify biological processes in living systems. Different imaging techniques in terms of sensitivity, resolution and complexity often require specific contrast agents to obtain satisfactory contrast enhancement in visualization reconstruction. Here, we will give a brief description on the properties of various imaging modalities, the nature of contrast enhancement and selected efficacious contrast agents commonly used at present. A detailed summary of characteristics, comparisons, and practical applications of these common imaging modalities is given in Table 2.

Table 2: Comparison of various imaging modalities in preclinical use.

Imaging modality	Optical imaging	Computed tomography	Magnetic resonance imaging	Ultrasound	Radionuclide imaging
Source of detection	visible or near-infrared light	X-ray	magnetic field, radiowaves	ultrasonic waves	γ-ray
Common imaging probes	fluorescent dyes, quantum dots	heavy atom-containing contrast agents, for example, iodine, barium and gadolinium salts	paramagnetic or superparamagnetic contrast agents, for example, gadolinium, manganese compounds and magnetofluids (Fe_3O_4)	microbubbles	Radionuclides ($^{18}F, ^{11}C, ^{64}Cu, ^{99m}Tc, ^{111}In$, etc.)-labeled probes
Advantages	inexpensive, low-cost, easy operation	anatomical imaging, applicable for humans	high spatial resolution, combines morphological and functional imaging, no tissue penetrating limit, applicable for humans	safety, real-time, low cost, wide availability, easy handling	high sensitivity, quantitative, no penetration limit

Disadvantages	photobleaching, limited tissue penetrating depth, surface-weighted, relatively low spatial resolution, autofluorescence disturbance	radiation risks, limited soft tissue resolution, not quantitative	relatively low sensitivity, time-consuming scan and processing, high cost	limited resolution and sensitivity, low data reproducibility	low spatial resolution, radiation risks, high cost
Some practical applications	cellular/intracellular expression, trafficking or movement monitoring of reporter/gene	bone and tumor imaging, fused image with other modalities	cell trafficking, morphological reporter/gene expression, cerebral and coronary angiography in clinics	potential application in drug delivery and controlled release, echocardiography and intracranial neoplasm in clinics	noninvasive evaluation of pharmacokinetics and metabolism of drugs, cerebral, cardiac and tumor imaging in clinics

Optical imaging is a relatively lowcost, noninvasive and extensively used imaging technique, in that light is taken as the most readily available and versatile imaging radiation source in nature. Fluorescence imaging (FLI) and bioluminescence imaging (BLI) are two major techniques in optical imaging to analyze the propagation of nonionizing radiation, light photons through a medium such as tissue. Proteic fluorophores (green fluorescent protein/ GFP, red fluorescent protein/RFP) and organic/inorganic fluorescent dyes are popular imaging agents in fluorescence imaging [27]. Besides that, quantum dots and iron oxide nanoparticles with fluorescent shell structure can also be employed as fluorescent imaging building blocks for efficient molecular fluorescence imaging in vitro and in vivo.

Computed tomography (CT) is founded on the exploitation of X-ray scanning, its attenuation in tissues and computed image reconstruction to obtain morphologic and vascular information within the body. Different components inside the body, such as soft tissue, fat, water and air, have different capability of X-ray absorption and attenuation. In this way, an anatomical visualization of body structures (lung, bones, tumor, and others) can be imaged with high contrast performance. Iodine, barium salt and gastrografin have been common CT contrast agents in clinics to highlight blood vessels, stomach and gastrointestinal organs. Theoretically, CT is not a molecular imaging modality as the sensitivity for contrast material is not sufficient to detect as low amounts as would be bound to a molecular target. However, this will not be an obstacle in the application of multimodal imaging approaches if CT contrast agents are structurally modified to meet other imaging requirements such as contrasting the blood pool or organs of the reticulo-endothelial system (RES).

Magnetic resonance imaging (MRI) is a noninvasive medical diagnostic technique. Although certain endogenous contrast can be achieved in the process of nuclei excitation and relaxation of magnetic spins, specific exogenous contrast agents are often required to give an acceptable MRI image with high spatiotemporal resolution, sensitivity, specificity, and volumetric coverage. The most commonly used contrast agents for MRI include paramagnetic complexes (Ga^{3+} or Mn^{2+} based chelates), paramagnetic ion nanoparticles (Gd_2O_3 and MnO), and superparamagnetic iron oxide nanoparticles (Fe_3O_4, FeCO, and $MnFe_2O_4$) [28, 29]. Novel MRI contrast agents have been produced as derivatives of these three species, such as magnetic dendrimers, liposomes, and micelles. It is notable that the major advantage of MRI is its high spatial resolution (25–100 μm level) and the excellent tissue contrast. In this context, MRI overruns other molecular imaging approaches up to date, and it is available for both morphological and functional assessments, while a certain quantity of contrast agents (approximately μg to mg) is necessary for the relatively time-consuming imaging process.

Ultrasound (or ultrasonography) [30] is one of the most applied medical imaging techniques. The major advantage of US imaging is low cost, high safety and readily availability for portable devices. In experimental ultrasound imaging, gas-filled microbubbles, perfluorocarbon emulsions and colloidal suspensions are commonly used contrast agents for blood pool enhancement, lesion characterization and perfusion imaging. These contrast agents can resonate in an ultrasound beam, undergo shell oscillations, and expand and contract in response to acoustic pressure changes and thus achieve the ultrasound contrast enhancement. Many contrast agents are made from biocompatible and biodegradable materials, exhibit excellent vascular circulation properties, and have proven to be safe and stable in vivo, which make them suitable candidates for US imaging and medical usage.

Radionuclide imaging, as this name implies, implements radioactive nuclei as the source for detection and image reconstruction [31]. Positron emission tomography (PET) and single photon emission computed tomography (SPECT) are two major types of radionuclide imaging modalities. Various radionuclides, such as ^{16}F, ^{64}Cu, and ^{68}Ga (for PET) and ^{125}I, ^{111}In, and ^{99m}Tc (for SPECT) can be applied as radioisotopes, respectively. The major disadvantage of radionuclide imaging is poor spatial resolution (1–2 mm in clinical scanners). High sensitivity (nmol/L), relatively low dose (ng) of radiotracers for detection, quantitative analysis and no limitation of tissue penetration facilitate the possibility of preclinical and translational applications, as well as noninvasive evaluations of pharmacokinetics properties for new drugs. For example, 2-[^{18}F]fluoro-2-deoxy-D-glucose ([^{18}F]FDG) has been a widely used PET imaging probe for both cancer diagnosis and radiotherapy [32].

Among all the molecular imaging modalities, no single modality is perfect to meet all the requirements in medical applications. As a consequence there is a clear trend towards hybrid imaging modalities such as PET-CT, PET-MRI, SPECT-CT, as well as optical imaging and CT. Also the combination of ultrasound and MRI is attractive, which particularly holds true for the real time monitoring of high focused ultrasound therapy. Thus, developing dual-modal imaging probes for both ultrasound and MRI, and combining the advantages of these two modalities can offer synergistic advantages in providing more valuable diagnostic information and treatment strategies [33].

Although it is challenging to integrate dual or multimodal imaging properties into a single probe (engineered molecule), this concept is probably the most promising and convenient approach to overcome current drawbacks in one single imaging modality. In particular, as nanoparticles have large surface area and multiple functional groups on the shell, it is possible to employ appropriate structural modifications and engineering along with binding ligands conjugation to produce multifunctional nanoparticles for multimodal molecular imaging. Nowadays, more and more nanoparticle-based dual or multimodal imaging probes (or contrast agents) have been developed and applied to multimodal functional imaging in living subjects.

STRUCTURE DESIGN AND FUNCTIONALIZATION STRATEGIES OF NANOPARTICLES

To devise multimodal imaging agents or therapeutic probes with nanoparticles, several parameters have to be taken into considerations:

- Toxicity of nanoparticles for living subjects and humans;
- Any possible metabolites after vascular circulation or cell uptake;

- Biocompatibility and biodegradability to avoid harmful accumulations in organs, tissues or blood;
- Availability for chemical modifications of nanoparticles; and
- Comprehensive assessments of fabricated nanoparticles in vitro and in vivo before practical applications for living subjects or humans.

Generally, common methodologies for structure design and functionalization strategies of nanoparticles can be summarized as follows.

- ***Selection and Fabrication of Nanoparticles Core***: What kind of nanoparticle is eligible to be the core and how to fabricate the structure of imaging probes are determined by the imaging purposes and a specific imaging modality.
- ***Shell Structure Synthesis***: The shell structure usually serves more complicated purposes than the core of nanoparticles. It may protect the core from the external microenvironment and improve the core stability and physical property. Shell materials bearing good biocompatibility will reduce unexpected immunophysiological side effects in vivo and facilitate fast clearance of nanoparticles from the body. As far as drug delivery and therapy are concerned, the shell structure defines the efficiency of shell burst and controllable release of therapeutic drugs encapsulated in the core. On the other hand, the nanoparticle shell could be considered as a cargo for imaging probes (or contrast agents), drug payload or therapeutic agents incorporation so as to strengthen the integral functional efficacy.
- *Surface Modifications*: As the outer interface of the shell might be too sensitive when exposed to biomedium, such as blood, plasma or receptors at the binding sites, surface coatings with stabilizer or emulsions (surfactants, polymers, etc.) may be necessary to maintain the nanoparticles' stability. For targeted molecular imaging, drug delivery and therapy, specific targeting ligands on the nanoparticle surface are desired to be conjugated to biomolecules (small molecules, peptides, antibodies or, proteins). Apart from that, PEGylations (for pharmacokinetics adjustment) or spacer/linker incorporations (for conjugation or segregation of biomolecules) on the surface may be possibly involved for certain imaging or therapy purposes. Several important bioconjugation strategies for nanoparticle surface functionality are summarized in Figure 2 [34].

Although various functionalities, targeting ligands, imaging probes and therapeutic payloads could be incorporated within the inner structure of nanoparticles or conjugated on the surface by different modification protocols as illustrated in Figure 3, not every possible modification unit is all necessary for an individual nanoparticle. Specific imaging modalities or defined therapeutic purposes will direct to necessary strategies for structure design and functionalization and thus lead to an optimal performance in practical medical applications.

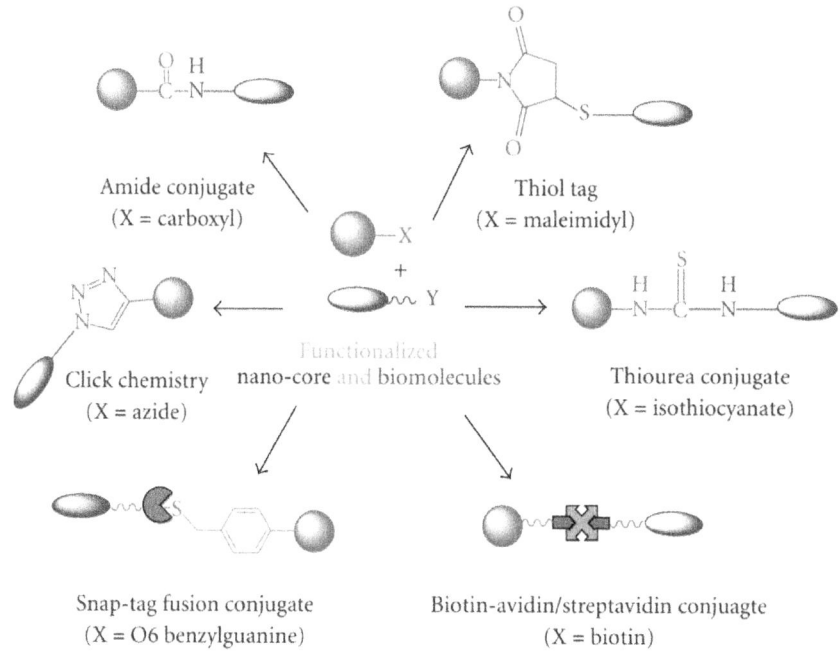

Figure 2: Bioconjugation strategies for chemical modification on nanoparticle surface.

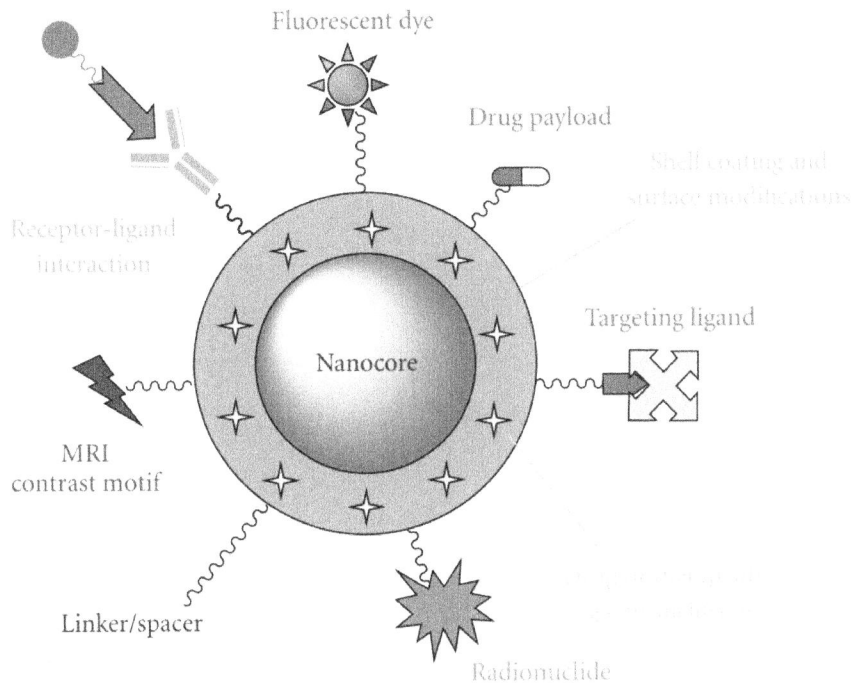

Figure 3: Scheme of multifunctional nanoparticle for molecular imaging, drug delivery and therapy. Optionally functionalized and devised nanoparticles could be achieved for individualized diagnosis and treatments.

BIOMEDICAL APPLICATIONS OF NANOPARTICLES IN IMAGING

The biomedical applications of nanoparticles (or nanomedicine) are rooted in the advanced functional design, and have been realized in preclinical experimental diagnosis. In the long run, they will contribute to personalized clinical treatment on the basis of molecular profiles of each individual patient. The development is rapid and multidirectional, but at present is still in its early stages (Figure 4). The main applications of nanoparticles can be divided into several major directions: diagnostic molecular imaging, delivery of drug and gene, and targeted therapy [35, 36]. In the following context, we intend to review the state-of-the-art imaging applications in recent years.

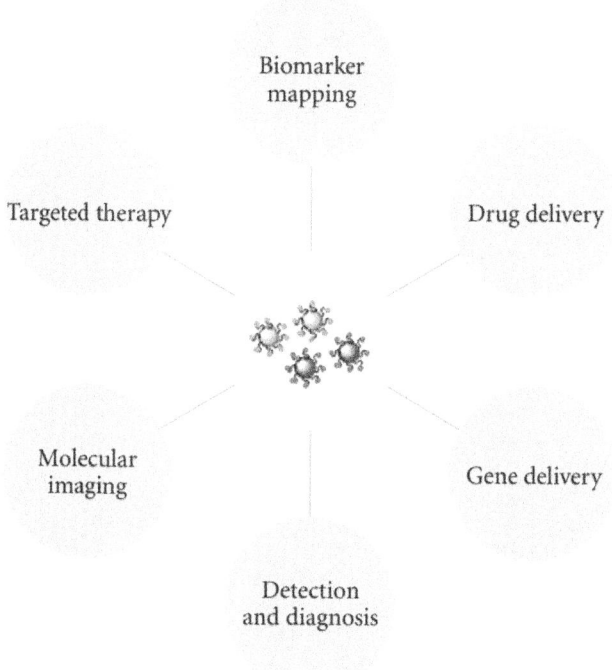

Figure 4: State of the art of multi-directional applications in nanomedicine.

Iron Oxide Nanoparticles

Iron oxide nanoparticles (IONPs) are representative contrast agents which can be used for T_2- and T_2^*- weighted MRI molecular imaging [37]. There are several methods for chemical synthesis of iron oxide nanoparticles. Among these methods, coprecipitation of Fe^{2+} and Fe^{3+} ions in a basic aqueous media (NaOH or NH_4OH solutions) is the simplest way, but usually polydispersed, poorly crystallized nanoparticles are obtained. To avoid these disadvantages, thermal decomposition method was employed to produce iron oxide nanoparticles with monodispersity and uniform crystalline. Subsequently, the hydrophobic iron oxide nanoparticles can be coated with phospholipids, silica, or amphiphilic polymers as shells to display good solubility and biocompatibility in vivo. Iron oxide nanoparticles can be classified into three subtypes in terms of different size distributions of particle population: VSOP (very small superparamagnetic iron oxide nanoparticles, diameter <10 nm), USPIO (ultrasmall superparamagnetic iron oxide nanoparticles, diameter ~ 20 nm) and SPIO (superparamagnetic iron oxide nanoparticles, diameter >30 nm). Table 3 has listed recent developments of IONP-based contrast agents that are intended for multimodal imaging and drug delivery.

Table 3: A chronological list of various IONP-based nanoparticles as MR imaging probes*.

SPIO-based nanoparticles	Application	Particle size (nm)	Relaxivity (mM^{-1}s^{-1})	Reference
Cy5.5-R$_4$-SS-CLIO	MR/optical in vivo imaging	68±13	r^1=27.8; r^2=91.2	[38]
Cy5.5-CLIO-EPPT-FITC	MR/optical in vivo imaging for uMUC-1 tumor	35.8	r^1=26.43; r^2=53.44	[39]
Cy5.5-Cltx-PEG-USPIO	MR/optical in vitro imaging for glioma brain tumor	10.5±1.5	n.a.	[40]
CoFe$_2$O$_4$@SiO$_2$(FITC)	MR/optical in vitro imaging	60	n.a.	[41]
DySiO$_2$-(Fe3O4)n	MR/optical in vitro imaging	46	r^2=397	[42, 43]
FePt-Au	MR in vitro imaging	6(FePt)–10(Au)	r^2=58.7	[44]
PVLA-USPIO	MR in vivo imaging for hepatocytes	25.8±6.1	n.a.	[45]
SPIO@SiO$_2$(FITC)	MR in vivo imaging and human stem cell labeling	50	r^2=128	[46, 47]
USPIO@SiO$_2$	MR in vitro imaging and cell labeling	9.9±1.6	r^1=0.55; r^2=339.80	[48]
APTMS-USPIO-cRGD	MR in vitro imaging for α$_v$β$_3$ integrin	10±3	r^1=1.0; r^2=134.0	[49]
USPIO-4-MC-cRGD	MR in vivo imaging for α$_v$β$_3$ integrin	8.4±1.0	r^2=165.0	[50]
VSOP-Cys-PEG-LCP	MR in vitro imaging for lung tumors	6.2±0.9	r^1=11.5; r^2=142.0	[51]
SPIO-Doxo@LCP-MFM	MR in vitro imaging and drug delivery	48~60	r^2=400	[52]

*Abbreviations: CLIO: Cross-linked iron oxide; FITC: fluorescein isothiocyanate; Cltx: Chlorotoxin; PVLA: Polyvinylbenzyl-O-β-D-galactopyranosyl-D-gluconamide; APTMS: Aminopropyl trimethoxysilane; MC: Methylcatechol; Doxo: Doxorubicin; LCP: Lung cancer-targeting peptide; MFM: Multifunctional micelle.

Synergistic effects of different imaging modalities will lead to multimodal imaging developments and applications in in vitro assay, ex vivo assessment, and in vivo diagnosis of living subjects including humans. Among multimodal imaging strategies, MRI-optical dual modal imaging can be relatively easily achieved [53]. In 2002, Weissleder's group developed a so-called smart crosslinked iron oxide (CLIO)-based Cy5.5-arginyl peptide conjugated nanoparticle for in vivo MR and near-infrared fluorescence (NIRF) dual imaging [38]. This probe could be prepared by conjugation of the arginyl peptides (R_4) to CLIO amine via either a disulfide linkage or a thioether linker, followed by the attachment of the indocyanine dye Cy5.5 as a NIR fluorescent probe. Since the absorbance spectra for all biomolecules (hemoglobin, water and lipids in living systems) reach a minimal absorption coefficient in the NIR wavelength range of 650–900 nm, this could minimize tissue autofluorescence and provide a clear imaging window for successful image visualization. After the administration of this probe via subcutaneous injection into nude mice, the lymph nodes were negatively enhanced in MR images and simultaneously visualizable in NIR fluorescent imaging.

Moore et al. designed a novel imaging probe consisting of CLIO-Cy5.5 conjugated to peptide EPPT which was specifically recognizing underglycosylated mucin-1 antigen (uMUC-1) on various tumor cells [39]. The synthesized probe displayed high specificity toward a variety of uMUC-1-positive human adenocarcinomas in vitro. Additionally, in vivo MR and NIRF imaging showed specific accumulation of this probe on uMUC-1-positive tumors and virtually no signal in control tumors. Thus, this imaging probe would be applied to not only early detection and staging of the recurrence of tumors, but also for monitoring of therapeutic efficacy. Apart from small peptides, Chlorotoxin, a larger 36-amino acid peptide purified from the venom of the giant Israeli scorpion and with high affinity to MMP-2 endopeptidase, could be conjugated to USPIO for imaging glioma tumors. Veiseh et al. synthesized a multifunctional nanoprobe consisting of Cy5.5-Cltx motifs on the surface of USPIO via PEGylation linkage for both MR imaging and fluorescence microscopy [40]. This nanoprobe has been proved to be highly stable and showed prolonged retention within targeted cells (at least 24 h), which is advantageous in intraoperative imaging applications compared to conventional optical fluorophores. A preferential uptake and a high degree of internalization of the nanoprobe conjugates by glioma cells versus control normal cells were also observed in in vitro experiments.

Cheon's group developed a "core-satellite" structured $DySiO_2$-$(Fe_3O_4)_n$ nanoparticle which demonstrated high performance of T_2-weighted contrast-enhanced MR images as well as good fluorescent properties for the detection of polysialic acids (PSAs) expressed on neuroblastoma and many other cell lines [42]. A similar concept could also be applied to MR imaging of targeted specific adenovirus gene delivery [43]. Lys residue-containing adenovirus and MnMEIO (manganese-doped magnetism engineered iron oxide) could be incorporated into the same cross-linker sulfo-SMCC. When the hybrid nanoparticles were applied to target specific delivery to coxsackie and adenovirus receptor (CAR)-expressed U251 cell lines, a dark MR contrast was apparently observed compared to control nontreated and free MnMEIO-treated cell lines. Targeted infection and gene delivery have also been proved successful by utilizing these virus-magnet hybrid nanoparticles.

Lu and coworkers reported a silica-coated core-shell SPIO nanoparticle with incorporation of fluorescein isothiocyanate (FITC), namely SPIO@SiO_2(FITC), for MRI-detectable human mesenchymal stem cells (hMSCs) labeling [46]. The SPIO nanoparticles were synthesized by thermal decomposition and stabilized with oleic acid and oleyl amine. Afterwards, the hydrophobic nanoparticles were coated with dye-doped silica shells via a reversed microemulsion system. This magnetic vector displayed perfect stability (7 days) after differentiations and could efficiently label hMSCs via clathrin- and actin-dependent endocytosis. This was the first report that hMSCs could be labeled with MRI contrast agents and also monitored in vivo with a clinical 1.5, T MRI scanner.

Recently, our group developed the silica- and alkoxysilane-coated USPIO nanoparticles and successfully applied them to MRI-guided cell-labeling and cell-tracking in vitro [48]. After surface coating of USPIO with silica, APTMS and AEAPTMS, respectively, a narrow size distribution with mean diameter of about 10 nm was observed, which indicated no significant size difference compared to uncoated USPIO nanoparticles. In both MR phantom imaging and cell uptake experiments, silica-coated USPIOs exhibited the highest T_2 relaxivity and maximum cellular iron concentration at same incubation conditions, compared to APTMS- and AEAPTMS-coated ones. Interestingly, a process of internalization of USPIO particle uptake into the cells could be observed by TEM as shown in Figure 5. After attachment of the USPIO particles to the cell plasma membrane, they were incorporated by small pinocytotic vesicles with membrane specifications (arrows in Figure 5(d)).

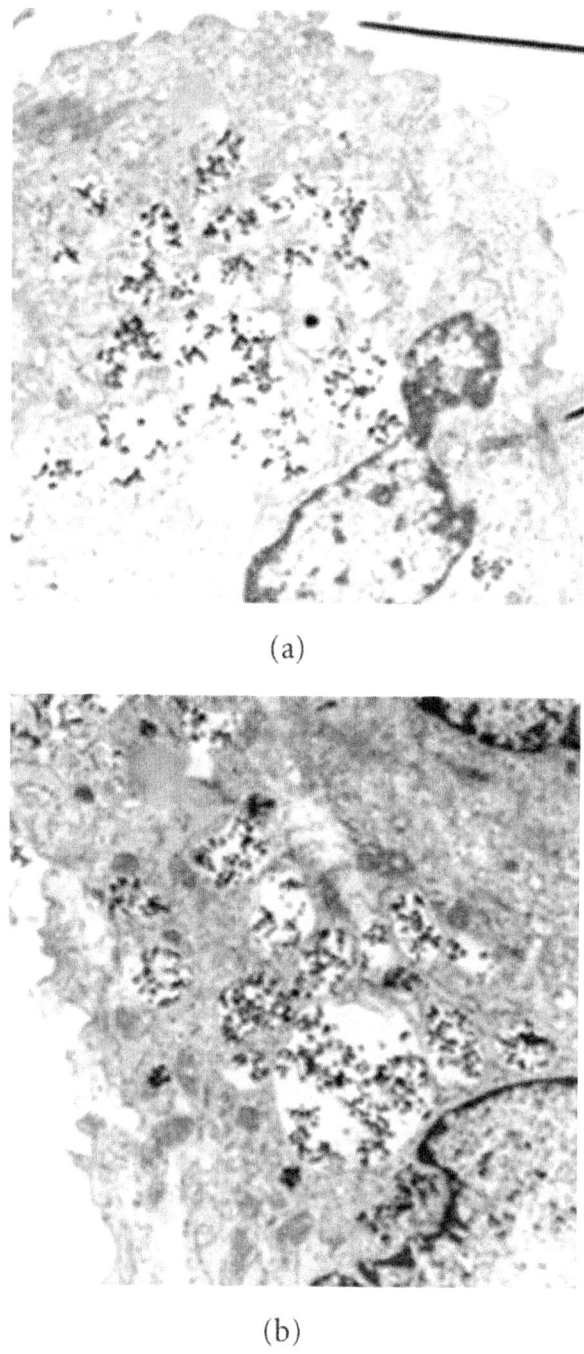

(a)

(b)

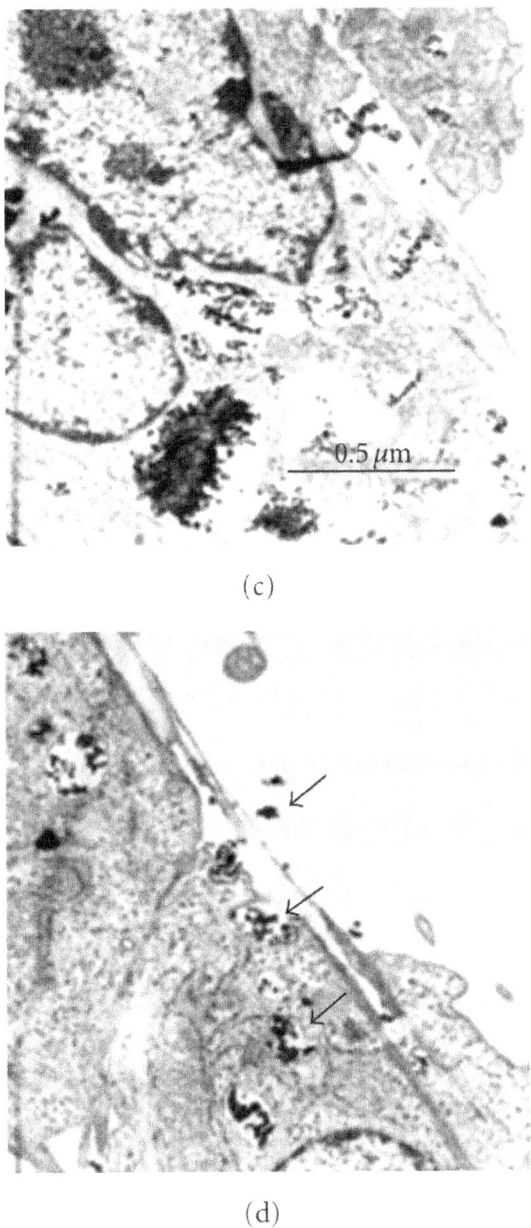

Figure 5: TEM images of immortalized progenitor cells 6 h after incubation with (a) silica-, (b) APTMS- and (c) AEAPTMS-coated USPIO particles at an iron concentration of 0.3 μmol/mL. The particles were located intracellularly and most of them were aggregated in large lysosomes. Image (d) shows the process of USPIO nanoparticle uptake into the cells. (Reprinted with permission from [48]).

On basis of these silica- and alkoxysilane-coated USPIO nanoparticles and as an extension of this research, Jugold et al. developed a kind of MRI contrast agent for selectively targeting urokinase-type plasminogen activator receptors (uPARs) when USPIOs were conjugated with polypeptides as binding-directed biomolecules [54]. This approach would be promisingly applied as a useful diagnostic strategy to more diseases if associated with PET and SPECT as synergistic imaging modalities.

Some biological and physiological processes are essential for development of malignant tumors and can be important targets for cancer diagnosis, prognosis and therapy. Angiogenesis is a crucial step for tumor growth, extension and even metastasis [55]. After surface conjugation with receptors or specific biomolecules, USPIO-based tumor-targeted nanoparticles could also be used for molecular MR imaging. Zhang et al. reported the fabrication of RGD-conjugated APTMS-coated USPIO nanoparticles and their applications to specifically targeting $\alpha_v\beta_3$ integrins, an important receptor family for tumor angiogenesis, by a clinical 1.5 T magnetic resonance scanner [49]. Compared to plain USPIO control, RGD-USPIO particles had a higher cell uptake capability for both internalization through the cell membrane and accumulation within endosomes. When incubated with HUVECs in vitro in the phantom imaging experiments, RGD-coupled USPIO particles exhibited a strong T_2^* contrast enhancement in gelatin gels, and the signal intensity (SI) for control USPIO and RGD-USPIO plus free RGD showed relatively lower darkness, which indicated that the RGD-USPIO particles could efficiently reduce the relaxivity, decrease the grey scale and therefore resulted in improved MR imaging contrast in vitro. In addition, these fabricated RGD-specific APTMS-coated USPIOs could be used to visualize $\alpha_v\beta_3$ integrin expression and distinguish tumors with different angiogenic profiles in nude mice in vivo MR imaging (Figure 6). These binding specificities and imaging differences between RGD-USPIO and plain USPIO particles facilitated high efficient image contrast and provided a useful tool for monitoring molecular profiles of tumor vessels in angiogenesis by MRI.

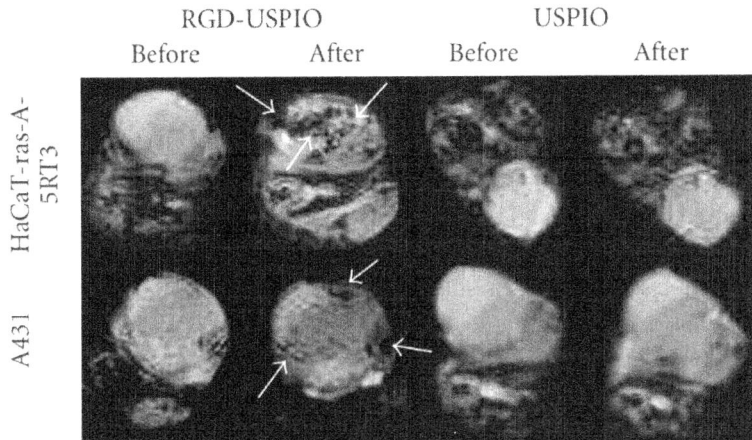

Figure 6: T_2^*-weighted MR images of nude mice bearing s.c. HaCaT-ras-A-5RT3 (top) and A431 (bottom) tumors before and 6 h after i.v. injection of RGD-USPIO and US-PIO, respectively. (Reprinted with permission from [49]).

SPECT/PET imaging has a major disadvantage of poor spatial resolution. In comparison, MRI images display satisfactory and anatomical structures in soft tissues and bones. Complementary SPECT-MRI/PET-MRI fused imaging modalities could potentially afford sufficient resolutions and high sensitivity. Weissleder et al. recently have developed trimodal imaging probes for PET, MRI, and fluorescence microscopy modalities [56, 57]. Multiple functionalities, including ^{18}F-PEG$_3$ radiotracer motif as a PET reporter and near-infrared fluorochrome, and azide group for "click" chemistry as a convenient approach for further modifications, could be incorporated on the surface of aminated polysaccharide layered cross-linked iron oxide (CLIO) nanoparticles for multiple imaging detections. Fused PET-CT dynamic images displayed a clearly resolved anatomical registration and radiotracer distribution information in BALB/C mice. This multifunctional nanoplatform could be applied to imaging of macrophages in inflammatory atherosclerosis after slight modifications of radionuclide chelators and coating materials.

Quantum Dots

Quantum dots (QDs) are a kind of versatile optical imaging agents that are readily stabilized with surface modifications, and conjugated with targeting ligands along with magnetic coatings serving as bimodal imaging probes for efficacious imaging of tumor angiogenesis. Mulder et al. showed the synthesis of PEGylated, cyclic RGD-functionalized QDs with a water-soluble and paramagnetic micellular coating as a molecular imaging probe for MRI and fluorescence microscopy [58]. The hydrophobic CdSe/ZnS core/shell QDs were PEGylated and then immobilized with Gd^{3+}-ion phospholipid micelles to render them water-soluble, biocompatible and MR-detectable. Maleimide functionalization ensured the following cyclic RGD peptide conjugation on the QDs surface for specifically targeting $\alpha_v\beta_3$-integrins overexpressed on the angiogenic endothelial cells and tumors. Satisfactory fluorescent imaging capability and prominent T_1-weighted contrast enhancement were observed in in vitro test assessment for angiogenic human umbilical vein endothelial cells (HUVECs).

Apart from that, quantum dots can be serving as useful carriers for drug and gene delivery. This suggests a bilateral character for quantum dots to transport drugs and simultaneously be detected optically. Bagalkot reported quantum dot-aptamer-doxorubicin (QD-Apt-DOX) conjugates for synchronous FRET-mediated imaging, therapy and sensing of drug delivery process [59]. The CdSe-ZnS core-shell QDs as optical imaging agents are surface functionalized with A10 RNA aptamer for specifically targeting prostate specific membrane antigen (PSMA) cancer cells. The intercalation of DOX within the A10 DNA aptamer resulted in the formation of QD-Apt-DOX conjugates. Due to a fluorescence resonance energy transfer (FRET) mechanism, the QD-Apt-DOX conjugates turned to be on "OFF" state and no fluorescence was detected. When internalized into the cancer cells, DOX, a widely used anthracycline drug, was gradually released and therefore recovered fluorescence detection by optical techniques, as well as inducing the therapeutic effect in the target cancer cells. By using this smart conjugates, molecular imaging of DOX delivery, location and simultaneous cancer therapy could be achieved.

Additionally, as quantum dots may be photosensitized for production of radicals upon absorption of visible light, they can also be used to in photodynamic therapy (PDT) and radiation therapy against various cancers. Juzenas reviewed several fundamental concepts and recent developments in these fields [60].

Gold Nanoparticles

Gold nanoparticles [61, 62] are a subtype in the nanoparticle family and have been studied over years. A wide range of gold nanoparticles, including nanorods, nanoshells, nanocages, and surface-enhanced Raman scattering (SERS) nanoparticles, have been well-investigated and found broad applications in nanomedicine. In particular, gold nanoparticles can be used as photothermal therapeutic agents and trigger controlled release of chemical drugs. The Hyeon group proposed using magnetic gold nanoshells (Mag-GNS) as a platform for

MRI imaging and photothermal therapy of breast cancers [63]. The starting silica sphere provided amino groups for conjugation to BMPA via direct nucleophilic substitution. Gold seed nanoparticles were attached to the residue amino groups on the silica sphere. After gold shell growth, magnetic gold shell nanoparticles were synthesized following the linkage of Anti-HER2/neu on the surface of Mag-GNS for targeting cancer cells. The MR imaging could be achieved and selectively killing of breast cancer cells were accomplished by NIR radiation.

The Yoo group also developed a rhodamine-encapsulated PLGA-Au-Mn nanocomposite for photothermal therapy and T_2 MRI contrast imaging [64]. Gold and manganese shells enabled the photothermal property and MR contrast enhancement. And the burst release of rhodamine, a model drug, could be activated upon NIR radiation.

Nanotubes

Graphitic carbon nanotubes exhibit potential medical imaging capabilities and also integrated diagnostic/therapeutic efficacies. Dai and coworkers reported the fabrication of single wall carbon nanotubes (SWCNTs) as photothermal therapeutic (PTT) agents [65]. Selective cancer cell destruction and intrinsic NIR-triggered cell killing capacity could be realized without any harmful effects to normal cells. Thus, this class of nanomaterials can be applied to drug delivery and targeted therapy for various cancers. Recently, their group engineered a new type of FeCO/single-graphitic-shell nanocrystals for both MRI-NIR imaging and therapy [66]. A scalable chemical vapour deposition method was utilized to synthesize the nanocrystals, and multiple functionalities of these coreshell materials were made for magnetism and near-infrared optical absorbance. The fabricated nanocrystals provided ultrahigh saturation magnetization and remarkable r_1 and r_2 relaxivities in both T_1-weighted (positive contrast) and T_2-weighted (negative-contrast) MR images, compared to commercial MRI contrast agents.

In in vivo animal imaging experiments, FeCO/single-graphitic-shell nanocrystals were injected into a rabbit. Longer circulation and better positive contrast enhancement at lower dosages were obtained. The near-infrared optical properties of FeCO/single-graphitic-shell nanocrystals under $\lambda = 808\,nm$ near-infrared laser radiation demonstrated significant temperature increase over time, which implied good capability to convert NIR photon energy into thermal energy. These characteristic profiles afford FeCO nanocrystals potential applications for in vivo multimodal diagnostic imaging, controlled drug release and effective NIR-triggered tumor destruction.

Dendrimers

Dendrimer is one kind of advanced materials used as MRI imaging contrast agents [67]. Polyamidoamine (PAMAM) is one of most extensively studied and widely used commercial dendrimeric compounds. Relatively easy chemical synthesis of different sizes of dendrimers, termed "generations" will permit a tuneable molecular structure for adjustable MR imaging applications. Boswell et al. synthesized a G3 PAMAM-based multimodal dendrimeric imaging nanoplatform and made tentative biological investigations for fluorescence, MRI and SPECT molecular imaging in melanoma xenografted nude mice [68]. Fluorescent dye (Alexa Fluor 594) and MR imaging chelator (1B4M DTPA) were consecutively conjugated to the amino groups on the external surface of PAMAM dendrimers. Cyclic RGD peptides were then bonded to render the multicovalency binding effect for $\alpha_v\beta_3$-integrins. Different building blocks in the dendrimeric nanoparticles could be modified according to practical imaging requirements. In vitro studies revealed appreciable capability of this macromolecule agent in both MRI and optical imaging. However, in in vivo experiments, unsatisfactory tumor uptake implied poor extravasation capability of these particles into tumors. This requires further structural modifications to meet an improved binding affinity for the targeted tumor cells.

Enormous efforts on development of dendrimer-based nanoparticles have been also made by Kobayashi and his colleagues [69, 70]. They fabricated a type of G6 PAMAM-based nanoprobes with multimodal and multicolor potentials. The PAMAM dendrimer was designed as a platform to be linked to both radionuclides and optical fluorophores, and therefore allowed for dual-modality scintigraphic and five-color near-infrared optical lymphatic imaging using a multiple-excitation spectrally resolved fluorescence imaging technique. The incorporation of ^{111}In radionuclide made it possible to provide semiquantitative distribution information, while optical imaging could provide qualitative visualization with acceptable spatial resolution. Although each nanoprobe has a similar structure and an identical chemical characteristics, multimodal imaging and multicolour resolution could be simultaneously achieved in vivo after one single injection of these nanoprobes were administered in nude mice. This powerful imaging tool could be encouraging people to explore more advanced multifunctional imaging probes.

Polymeric Micro-/Nanobubbles

Polymeric nanosized vectors are increasingly attractive for effective drug and gene delivery. Common small pharmaceutical compounds, therapeutic peptides and nucleic acids could be encapsulated as payloads into the polymeric nanocarriers. Compared to conventional viral vectors, polymeric nanoparticles might be biocompatible and biodegradable in vivo, and could avoid undesireable side effects, such as mutational virus insertion into host genome and development of replication competent viruses [71, 72]. However, as polymeric nanovectors may be cleared by the reticuloendothelial system (RES) due to the opsinization interactions in the blood stream, necessary surface modifications on the nanovectors will improve pharmacokinetics and lead to long circulating property. Figure 7 shows a series of representative intercellular delivery process of polymeric nanoparticles.

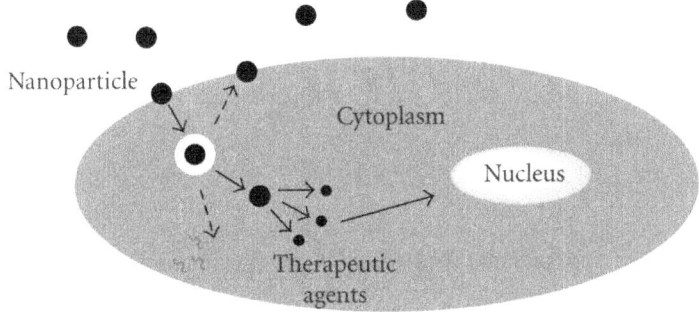

Figure 7: Cartoon for cytosolic delivery process of polymeric nanoparticles (NPs). (a) Cellular internalization of NPs; (b) Formation of lysosome-NPs by endocytosis; (c) Release of therapeutic agents in cytoplasm; (d) Degradation of agents; (e) Exocytosis of NPs; (f) Cytosolic transport of agents to targets.

In ultrasonography, microbubbles can be used as imaging/diagnostic agents and make themselves as drug-loaded vehicles in the blood vessels for therapeutic purposes [73–75]. In comparison to traditional soft shell microbubbles, hard shell polymeric microbubbles could provide higher stability to prevent gas diffusion and thus can be loaded with regular air instead of heavy gases. Palmowski et al. fabricated polymeric cyanoacrylate streptavidin-coated microbubbles, discussed the pharmacodynamics and found imaging and therapy applications in antiangiogenesis [76]. Following the polymerization of cyanoacrylate, streptavidin was coated on the surface to form polymer-stabilized microbubbles. Before administration, biotinylated antibodies (VEGFR2 and IgG) or peptides (RGD and RAD) were added to generate target-specific microbubbles as ultrasound contrast agents (Figure 8). Sensitive particle acoustic quantification (SPAQ)-based quantitative analysis proved significant accumulation of RGD- and VEGFR2-conjugated microbubbles, compared with uncoated microbubbles and unspecific RAD- and IgG-controls. As for the imaging property, although the thicker and more robust polymer-stabilized shell would reduce the backscatter signal to a certain degree, this signal is strong enough to detect single microbubble in tissues and therefore enabled imaging with superb sensitivity.

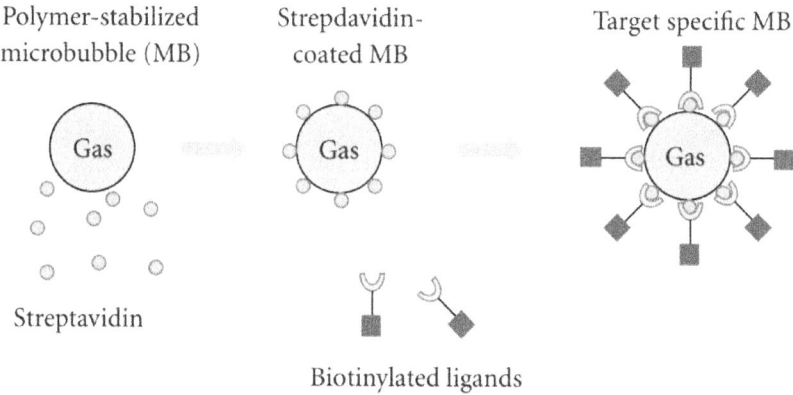

Figure 8: Schematic synthetic approach of target-specific microbubbles.

Afterwards two groups of animals were administered with these polymeric microbubbles to assess their antiangiogenic therapy effects. Before therapy started, no significant difference in the mean number of bound VEGFR2-specific and $\alpha_v\beta_3$ integrin-specific microbubbles was found between tumors of control and therapy group. Seven days after treatment, significantly increased binding of VEGFR2-specific microbubbles and $\alpha_v\beta_3$ integrin-specific microbubbles was observed in untreated tumors. In contrast, in treated tumors, accumulation of VEGFR2-specific microbubbles and $\alpha_v\beta_3$ integrin-specific microbubbles was reduced and significantly lower than that in control group. Thus these targeted microbubbles proved applicable to both effective molecular profiling of tumor angiogenesis and sensitive assessment of therapy in vivo.

Encouraged by these experimental findings, our group then made further explorations with these specific polymeric microbubble ultrasound contrast agents on imaging and therapy of early vascular response in prostate tumors irradiated with carbon ions (16 Gy) [77]. After injection of unspecific, RGD-coated and anti-ICAM-1-coated microbubbles, more significant accumulations were observed in tumor vessels bearing $\alpha_v\beta_3$ integrin and ICAM-1 receptors that for unspecific microbubbles. Retention of $\alpha_v\beta_3$ integrin- and ICAM-1-specific microbubbles was evaluated by treating these microbubbles with the corresponding tumor cells after carbon ion irradiations. Significant increase of specific microbubbles within tumor vessels was found after 3-day carbon ion irradiation due to the induction of up-regulation of $\alpha_v\beta_3$ integrin and ICAM-1.

Microbubbles are regarded as the most popular contrast agents in ultrasound due to excellent acoustic backscattering properties of the gas core. Besides that, the encapsulation of paramagnetic MR imaging probes into the shell structure could facilitate a potential multimodal imaging capacity. Recently, Gu and his colleagues developed PLA-PVA double-layered polymeric microbubbles with the encapsulation of superparamagnetic iron oxide (Fe_3O_4, SPIO) nanoparticles in the bubble shell [78, 79]. The microbubble shell was 50–70 nm thickness and could afford the successful heterogeneous encapsulation of approximately 12 nm-sized SPIO nanoparticles. The engineered double-layered SPIO-encapsulated microbubbles (EMBs) showed improved r_2 relaxivity and better contrast enhancement than SPIO-free microbubbles or SPIO-included microbubbles on the surface. The in vitro MRI experiments exhibited a gradient decrease of gray scale associated with a corresponding increase of the SPIO concentrations. And the transverse relaxation fitted well to a linear relationship with different SPIO-inclusion amounts in the microbubbles. Then in vitro ultrasound imaging was performed to observe a distinct "brightening" contrast enhancement in the region of interest (ROI) with a certain SPIO-EMBs concentration. After the injection of SPIO-EMBs into the liver of living rats, real-time MRI anatomical images demonstrated a clear negatively enhanced contrast subsequently over several time intervals. These exiting findings revealed a promising ultrasound-MRI dual modal imaging modality in medical applications.

To utilize microbubble-nanoparticle hybrid vehicles for drug release and therapy is another promising strategy for medical treatment applications [80]. Combination of microbubbles with nanoparticles into one single drug payload entity facilitates the possibility of effective drug transport from extracellular microenvironment to cell membrane, and controllable release at the diseased sites. A train of ultrasound pulses or radiation forces (RF) with certain pressure and frequency will lead to an enhanced permeability and retention effect (EPR) and significantly contribute to close proximity to vessel wall, internalization into the membrane, controllable burst of the microbubble shell, and targeted drug deposit site-specifically at the diseased regions. The Rapoport group described the applications of drug-loaded nano/microbubbles for combined ultrasonography and targeted chemotherapy of breast cancer [81]. In this nonthermal therapeutic technique, perfluorocarbon nanodroplets with stabilization of block copolymers could be converted into nano/microbubbles upon heating. The phase state and size of bubbles could be controlled by different ratio of copolymer/gas core component. After the intravenous administration, strong and selective contrast was observed, and these bubbles could extravasate through leaky tumor vascular spaces. With oscillation and collapse of bubbles, encapsulated drugs could be controllably released, which consequently led to enhanced uptake by the targeted tumor cells.

Fullerene-Based Nanoparticles

Fullerene-based nanaoparticles are playing a more active role in modern medical imaging and therapeutic applications. For example, gadolinium-doped fullerenes and derivatives, such as $Gd@C_{82}$, $Gd@C_{82}(OH)_{22}$, $Gd3N@C_{80}$, and $Gd@C_{60}[C(COOH)_2]_{10}$ could be used as MR imaging contrast agents, and polonium-doped fullerenes act as candidate radiotracers in nuclear medicine [82, 83]. On the other hand, after readily multifunctional modifications on the surface of fullerenes, they could be potential nanosized agents for medical gene therapies. Sitharaman and coworkers developed a new class of water-soluble C_{60} transfecting agents as gene-delivery vectors in vitro [84]. Three types of fullerenes with different chemical modifications via Hirsch-Bingel chemistry on the surface displayed positively, neutrally and negatively charged properties, respectively. Although all these C_{60} derivatives showed certain efficiency to transfect cells with DNA, only positively charged fullerenes exhibited efficient in vitro transfection. After reducing the aggregation and lowering the toxicity levels of these C_{60} fullerenes, it will offer great opportunity for them to serve as simultaneously diagnostic and therapeutic agents.

Rare-Earth Doped Nanoparticles

Recently, rare-earth doped nanoparticles have emerged as a fast-growing platform in cell trafficking and imaging due to low background noise for their near-infrared (NIR, $\lambda > 760nm$) emission. In particular, rare-earth oxide (REO) phosphor system has been widely studied so far. Meiser reported that $LaPO_4$ nanoparticles were biofunctionalized via biotin-avidin chemistry with good photostability and fluorescent properties [85]. Setua produced highly monodispersed Eu^{3+} and Gd^{3+} doped Y_2O_3 nanocrystals, and presented bi-modal imaging applications of both paramagnetism that enabled magnetic resonance imaging and bright red-fluorescence, aiding optical imaging of cancer cells, targeted specifically to their molecular receptors [86].

Rare-earth upconverting nanoparticles (UCNPs) have been developed as a new generation of luminescent labels due to their superb optical features, long lifetimes and excellent photostability [87]. Among various kinds of upconverting nanoparticles, $NaYF_4$ is the most well-known system that has been employed in cellular and in vivo animal imaging [88–90]. As this field is rapidly developing, we can expect that rare-earth doped nanoparticles will find their way into even more elaborate biotechnological applications in the coming future due to their relatively simple nanocomposition, deep penetration depth of NIR and other advantageous physical features. However, due to potentially high toxicity of lanthanides, more detailed investigations will be probably required to evaluate their biochemical and physiological behaviours before rare-earth doped nanoparticles are eventually translated for biomedical applications.

CONCLUSION AND PERSPECTIVE

Nanotechnology has been witnessing explosive growth, and the field of nanomedicine is undergoing revolutionary developments from traditional strategies to modern applications. However, several challenging issues still circumvent widespread biomedical uses of advanced nanotechnology. For example, novel multifunctional nanomaterials and improved treatment strategies are required to meet the needs of real-time, noninvasive imaging in living subjects or humans, and those of satisfactory drug delivery and therapy efficiency in vivo. Secondly, standardized nanoplatforms to be applied to diagnostic or therapeutic investigations of various diseases still have to be developed and formulated. More importantly, limited information has been obtained about the possible toxicity of nanoparticles and potential risks for the environment and human health up to date. It is quite urgent to evaluate the safety and the fate of nanomaterials in the body, so that rational and sufficient biological assessments could be concluded prior to ultimately translations into clinics. This will also be helpful for us to get a down-to-earth understanding and well-organized interpretation for biological and physiological processes. Considering the vast potential of nanoparticles in medicine, it is believed that nanomedicine will have a high impact on human life and contribute to the concept "small stuff makes big sense" in the coming future.

ACKNOWLEDGMENT

This work was financially supported by a grant of the BMBF (no. 0315481).

REFERENCES

1. G. Cao, Nanostructures & Nanomaterials: Synthesis, Properties & Applications, Imperial College Press, London, UK, 2004.
2. M. Ferrari, "Cancer nanotechnology: opportunities and challenges," Nature Reviews Cancer, vol. 5, no. 3, pp. 161–171, 2005.
3. National Nanotechnology Initiative, "What is nanotechnology?"http://www.nano.gov/html/facts/whatIsNano.html.
4. D. Thassu, M. Deleers, and Y. Pathak, Nanoparticulate Drug Delivery Systems, Informa Healthcare, New York, NY, USA, 2007.

5. V. E. Kagan, H. Bayir, and A. A. Shvedova, "Nanomedicine and nanotoxicology: two sides of the same coin," Nanomedicine: Nanotechnology, Biology, and Medicine, vol. 1, no. 4, pp. 313–316, 2005.
6. D. A. Mankoff, "A definition of molecular imaging," Journal of Nuclear Medicine, vol. 48, no. 6, pp. 18N–21N, 2007.
7. R. P. Choudhary, V. Fuster, and Z. A. Fayad, "Molecular, cellular and functional imaging of atherothrombosis," Nature Reviews Drug Discovery, vol. 3, no. 11, pp. 913–925, 2004.
8. F. A. Jaffer, P. Libby, and R. Weissleder, "Molecular and cellular imaging of atherosclerosis: emerging applications," Journal of the American College of Cardiology, vol. 47, no. 7, pp. 1328–1338, 2006.
9. W. J. M. Mulder, D. P. Cormode, S. Hak, M. E. Lobatto, S. Silvera, and Z. A. Fayad, "Multimodality nanotracers for cardiovascular applications," Nature Clinical Practice Cardiovascular Medicine, vol. 5, no. 2, pp. S103–S111, 2008.
10. V. P. Torchilin, "Multifunctional nanocarriers," Advanced Drug Delivery Reviews, vol. 58, no. 14, pp. 1532–1555, 2006.
11. P. Alivisatos, "The use of nanocrystals in biological detection," Nature Biotechnology, vol. 22, no. 1, pp. 47–52, 2004.
12. E. Yaghini, A. M. Seifalian, and A. J. MacRobert, "Quantum dots and their potential biomedical applications in photosensitization for photodynamic therapy," Nanomedicine, vol. 4, no. 3, pp. 353–363, 2009.
13. I. J. Majoros, C. R. Williams, and J. R. Baker Jr., "Current dendrimer applications in cancer diagnosis and therapy," Current Topics in Medicinal Chemistry, vol. 8, no. 14, pp. 1165–1179, 2008.
14. G. A. Koning and G. C. Krijger, "Targeted multifunctional lipid-based nanocarriers for image-guided drug delivery," Anti-Cancer Agents in Medicinal Chemistry, vol. 7, no. 4, pp. 425–440, 2007.
15. X. Huang, P. K. Jain, I. H. El-Sayed, and M. A. El-Sayed, "Gold nanoparticles: Interesting optical properties and recent applications in cancer diagnostics and therapy," Nanomedicine, vol. 2, no. 5, pp. 681–693, 2007.
16. A. A. Shvedova, E. R. Kisin, D. Porter et al., "Mechanisms of pulmonary toxicity and medical applications of carbon nanotubes: two faces of Janus?" Pharmacology and Therapeutics, vol. 121, no. 2, pp. 192–204, 2009.
17. S. Qin, C. F. Caskey, and K. W. Ferrara, "Ultrasound contrast microbubbles

in imaging and therapy: physical principles and engineering," Physics in Medicine and Biology, vol. 54, no. 6, pp. R27–R57, 2009.

18. M. Liong, J. Lu, M. Kovochich et al., "Multifunctional inorganic nanoparticles for imaging, targeting, and drug delivery," ACS Nano, vol. 2, no. 5, pp. 889–896, 2008.

19. E. Blanco, C. W. Kessinger, B. D. Sumer, and J. Gao, "Multifunctional micellar nanomedicine for cancer therapy," Experimental Biology and Medicine, vol. 234, no. 2, pp. 123–131, 2009.

20. R. Singh and K. Kostarelos, "Designer adenoviruses for nanomedicine and nanodiagnostics," Trends in Biotechnology, vol. 27, no. 4, pp. 220–229, 2009.

21. I. I. Slowing, J. L. Vivero-Escoto, C.-W. Wu, and V. S.-Y. Lin, "Mesoporous silica nanoparticles as controlled release drug delivery and gene transfection carriers," Advanced Drug Delivery Reviews, vol. 60, no. 11, pp. 1278–1288, 2008.

22. A. D. Durnev, "Toxicology of nanoparticles," Bulletin of Experimental Biology and Medicine, vol. 145, no. 1, pp. 72–74, 2008.

23. R. Weissleder and U. Mahmood, "Molecular imaging," Radiology, vol. 219, no. 2, pp. 316–333, 2001.

24. T. F. Massoud and S. S. Gambhir, "Molecular imaging in living subjects: seeing fundamental biological processes in a new light," Genes and Development, vol. 17, no. 5, pp. 545–580, 2003.

25. X. Michalet, F. F. Pinaud, L. A. Bentolila et al., "Quantum dots for live cells, in vivo imaging, and diagnostics," Science, vol. 307, no. 5709, pp. 538–544, 2005.

26. J. W. M. Bulte and M. M. J. Modo, Nanoparticles in Biomedical Imaging: Emerging Technologies and Applications, Springer, New York, NY, USA, 2008.

27. V. Ntziachristos, "Fluorescence molecular imaging," Annual Review of Biomedical Engineering, vol. 8, pp. 1–33, 2006.

28. Z. Zhang, S. A. Nair, and T. J. McMurry, "Gadolinium meets medicinal chemistry: MRI contrast agent development," Current Medicinal Chemistry, vol. 12, no. 7, pp. 751–778, 2005.

29. V. K. Varadan, L. Chen, and J. Xie, Nanomedicine: Design and Applications of Magnetic Nanomaterials, Nanosensors and Nanosystems, Wiley, London, UK, 2008.

30. S. H. Bloch, P. A. Dayton, and K. W. Ferrara, "Targeted imaging using ultrasound contrast agents," IEEE Engineering in Medicine and Biology

Magazine, vol. 23, no. 5, pp. 18–27, 2004.
31. H. N. Wagner, Z. Szabo, and J. W. Buchanan, Principles of Nuclear Medicine, W. B. Saunders, Philadelphia, Pa, USA, 2nd edition, 1995.
32. R. Essner, F. Daghighian, and A. E. Giuliano, "Advances in FDG PET probes in surgical oncology,"Cancer Journal, vol. 8, no. 2, pp. 100–108, 2002.
33. J. Cheon and J.-H. Lee, "Synergistically integrated nanoparticles as multimodal probes for nanobiotechnology," Accounts of Chemical Research, vol. 41, no. 12, pp. 1630–1640, 2008.
34. G. T. Hermanson, Bioconjugate Techniques, Elsevier, New York, NY, USA, 2nd edition, 2008.
35. R. Sinha, G. J. Kim, S. Nie, and D. M. Shin, "Nanotechnology in cancer therapeutics: bioconjugated nanoparticles for drug delivery," Molecular Cancer Therapeutics, vol. 5, no. 8, pp. 1909–1917, 2006.
36. Y. Liu, H. Miyoshi, and M. Nakamura, "Nanomedicine for drug delivery and imaging: a promising avenue for cancer therapy and diagnosis using targeted functional nanoparticles," International Journal of Cancer, vol. 120, no. 12, pp. 2527–2537, 2007.
37. S. Laurent, D. Forge, M. Port et al., "Magnetic iron oxide nanoparticles: synthesis, stabilization, vectorization, physicochemical characterizations and biological applications," Chemical Reviews, vol. 108, no. 6, pp. 2064–2110, 2008.
38. L. Josephson, M. F. Kircher, U. Mahmood, Y. Tang, and R. Weissleder, "Near-infrared fluorescent nanoparticles as combined MR/optical imaging probes," Bioconjugate Chemistry, vol. 13, no. 3, pp. 554–560, 2002.
39. A. Moore, Z. Medarova, A. Potthast, and G. Dai, "In vivo targeting of underglycosylated MUC-1 tumor antigen using a multimodal imaging probe," Cancer Research, vol. 64, no. 5, pp. 1821–1827, 2004.
40. O. Veiseh, C. Sun, J. Gunn et al., "Optical and MRI multifunctional nanoprobe for targeting gliomas,"Nano Letters, vol. 5, no. 6, pp. 1003–1008, 2005.
41. T.-J. Yoon, K. N. Yu, E. Kim et al., "Specific targeting, cell sorting, and bioimaging with smart magnetic silica core-shell nanomaterials," Small, vol. 2, no. 2, pp. 209–215, 2006.
42. J.-H. Lee, Y.-W. Jun, S.-I. Yeon, J.-S. Shin, and J. Cheon, "Dual-mode nanoparticle probes for high-performance magnetic resonance and fluorescence imaging of neuroblastoma," Angewandte Chemie -

International Edition, vol. 45, no. 48, pp. 8160–8162, 2006.

43. Y.-M. Huh, E.-S. Lee, J.-H. Lee et al., "Hybrid nanoparticles for magnetic resonance imaging of target-specific viral gene delivery," Advanced Materials, vol. 19, no. 20, pp. 3109–3112, 2007.

44. J.-S. Choi, Y.-W. Jun, S.-I. Yeon, H. C. Kim, J.-S. Shin, and J. Cheon, "Biocompatible heterostructured nanoparticles for multimodal biological detection," Journal of the American Chemical Society, vol. 128, no. 50, pp. 15982–15983, 2006.

45. M. K. Yoo, I. Y. Kim, E. M. Kim et al., "Superparamagnetic iron oxide nanoparticles coated with galactose-carrying polymer for hepatocyte targeting," Journal of Biomedicine and Biotechnology, vol. 2007, Article ID 94740, 9 pages, 2007.

46. Y.-S. Lin, S.-H. Wu, Y. Hung et al., "Multifunctional composite nanoparticles: magnetic, luminescent, and mesoporous," Chemistry of Materials, vol. 18, no. 22, pp. 5170–5172, 2006.

47. C.-W. Lu, Y. Hung, J.-K. Hsiao et al., "Bifunctional magnetic silica nanoparticles for highly efficient human stem cell labeling," Nano Letters, vol. 7, no. 1, pp. 149–154, 2007.

48. C. Zhang, B. Wängler, B. Morgenstern et al., "Silica- and alkoxysilane-coated ultrasmall superparamagnetic iron oxide particles: a promising tool to label cells for magnetic resonance imaging,"Langmuir, vol. 23, no. 3, pp. 1427–1434, 2007.

49. C. Zhang, M. Jugold, E. C. Woenne et al., "Specific targeting of tumor angiogenesis by RGD-conjugated ultrasmall superparamagnetic iron oxide particles using a clinical 1.5-T magnetic resonance scanner,"Cancer Research, vol. 67, no. 4, pp. 1555–1562, 2007.

50. J. Xie, K. Chen, H.-Y. Lee et al., "Ultrasmall c(RGDyK)-coated Fe3O4 nanoparticles and their specific targeting to integrin $\alpha v \beta 3$-rich tumor cells," Journal of the American Chemical Society, vol. 130, no. 24, pp. 7542–7543, 2008.

51. G. Huang, C. Zhang, S. Li et al., "A novel strategy for surface modification of superparamagnetic iron oxide nanoparticles for lung cancer imaging," Journal of Materials Chemistry, vol. 19, no. 35, pp. 6367–6372, 2009.

52. J. S. Guthi, S.-G. Yang, G. Huang et al., "MRI-visible micellar nanomedicine for targeted drug delivery to lung cancer cells," Molecular Pharmaceutics, vol. 7, no. 1, pp. 32–40, 2010.

53. W. J. M. Mulder, A. W. Griffioen, G. J. Strijkers, D. P. Cormode, K. Nicolay, and Z. A. Fayad, "Magnetic and fluorescent nanoparticles for

multimodality imaging," Nanomedicine, vol. 2, no. 3, pp. 307–324, 2007.

54. M. Jugold, F. Kiessling, C. Zhang, B. Wangler, R. Pipkorn, and W. Semmler, "uPAR Selective Contrast Agent for Magnetic Resonance Imaging," European Patent no. EP1902734A1.

55. J. Folkman, "Angiogenesis in cancer, vascular, rheumatoid and other disease," Nature Medicine, vol. 1, no. 1, pp. 27–31, 1995.

56. M. Nahrendorf, H. Zhang, S. Hembrador et al., "Nanoparticle PET-CT imaging of macrophages in inflammatory atherosclerosis," Circulation, vol. 117, no. 3, pp. 379–387, 2008.

57. N. K. Devaraj, E. J. Keliher, G. M. Thurber, M. Nahrendorf, and R. Weissleder, "18F labeled nanoparticles for in vivo PET-CT imaging," Bioconjugate Chemistry, vol. 20, no. 2, pp. 397–401, 2009.

58. W. J. M. Mulder, R. Koole, R. J. Brandwijk et al., "Quantum dots with a paramagnetic coating as a bimodal molecular imaging probe," Nano Letters, vol. 6, no. 1, pp. 1–6, 2006.

59. V. Bagalkot, L. Zhang, E. Levy-Nissenbaum et al., "Quantum dot-aptamer conjugates for synchronous cancer imaging, therapy, and sensing of drug delivery based on Bi-fluorescence resonance energy transfer," Nano Letters, vol. 7, no. 10, pp. 3065–3070, 2007.

60. P. Juzenas, W. Chen, Y.-P. Sun et al., "Quantum dots and nanoparticles for photodynamic and radiation therapies of cancer," Advanced Drug Delivery Reviews, vol. 60, no. 15, pp. 1600–1614, 2008.

61. M.-C. Daniel and D. Astruc, "Gold nanoparticles: assembly, supramolecular chemistry, quantum-size-related properties, and applications toward biology, catalysis, and nanotechnology," Chemical Reviews, vol. 104, no. 1, pp. 293–346, 2004.

62. W. Cai, T. Gao, H. Hong, and J. Sun, "Applications of gold nanoparticles in cancer nanotechnology,"Nanotechnology, Science and Applications, vol. 1, pp. 17–32, 2008.

63. J. Kim, S. Park, J. E. Lee et al., "Designed fabrication of multifunctional magnetic gold nanoshells and their application to magnetic resonance imaging and photothermal therapy," Angewandte Chemie - International Edition, vol. 45, no. 46, pp. 7754–7758, 2006.

64. H. Park, J. Yang, S. Seo et al., "Multifunctional nanoparticles for photothermally controlled drug delivery and magnetic resonance imaging enhancement," Small, vol. 4, no. 2, pp. 192–196, 2008.

65. W. S. Seo, J. H. Lee, X. Sun et al., "FeCo/graphitic-shell nanocrystals as advanced magnetic-resonance-imaging and near-infrared agents," Nature

Materials, vol. 5, no. 12, pp. 971–976, 2006.

66. J. H. Lee, S. P. Sherlock, M. Terashima et al., "High-contrast in vivo visualization of microvessels using novel FeCo/GC magnetic nanocrystals," Magnetic Resonance in Medicine, vol. 62, no. 6, pp. 1497–1509, 2009.

67. V. J. Venditto, C. A. S. Regino, and M. W. Brechbiel, "PAMAM dendrimer based macromolecules as improved contrast agents," Molecular Pharmaceutics, vol. 2, no. 4, pp. 302–311, 2005.

68. C. A. Boswell, P. K. Eck, C. A. S. Regino et al., "Synthesis, characterization, and biological evaluation of integrin αvβ3-targeted PAMAM dendrimers," Molecular Pharmaceutics, vol. 5, no. 4, pp. 527–539, 2008.

69. T. Barrett, G. Ravizzini, P. L. Choyke, and H. Kobayashi, "Dendrimers in medical nanotechnology," IEEE Engineering in Medicine and Biology Magazine, vol. 28, no. 1, pp. 12–22, 2009.

70. H. Kobayashi, Y. Koyama, T. Barrett et al., "Multimodal nanoprobes for radionuclide and five-color near-infrared optical lymphatic imaging," ACS nano, vol. 1, no. 4, pp. 258–264, 2007.

71. A. E. Nel, L. Mädler, D. Velegol et al., "Understanding biophysicochemical interactions at the nano-bio interface," Nature Materials, vol. 8, no. 7, pp. 543–557, 2009.

72. S. Kommareddy, S. B. Tiwari, and M. M. Amiji, "Long-circulating polymeric nanovectors for tumor-selective gene delivery," Technology in Cancer Research and Treatment, vol. 4, no. 6, pp. 615–625, 2005.

73. J. R. Lindner, "Microbubbles in medical imaging: current applications and future directions," Nature Reviews Drug Discovery, vol. 3, no. 6, pp. 527–532, 2004.

74. P. V. Kulkarni, C. A. Roney, P. P. Antich, F. J. Bonte, A. V. Raghu, and T. M. Aminabhavi, "Quinoline-n-butylcyanoacrylate-based nanoparticles for brain targeting for the diagnosis of Alzheimer's disease," Wiley Interdisciplinary Reviews: Nanomedicine and Nanobiotechnology, vol. 2, no. 1, pp. 35–47, 2009.

75. Q. Zhang, Z. Shen, and T. Nagai, "Prolonged hypoglycemic effect of insulin-loaded polybutylcyanoacrylate nanoparticles after pulmonary administration to normal rats," International Journal of Pharmaceutics, vol. 218, no. 1-2, pp. 75–80, 2001.

76. M. Palmowski, J. Huppert, G. Ladewig et al., "Molecular profiling of angiogenesis with targeted ultrasound imaging: early assessment of

antiangiogenic therapy effects," Molecular Cancer Therapeutics, vol. 7, no. 1, pp. 101–109, 2008.

77. M. Palmowski, P. Peschke, J. Huppert et al., "Molecular ultrasound imaging of early vascular response in prostate tumors irradiated with carbon ions," Neoplasia, vol. 11, no. 9, pp. 856–863, 2009.

78. F. Yang, L. Li, Y. Li, Z. Chen, J. Wu, and N. Gu, "Superparamagnetic nanoparticle-inclusion microbubbles for ultrasound contrast agents," Physics in Medicine and Biology, vol. 53, no. 21, pp. 6129–6141, 2008.

79. F. Yang, Y. Li, Z. Chen, Y. Zhang, J. Wu, and N. Gu, "Superparamagnetic iron oxide nanoparticle-embedded encapsulated microbubbles as dual contrast agents of magnetic resonance and ultrasound imaging," Biomaterials, vol. 30, no. 23-24, pp. 3882–3890, 2009.

80. S. Hernot and A. L. Klibanov, "Microbubbles in ultrasound-triggered drug and gene delivery," Advanced Drug Delivery Reviews, vol. 60, no. 10, pp. 1153–1166, 2008.

81. Z. Gao, A. M. Kennedy, D. A. Christensen, and N. Y. Rapoport, "Drug-loaded nano/microbubbles for combining ultrasonography and targeted chemotherapy," Ultrasonics, vol. 48, no. 4, pp. 260–270, 2008.

82. R. D. Bolskar, "Gadofullerene MRI contrast agents," Nanomedicine, vol. 3, no. 2, pp. 201–213, 2008.

83. J. Liu, S.-I. Ohta, A. Sonoda et al., "Preparation of PEG-conjugated fullerene containing Gd^{3+} ions for photodynamic therapy," Journal of Controlled Release, vol. 117, no. 1, pp. 104–110, 2007.

84. B. Sitharaman, T. Y. Zakharian, A. Saraf et al., "Water-soluble fullerene (C60) derivatives as nonviral gene-delivery vectors," Molecular Pharmaceutics, vol. 5, no. 4, pp. 567–578, 2008.

85. F. Meiser, C. Cortez, and F. Caruso, "Biofunctionalization of fluorescent rare-earth-doped lanthanum phosphate colloidal nanoparticles," Angewandte Chemie, vol. 116, no. 44, pp. 6080–6083, 2004.

86. S. Setua, D. Menon, A. Asok, S. Nair, and M. Koyakutty, "Folate receptor targeted, rare-earth oxide nanocrystals for bi-modal fluorescence and magnetic imaging of cancer cells," Biomaterials, vol. 31, no. 4, pp. 714–729, 2010.

87. F. Wang and X. Liu, "Recent advances in the chemistry of lanthanide-doped upconversion nanocrystals,"Chemical Society Reviews, vol. 38, no. 4, pp. 976–989, 2009.

88. Z. Li, Y. Zhang, and S. Jiang, "Multicolor core/shell-structured upconversion fluorescent nanoparticles,"Advanced Materials, vol. 20, no. 24, pp. 4765–4769, 2008.
89. Y. I. Park, J. H. Kim, K. T. Lee et al., "Nonblinking and nonbleaching upconverting nanoparticles as an optical imaging nanoprobe and T1 magnetic resonance imaging contrast agent," Advanced Materials, vol. 21, no. 44, pp. 4467–4471, 2009.
90. L.-Q. Xiong, Z.-G. Chen, M.-X. Yu, F.-Y. Li, C. Liu, and C.-H. Huang, "Synthesis, characterization, and in vivo targeted imaging of amine-functionalized rare-earth up-converting nanophosphors," Biomaterials, vol. 30, no. 29, pp. 5592–5600, 2009.

Chapter 7

NANOPHOTONICS FOR MOLECULAR DIAGNOSTICS AND THERAPY APPLICATIONS

João Conde,[1,2] João Rosa,[1,3] João C. Lima[3], and Pedro V. Baptista[1]

[1]CIGMH, Departamento de Ciências da Vida, Faculdade de Ciências e Tecnologia, Universidade Nova de Lisboa, Campus de Caparica, 2829-516 Caparica, Portugal

[2]Instituto de Nanociencia de Aragón, Universidad de Zaragoza, Campus Río Ebro, Edifício I+D, Mariano Esquillor s/n, 50018 Zaragoza, Spain

[3]REQUIMTE, Departamento de Química, Faculdade de Ciências e Tecnologia, Universidade Nova de Lisboa, Campus de Caparica, 2829-516 Caparica, Portugal

ABSTRACT

Light has always fascinated mankind and since the beginning of recorded history it has been both a subject of research and a tool for investigation of other phenomena. Today, with the advent of nanotechnology, the use of light has reached its own dimension where light-matter interactions take place at wavelength and subwavelength scales and where the physical/chemical nature of nanostructures controls the interactions. This is the field of nanophotonics which allows for the exploration and manipulation of light in and around nanostructures, single molecules, and molecular complexes. What is more is the use of nanophotonics in biomolecular interactions—nanobiophotonics—has prompt for a plethora of molecular diagnostics and therapeutics making use of the remarkable nanoscale properties. In this paper, we shall focus on the uses of nanobiophotonics for molecular diagnostics involving specific sequence characterization of nucleic acids and for gene delivery systems of relevance for therapy strategies. The use of nanobiophotonics for the combined diagnostics/therapeutics (theranostics) will also be addressed, with particular focus on those systems enabling the development of safer, more efficient, and specific platforms. Finally, the translation of nanophotonics for theranostics into the clinical setting will be discussed.

INTRODUCTION

Nanophotonics deals with the interaction of light with matter at a nanometer scale, providing challenges for fundamental research and opportunities for new technologies, encompassing the study of new optical interactions, materials, fabrication techniques, and architectures, including the exploration of natural and synthetic, or artificially engineered, structures such as photonic crystals, holey fibers, quantum dots, subwavelength structures, and plasmonics [1, 2]. The use of photonic nanotechnologies in medicine is a rapidly emerging and potentially powerful approach for disease protection, detection, and treatment. The high speed of light manipulation and the remote nature of optical methods suggest that light may successfully connect diagnostics, treatment, and even the guidance of the treatment in one theranostic procedure combination of therapeutics with diagnostics (including patient prescreening and therapy monitoring).

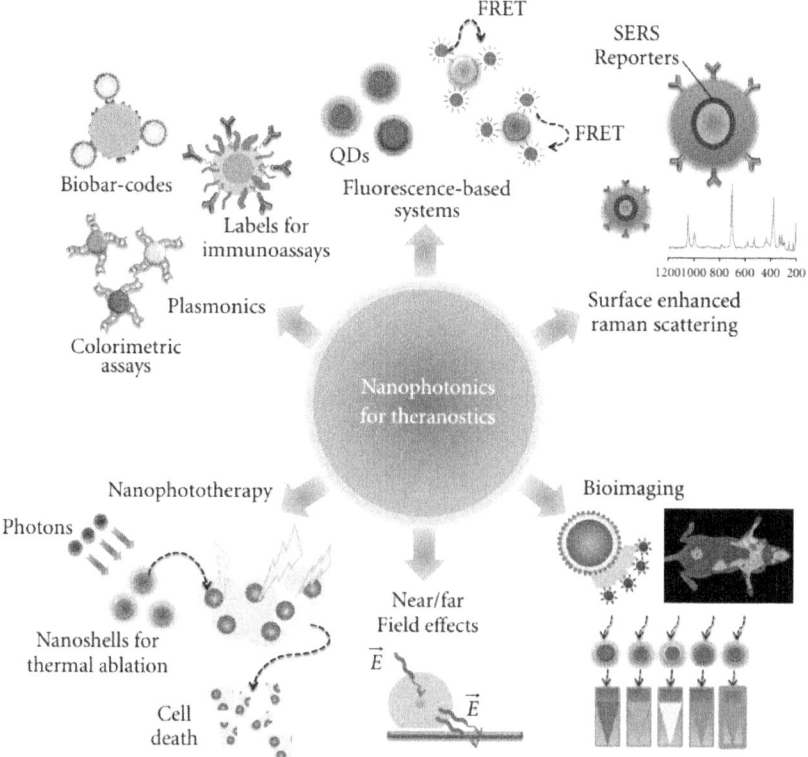

Figure 1: Nanophotonics for theranostics. Nanoparticles-based strategies can be used for biosensing using plasmonic nanosensors, such as metal nanoparticles functionalized with nucleic acid strand for colorimetric assays and biobar codes for protein

detection or intense labels for immunoassays. Some nanoparticle systems can also be used for sensing by exploring a typical FRET system or can be surrounded with Raman reporters in order to provide in vivo detection and tumour targeting. In fact, NPs symbolize an important class of materials with unique features suitable for biomedical imaging applications such as increased sensitivity in detection and high quantum yields for fluorescence. Alternatively, NPs can survey near/far field enhancing qualities that hold promise for a bounty of novel applications in optics and photonics. Engineered NPs can also act as phototherapeutic agents that can be attached to specific targets for selective damage to cancer cells.

Limitations in medical practice are closely associated with the fact that diagnostics, therapy, and therapy guidance are three discrete and isolated stages. In order to overcome some of the sensitivity and specificity of current medicines, theranostics unites the three above stages in one single process, supporting early-stage diagnosis and treatment [3, 4]. Nowadays, there is an ever increasing need to enhance the capability of theranostics procedures where nanophotonics-based sensors may provide for the simultaneous detection of several gene-associated conditions and nanodevices utilizing light-guided and light-activated therapy with the ability to monitor real-time drug action (see Figure 1).

NANOPHOTONICS FOR DIAGNOSTICS

Surface Plasmons on Nanoparticles and Surfaces

Surface plasmons are collective charge oscillations that occur at the interface between conductors and dielectrics. They can take various forms, ranging from freely propagating electron density waves along metal surfaces to localized electron oscillations on metal nanoparticles (NPs) [5, 6]. When light passes through a metal nanoparticle, it induces dipole moments that oscillate at the respective frequency of the incident wave, consequently dispersing secondary radiation in all directions. This collective oscillation of the free conduction electrons is called localized surface plasmon resonance (LSPR). Light on NP induces the conduction electrons to oscillate collectively with a resonant frequency that depends on the nanoparticles' size, shape, composition, interparticle distance, and environment (dielectric properties) [7–10]. As a result of these SPR modes, the nanoparticles absorb and scatter light so intensely that single NPs are easily observed by eye using dark-field (optical scattering) microscopy. Plasmonic NPs provide a nearly unlimited photon resource for observing molecular binding for longer periods of time, once they do not blink or bleach like fluorophores [11].

Nanoparticle-based colorimetric assays for diagnostics have been a subject of intensive research, where LSPR can be used to detect DNA or proteins by the changes in the local index of refraction upon adsorption of the target molecule to the metal surface. Due to the intense SPR in the visible yielding extremely bright colors, gold nanoparticle colloids have been widely used of molecular diagnostics. In fact, gold nanoparticles (AuNPs) functionalized with ssDNA capable of specifically hybridizing to a complementary target for the detection of specific nucleic acid sequences in biological samples have been extensively used [12–27]. Other approaches use the AuNPs' plasmonic as a core/seed that can be tailored with a wide variety of surface functionalities to provide highly selective nanoprobes for diagnostics [28] or the SPR scattering imaging or SPR absorption spectroscopy generated from antibody-conjugated AuNPs in molecular biosensor techniques for the diagnosis of oral epithelial living cancer cells in vivo and in vitro [29] and the use of multifunctional AuNPs which incorporate both cytosolic delivery and targeting moieties on the same particle functioning as intracellular sensors to monitor actin rearrangement in live fibroblasts [30].

Plasmonic NPs have also been used as extremely intense labels for immunoassays [31–34] and biochemical sensors [19, 35–37]. Also, the use of colloidal silver plasmon resonant particles (PRPs) coated with standard ligands as target-specific labels has been reported for in situ hybridization and immunocytology assays [34]. Most notably, a nanoparticle-based Biobar code has been developed for the detection of proteins that relies on magnetic microparticle probes with antibodies that specifically bind a target of interest and nanoparticle probes that are encoded with DNA that is unique to the protein target of interest and antibodies that can sandwich the target captured by the microparticle probes [33]. Haes and coworkers have reported on an optical biosensor based on localized surface plasmon resonance spectroscopy developed to monitor the interaction between the antigen, amyloid-β-derived diffusible ligands (ADDLs), and specific anti-ADDL antibodies, used in the detection of a biomarker for Alzheimer's Disease [35].

Raman-Spectroscopy-Based Systems

When light interacts with a substance, it can be absorbed, transmitted, or scattered. Scattered radiation can result from an elastic collision (Rayleigh scattering) or inelastic (Raman scattering). Raman spectroscopy is based on a change of frequency when light is inelastically scattered by molecules or atoms resulting in a molecular fingerprint information on molecular structure or intermolecular interaction of a specific process or molecule. The potential of Raman spectroscopy as biomedical diagnostics tool is rather low due to its low cross-section (~10^{-30} cm^2) that results in low sensitivity [38]. However, in 1977, two groups independently described the use of noble metal surfaces to enhance the Raman scattering signal of target molecules [39, 40]—Surface enhancement raman spectroscopy (SERS). Jeanmaire and Van Duyne proposed a twofold electromagnetic field enhancement that was later associated with the interaction between the incident and scattered photons with the nanostructure's LSPR [41]. Simultaneously, Albrecht and Creighton suggested the source of the enhancement to be caused by a specific interaction between an adsorbate and the nanoparticle surface, briefly, a charge transfer from the adsorbate into the empty energetic levels on the metal surface or from the occupied levels of the nanoparticle's surface to the adsorbate [42–44].

Generally, SERS requires that the biological analyte reaches a suitable surface where the substrates are treated as two-dimensional macroscopic surfaces onto which adsorbed molecules suffer a local-field enhancement. Despite direct adsorption not being a good solution because of its dependence on the affinity between substrate and analyte, a method to identify and distinguish different strains of virus based on signal differences generated by the surface aminoacids using silver nanorods has been successfully developed [45]. Using a similar approach of direct adsorption, Pînzary et al. used naked silver nanoparticles to differentiate in situ healthy colon from carcinoma colon tissue [46]. Nanotags have been widely employed to address the lack of specificity [47, 48]. These nanotags usually possess a metallic colloidal core functionalized with a Raman reporting molecule and the specific molecule used to capture the analyte and have been used to directly detect DNA sequences [49, 50] and amplified DNA products of epizootic pathogens using complementary DNA strands so that only the complementary target hybridizes with the probes [51]. Using a similar system, but exploring the distance-dependent enhancement of the electromagnetic field with a hairpin probe molecule, Wabuyele has also been used to distinguish single nucleotide polymorphisms in cancer-related genes [52]. Combining nanotags with other nanoparticles or binding surfaces that target the same analyte in a sandwich conformation proved useful to detect antibodies in serum [53]. A similar approach using a flat substrate instead of NPs had already been proposed to detect DNA, RNA, and proteins [54, 55]. However, in this approach, the substrate is used only to immobilize the analyte; a gold-nanoparticle-based nanotag is used to identify the analyte and the surface enhancement is obtained by silver coating of the nanotag. miRNA profiling has also been pursued via a slightly different approach based on the hybridization of the target molecules with a thiolated oligonucleotide and subsequent functionalization on a silver substrate [56]. SERS have also been explored to identify changes in the analyzed system such as interaction between DNA and xenobiotic molecules like cisplatin [57] or DNA-binding proteins [58,

59]. The combination of magnetic iron/gold core-shell nanoparticles with gold nanorods has also been used to specifically enumerate E. coli in water samples in a rapid and sensitive test [60]. In this case, the magnetic nanoparticles are used to concentrate the bacteria, improving the Raman signal by concentration and the posteriorly added gold nanorods serve as Raman signal enhancers.

SERS can also be used in conjunction with colloidal gold to detect and target tumors in vivo, where the AuNPs are surrounded with Raman reporters that provide light emission 200 times brighter than quantum dots [61,62]. It was also found that the Raman reporters became more stable and yielded larger optical enhancements when NPs were encapsulated with a thiol-modified polyethylene glycol coat, which also allows for increased biocompatibility and circulation times in vivo. When conjugated to tumor-targeting ligands, these conjugated SERS-NPs were able to target tumor markers at surface of malignant cells, such as epidermal growth factor receptor (EGFR) that is sometimes overexpressed in cells of certain cancer types [29] and used to locate the tumor in xenograft tumor models [50].

Fluorescence-Based Systems

Quantum dots (QDs) are semiconductor nanoparticles with narrow, tunable, symmetrical emission spectra, and high quantum yields [63–65], and together with compatibility with DNA and proteins, make QDs exceptional substitutes as fluorescence labels. The use of QDs for nucleic acid characterization has long been proposed, for example, CdSe/ZnS QDs for SNP identification on human TP53 gene, multiallele detection of hepatitis B and C viruses [66], and in situ detection of chromosome abnormalities and mutations [67]. QDs have also been used as chemical sensors by exploring a typical FRET system where a dark quencher is placed at a protein-binding site attached to a QD surface. The quantum dots emission is quenched in presence of the analyte and upon analyte displacement the emission is restored [68]. A simpler approach was used to detect adenine using fluorescent ZnS nanoparticles at pH7, making use of capability of adenine itself to quench emission of the quantum-dot-like nanoparticles [69].

Several studies report on the modulation of fluorophores at the vicinity of nanoparticles (e.g., gold, silver, and quantum dots) [70, 71], an interaction that has found application in a variety of systems to detect biologically relevant targets with particular focus upon AuNPs due to their ease in functionalization with biomolecules [72–75]. Several methods based on the quenching of fluorescence have been proposed for DNA detection consisting of fluorophore-labeled ssDNA electrostatically adsorbed onto gold nanoparticles [76], carbon nanotubes [77], and carbon nanoclots [78], where the presence of a complementary target triggers desadsorption of the newly formed dsDNA from the nanostructures due to the electrostatic variation between ssDNA and dsDNA, and fluorescence emission is restored. Also, fluorescence quenching of fluorophores close to metal nanoparticles functionalized with thiol-modified oligonucleotides has been explored in different conformations. Tang and co-workers proposed a method to probe hydroxyl radicals using an AuNP-oligonucleotide-FAM system where the hydroxyl radical promotes strand breakage and consequent release of FAM, restoring the previously quenched fluorescence [79]. The same quenching mechanism was used to detect specific DNA strands using two probes (one with an AuNP label and another labeled with TAMRA) that hybridize to two DNA sequences near each other [80], bringing the fluorophore and AuNP close enough to quench fluorescence emission.

Proteins have also been probed through nanoparticle-fluorescence-mediated systems, for example, human blood proteins have been let to interact with fluorescent AuNPs and detected through quenching [81]. In another example, a sandwich immunoassay using AuNPs quenching has been proposed for the detection of the protein cardiac troponin T by its interaction with two different antibodies, one attached to AuNPs and the other labeled with fluorescent dyes [82]. By means of an opposite modulation, infrared fluorescent nanoparticles showed enhanced fluorescence when interacting with protein [83].

A popular application of fluorescence modulation by nanoparticles has been specific ion sensing. Su and co-workers developed a copper sensor by covering fluorescent DNA-Cu/Ag nanoclusters with mercaptopropionic acid which quenches the intrinsic fluorescence of the nanoparticle; in the presence of Cu^{2+}, the capping agent is oxidized to form a disulfide compound resulting in release of the nanoparticle and restoration of emission suitable for quantification between 5 and 200 nM [84]. A very specific colorimetric and fluorimetric method to detect Hg^{2+} ions was developed with porphyrin-modified $Au@SiO_2$ nanoparticles, where the intensively fluorescent red complex turns green and weakly fluorescent in presence of Hg^{2+} [85]. Another examples include sensing of Pb^{2+} and adenosine by combining an adenosine aptamer and a DNAzyme with an abasic site where 2-amino-5,6,7-trimethyl-1,8-naphthyridine is trapped to quench its fluorescence [86]. When in solution, Pb^{2+} enables the DNAzyme to cleave its substrate thus removing the fluorescent compound from the abasic site restoring its fluorescence. Similarly, the presence of adenosine induces structural change of the aptamer, resulting in the release of the fluorescent molecule from the DNA duplex and a subsequent fluorescence enhancement.

Nanophotonics Bioimaging

Nanoparticles show unique features suitable for biomedical imaging applications, such as an increased sensitivity in detection through amplification of signal changes (e.g., magnetic resonance imaging); high fluorescence quantum yields and large magnetic moments; properties that induce phagocytosis and selective uptake by macrophages (e.g., liposomes); physicochemical manipulations of energy (i.e., quantum dots); among others [87]. Because light absorption from biologic tissue components is minimized at near infrared (NIR) wavelengths, most nanoparticles (e.g., noble metal and magnetic NPs, nanoshells, nanoclusters, nanocages, nanorods and quantum dots) for in vivo imaging and therapy have been designed to strongly absorb in the NIR and used for in vivo diagnostics [83, 88, 89]. Ex vivo and in vivo imaging applications of nanoparticles have included their use as contrast agents for magnetic resonance imaging (MRI) [90], optical coherence tomography (OCT) [91–93], photoacoustic imaging (PAI) [94], and two-photon luminescence (TPL) spectroscopy [95].

Magnetic Resonance Imaging

Magnetic resonance imaging (MRI) is based heavily on nuclear magnetic resonance (NMR), first described by R. Damadian. Magnetic resonance measurements cause no obvious deleterious effects on biological tissue, and the incident radiation consists of common radio frequencies at right angles to a static magnetic field [96]. Iron oxide nanoparticles show superparamagnetism, allowing for the facile alignment of the magnetic moments to an applied magnetic field, thus of great interest as contrast agents for MRI [97]. Presently, magnetic iron oxide nanoparticles are routinely used as contrast agents to enhance an MRI image, providing sharper contrast between soft and hard tissue in the body (e.g. liver and spleen or lymph nodes) [98]. Jun et al. presented a synthetically controlled magnetic nanocrystal model system that led to the improvement of high-performance nanocrystal—antibody probe systems for the diagnosis of breast cancer cells via magnetic resonance imaging [99]. Also, $MnFe_2O_4$ nanocrystals functionalized with an antibody conjugate (herceptin) capable of specific targeting of cancerous cells was successfully used for in vivo MRI in mice [88]. Driehuysb et al. developed an imaging method to detect submillimeter-sized metastases with molecular specificity by targeting cancer cells with iron oxide nanoparticles functionalized with cancer-binding ligands, demonstrating in vivo detection of pulmonary micrometastases in mice injected with breast adenocarcinoma cells [100]. Hybrid NPs with a superparamagnetic iron oxide/silica core and a gold nanoshell, with significant absorbance and scattering in the NIR region, have been used in vivo as contrast agents for MRI presenting a good MR signal in hepatoma, each moiety providing for a distinct signal that enhanced detection [101].

Optical Coherence Tomography

Optical Coherence Tomography (OCT) is an imaging modality that provides cross-sectional subsurface imaging of biological tissue with micrometer scale resolution which is based on a broadband light source and a fiber-optic interferometer. It captures three-dimensional images from within optical scattering media, typically employing near-infrared light. The use of relatively long wavelength light allows it to penetrate into the scattering medium [102–104]. The extra scattering provided by Au-nanoshells enhances optical contrast and brightness for improved diagnostic imaging of tumors in mice due to the preferential accumulation of the nanoshells in the tumor [105]. Tseng et al. developed nanorings with a localized surface plasmon resonance covering a spectral range of 1300 nm that produced both photothermal and image contrast enhancement effects in OCT when delivered into pig adipose samples [106]. Additionally, the image contrast enhancement effect could be isolated by continuously scanning the sample with a lower scan frequency, allowing to effectively control the therapeutic modality. In the same way, gold capped nanoroses have been used in photothermal OCT to detect macrophages in ex vivo rabbit arteries [107].

Photoacoustic Imaging

In photoacoustic imaging (PAI) and photoacoustic tomography (PAT), a pulse of NIR laser light, typically 757 nm, is used in resonance with the surface plasmon instead of a continuous NIR source. With this technique causing rapid thermal expansion of the surrounding media, the generated sound wave can be detected on the surface of the subject. NIR reduces the amount of absorption that occurs, but absorption of light by various other organs is unavoidable [108, 109]. Yang et al. demonstrated the feasibility of using poly(ethylene glycol)-coated Au nanocages as a new in vivo NIR contrast-enhancing agent for photoacoustic tomography and image their distribution in the vasculature of rat brain. These Au-nanocages enhanced the contrast between blood and the surrounding tissues by up to 81%, achieving a more detailed image of vascular structures at greater depths. Additionally, they were shown to present slight advantages over Au-nanoshells, being better suited for in vivoapplications, specially due to their more compact size (<50 nm compared to >100 nm for Au-nanoshells) and larger optical absorption cross-sections [110]. Due to the ability of gold-nanorods to have the maximum of the plasmon resonance tuned further into the NIR, Motamedi et al. reported a contrast agent for a laser optoacoustic imaging system for in vivo detection of gold nanorods and to enhance the diagnostic power of optoacoustic imaging [111]. Song et al. proposed a noninvasive in vivo spectroscopic photoacoustic sentinel lymph node mapping in a rat model using gold nanorods as lymph node tracers [112].

Two-Photon Luminescence

In two-photon luminescence (TPL) spectroscopy, an electron is excited from the conductance band to the valence band of the metal nanoparticles using two photons. As the electron relaxes to the conductance band, light is released and amplified due to a resonant coupling with localized surface plasmons, enhancing a variety of linear and nonlinear optical properties [113, 114]. TPL was first described by Boyd et al. that found that roughened metal surfaces exhibited much higher induced luminescence efficiency than smooth surfaces [115]. In fact, TPL is a potentially powerful technique for noninvasive imaging at the micron scale hundreds of microns deep into scattering tissue. This way, it ought to be possible to discriminate cancerous and healthy tissue based on two-photon imaging from endogenous fluorophores. For enhanced imaging, two-photon contrast agents have been developed showing the ability to increase signal-to-noise ratio and targeted to molecular signatures of interest that are not fluorescent. Because imaging of intrinsic fluorophores is often difficult due to their relatively weak signals, the use of such a bright contrast agent holds the promise to enablein vivo applications of two photon imaging in a clinical setting [113, 116, 117]. Wang et al. collected images of single gold nanorods flowing in the mouse ear blood vessels with luminescence three times stronger than background [114]. It is worth mentioning that the TPL signal from a single nanorod is 58 times that of the two-photon fluorescence signal from a single rhodamine molecule.

QDs for In Vivo Imaging

In the last decade, water soluble bioconjugated QDs have been increasingly applied for imaging [63, 64, 118] However, QD probes for imaging show poor stability once inside cytosolic environment and reduced biocompatibility in living organisms [119], which constitutes a serious drawback for widespread in vivoapplication.

Despite the serious concerns related to the in vivo use of QDs, these nanocrystals show remarkable imaging properties that may be judged of value for improved diagnostics. In fact, QDs have proven of great value when imaging vascular networks of mammals such as lymphatic and cardiovascular systems [120–124]. Also, Kim et al. demonstrated that quantum dots allowed a major cancer surgery to be performed in large animals (mice and pigs) under complete image guidance, by locating the position of sentinel lymph nodes [125]. With similar potential to that observed when imaging the lymph system, the imaging of cardiovascular systems has also been achieved using QDs [126, 127]. Larson et al. demonstrated that QDs retained their fluorescence after injection and could be detected in the capillaries of skin and adipose tissue of a mouse [128]. The fluorescent emission and multiplexing capabilities of QDs are being exploited to improve the sensitivity and selectivity in the early detection of tumors [61, 129, 130]. Åkerman et al. described for the first time the application of targeted cancer imaging by using ZnS-capped CdSe QDs coated with a lung-targeting peptide that accumulate in the lungs of mice, whereas two other peptides specifically direct QDs to blood vessels or lymphatic vessels in tumors [131]. Later, Gao et al. described the development of multifunctional nanoparticle probes based on QDs with a copolymer linked to tumor-targeting ligands and drug-delivery functionalities for cancer targeting and imaging in living animals [132]. Once the toxicological aspects associated with QDs have been clarified, such studies demonstrate the potential of QDs for ultrasensitive and multiplexed imaging of molecular targets in vivo.

NANOPHOTONICS FOR THERAPY

Nanophototerapy uses pulsed lasers and absorbing nanoparticles attached to specific targets for selective damage to cancer cells. Plasmonic photothermal therapy (PPTT) and photodynamic therapy (PDT) are two of the main techniques that take advantage of the selective absorbance of the surface plasmon resonance and the fact that the nanoparticles relax by liberating heat into their surrounding environment.

Plasmonic Photothermal Therapy

Plasmonic photothermal therapy is a less invasive experimental technique that holds great promise for the treatment of cell malignancies and, in particular, of cancer. It combines two key components: (i) light source, specifically lasers with a spectral range of 650–900 nm for deep tissue penetration and (ii) optical absorption of AuNPs which release the optical irradiation as heat in the picoseconds time scale, thereby inducing photothermal ablation [133–135]. Kirui et al. reported the use of gold and iron oxide hybrid nanoparticles in targeting, imaging, and selective thermal killing of colorectal cancer cells [136]. Huang and colleagues have demonstrated that gold nanorods have a longitudinal absorption band in the NIR on account of SPR oscillations and are effective as photothermal agents [137]. Gold nanorods aspect ratios allow tuning the SPR band from the visible to the NIR (transmits readily through human skin and tissue), making them suitable for photothermal converters of near infrared light for in vivo applications [138, 139]. Effective photothermal destruction of cancer cells and tissue have been demonstrated for other gold nanostructures, such as branched gold nanoparticles [140], gold nanoshells [141–143], gold nanocages [134], and gold nanospheres [144].

Photodynamic Therapy

Photodynamic therapy employs chemical photosensitizers that generate reactive oxygen species (ROS), such as a singlet oxygen ($^1O_2$), capable of tumor destruction [145, 146]. This technique is noninvasive and can be applied locally or systemically without noticeable cumulative toxicity effects without high costs. To attain maximal killing efficiency of tumor cells, the photosensitizer must be in close proximity to the tumor cells, thus requiring specific targeting when administered systemically. One of the major limitations is the poor tissue penetration of high-energy light and the systemic dispersal of the photosensitizer [147, 148].

Aiming at circumventing some of the limitations of photodynamic therapy, Zhang et al. reported a new type of photosensitizers based on photon upconverting nanoparticles (a process where low energy light, usually near-infrared (NIR) or infrared (IR), is converted to higher energies, ultraviolet (UV), or visible, via multiple absorptions or energy transfer steps) [149, 150]. One year later, Yong et al. reported the use of NPs modified with zinc phthalocyanin photosensitizer that produce green/red emission on near-infrared (NIR) excitation and is capable of singlet oxygen sensitization; upon targeted binding to cancer cells, significant cell destruction was induced [151]. Recently, Qian et al. published similar results with the use of zinc phthalocyanine nanocrystals coated with a uniform layer of mesoporous silica [152].

CONCLUSIONS

Light is an amazing intermediate with a gargantuan capacity for carrying multiple information and functions. Instinctively, we view light as rays, which propagate in a single direction, either being absorbed or reflected to some extent by any object on which it impinges. However, the propagation of light through a material is itself a quantum effect, involving the excitation and relaxation of electrons in the material. It is well known that light has the facility to act through biological, chemical, mechanical, and thermal pathways at molecular/cellular levels in diagnostic and therapeutic applications.

Currently, we are in the dawn of a new age in therapy driven by nanotechnology vehicles. Although there are technical challenges associated with the therapeutic application of nanodevices, the integration of therapy with diagnostic profiling has accelerated the pace of discovery of new nanotechnology methods. In addition to continuing to push forward on the above challenges, nanotechnology together with photonics can be used both for identifying useful target candidates and for validating their importance in disease states. Nanophotonics may present new opportunities for personalized medicine in which diagnosis and treatment are based on each individual's molecular profile. Further research into the fundamental mechanisms that efficiently control light using nanodevices could unveil new dimensions of nanoparticle-mediated theranostic systems.

Here, we have attempted to give the reader a limited overview of some aspects of the current state of research into the fascinating aspects and control over nanophotonics in molecular diagnostics and therapy applications.

ACKNOWLEDGMENT

The authors acknowledge FCT/MCTES—CIGMH(Portugal) for financial support.

REFERENCES

1. M. Ohtsu, K. Kobayashi, T. Kawazoe, T. Yatsui, and M. Naruse, Principles of Nanophotonics, Series in Optics and Optoelectronics, Taylor & Francis, CRC Press, 2008.
2. Y. Shen and P. N. Prasad, "Nanophotonics: a new multidisciplinary frontier," Applied Physics B, vol. 74, no. 7-8, pp. 641–645, 2002.
3. T. Lammers, S. Aime, W. E. Hennink, G. Storm, and F. Kiessling, "Theranostic nanomedicines," Accounts of Chemical Research. In press.
4. F. Pene, E. Courtine, A. Cariou, and J. P. Mira, "Toward theragnostics," Critical Care Medicine, vol. 37, no. 1, pp. S50–S58, 2009.
5. W. L. Barnes, A. Dereux, and T. W. Ebbesen, "Surface plasmon subwavelength optics," Nature, vol. 424, no. 6950, pp. 824–830, 2003.
6. U. Kreibig and M. Vollmer, Optical Properties of Metal Clusters, vol. 25 of Springer Series in Materials Science, Springer, Berlin, Germany, 1995.
7. M. C. Daniel and D. Astruc, "Gold nanoparticles: assembly, supramolecular chemistry, quantum-size-related properties, and applications toward biology, catalysis, and nanotechnology," Chemical Reviews, vol. 104, no. 1, pp. 293–346, 2004.
8. S. Eustis and M. A. El-Sayed, "Why gold nanoparticles are more precious than pretty gold: Noble metal surface plasmon resonance and its enhancement of the radiative and nonradiative properties of nanocrystals

of different shapes," Chemical Society Reviews, vol. 35, no. 3, pp. 209–217, 2006.

9. S. K. Ghosh and T. Pal, "Interparticle coupling effect on the surface plasmon resonance of gold nanoparticles: from theory to applications," Chemical Reviews, vol. 107, no. 11, pp. 4797–4862, 2007.

10. W. Zhao, M. A. Brook, and Y. Li, "Design of gold nanoparticle-based colorimetric biosensing assays,"ChemBioChem, vol. 9, no. 15, pp. 2363–2371, 2008.

11. J. N. Anker, W. P. Hall, O. Lyandres, N. C. Shah, J. Zhao, and R. P. Van Duyne, "Biosensing with plasmonic nanosensors," Nature Materials, vol. 7, no. 6, pp. 442–453, 2008.

12. P. Baptista, G. Doria, D. Henriques, E. Pereira, and R. Franco, "Colorimetric detection of eukaryotic gene expression with DNA-derivatized gold nanoparticles," Journal of Biotechnology, vol. 119, no. 2, pp. 111–117, 2005.

13. P. Baptista, E. Pereira, P. Eaton et al., "Gold nanoparticles for the development of clinical diagnosis methods," Analytical and Bioanalytical Chemistry, vol. 391, no. 3, pp. 943–950, 2008.

14. Y. C. Cao, R. Jin, C. S. Thaxton, and C. A. Mirkin, "A two-color-change, nanoparticle-based method for DNA detection," Talanta, vol. 67, no. 3, pp. 449–455, 2005.

15. M. M. C. Cheng, G. Cuda, Y. L. Bunimovich et al., "Nanotechnologies for biomolecular detection and medical diagnostics," Current Opinion in Chemical Biology, vol. 10, no. 1, pp. 11–19, 2006.

16. J. Conde, J. M. de la Fuente, and P. V. Baptista, "RNA quantification using gold nanoprobes - application to cancer diagnostics," Journal of Nanobiotechnology, vol. 8, article no. 5, 2010.

17. P. Costa, A. Amaro, A. Botelho, J. Inácio, and P. V. Baptista, "Gold nanoprobe assay for the identification of mycobacteria of the Mycobacterium tuberculosis complex," Clinical Microbiology and Infection, vol. 16, no. 9, pp. 1464–1469, 2010.

18. G. Doria, R. Franco, and P. Baptista, "Nanodiagnostics: fast colorimetric method for single nucleotide polymorphism/mutation detection," IET Nanobiotechnology, vol. 1, no. 4, pp. 53–57, 2007.

19. R. Elghanian, J. J. Storhoff, R. C. Mucic, R. L. Letsinger, and C. A. Mirkin, "Selective colorimetric detection of polynucleotides based on

the distance-dependent optical properties of gold nanoparticles," Science, vol. 277, no. 5329, pp. 1078–1081, 1997.

20. H. Li and L. Rothberg, "Colorimetric detection of DNA sequences based on electrostatic interactions with unmodified gold nanoparticles," Proceedings of the National Academy of Sciences of the United States of America, vol. 101, no. 39, pp. 14036–14039, 2004.

21. C. A. Mirkin, R. L. Letsinger, R. C. Mucic, and J. J. Storhoff, "A DNA-based method for rationally assembling nanoparticles into macroscopic materials," Nature, vol. 382, no. 6592, pp. 607–609, 1996.

22. W. J. Qin and L. Y. L. Yung, "Nanoparticle-based detection and quantification of DNA with single nucleotide polymorphism (SNP) discrimination selectivity," Nucleic Acids Research, vol. 35, no. 17, article no. e111, 2007.

23. K. Sato, K. Hosokawa, and M. Maeda, "Rapid aggregation of gold nanoparticles induced by non-cross-linking DNA hybridization," Journal of the American Chemical Society, vol. 125, no. 27, pp. 8102–8103, 2003.

24. K. Sato, K. Hosokawa, and M. Maeda, "Non-cross-linking gold nanoparticle aggregation as a detection method for single-base substitutions," Nucleic Acids Research, vol. 33, no. 1, article e4, 2005.

25. J. J. Storhoff, A. D. Lucas, V. Garimella, Y. P. Bao, and U. R. Müller, "Homogeneous detection of unamplified genomic DNA sequences based on colorimetric scatter of gold nanoparticle probes," Nature Biotechnology, vol. 22, no. 7, pp. 883–887, 2004.

26. T. A. Taton, C. A. Mirkin, and R. L. Letsinger, "Scanometric DNA array detection with nanoparticle probes," Science, vol. 289, no. 5485, pp. 1757–1760, 2000.

27. C. S. Thaxton, D. G. Georganopoulou, and C. A. Mirkin, "Gold nanoparticle probes for the detection of nucleic acid targets," Clinica Chimica Acta, vol. 363, no. 1-2, pp. 120–126, 2006.

28. C. C. You, O. R. Miranda, B. Gider et al., "Detection and identification of proteins using nanoparticle-fluorescent polymer 'chemical nose' sensors," Nature Nanotechnology, vol. 2, no. 5, pp. 318–323, 2007.

29. I. H. El-Sayed, X. Huang, and M. A. El-Sayed, "Surface plasmon resonance scattering and absorption of anti-EGFR antibody conjugated gold nanoparticles in cancer diagnostics: applications in oral cancer," Nano Letters, vol. 5, no. 5, pp. 829–834, 2005.

30. S. Kumar, N. Harrison, R. Richards-Kortum, and K. Sokolov, "Plasmonic nanosensors for imaging intracellular biomarkers in live cells," Nano Letters, vol. 7, no. 5, pp. 1338–1343, 2007.

31. W.-C. Law, K.-T. Yong, A. Baev, and P. N. Prasad, "Sensitivity improved surface plasmon resonance biosensor for cancer biomarker detection based on plasmonic enhancement," ACS Nano, vol. 5, no. 6, pp. 4858–4864, 2011.

32. J. S. Mitchell and T. E. Lowe, "Ultrasensitive detection of testosterone using conjugate linker technology in a nanoparticle-enhanced surface plasmon resonance biosensor," Biosensors and Bioelectronics, vol. 24, no. 7, pp. 2177–2183, 2009.

33. J. M. Nam, C. S. Thaxton, and C. A. Mirkin, "Nanoparticle-based bio-bar codes for the ultrasensitive detection of proteins," Science, vol. 301, no. 5641, pp. 1884–1886, 2003.

34. S. Schultz, D. R. Smith, J. J. Mock, and D. A. Schultz, "Single-target molecule detection with nonbleaching multicolor optical immunolabels," Proceedings of the National Academy of Sciences of the United States of America, vol. 97, no. 3, pp. 996–1001, 2000.

35. A. J. Haes, L. Chang, W. L. Klein, and R. P. Van Duyne, "Detection of a biomarker for Alzheimer's disease from synthetic and clinical samples using a nanoscale optical biosensor," Journal of the American Chemical Society, vol. 127, no. 7, pp. 2264–2271, 2005.

36. A. D. McFarland and R. P. Van Duyne, "Single silver nanoparticles as real-time optical sensors with zeptomole sensitivity," Nano Letters, vol. 3, no. 8, pp. 1057–1062, 2003.

37. G. Raschke, S. Kowarik, T. Franzl et al., "Biomolecular recognition based on single gold nanoparticle light scattering," Nano Letters, vol. 3, no. 7, pp. 935–938, 2003.

38. M. K. Hossain and Y. Ozaki, "Surface-enhanced Raman scattering: facts and inline trends," Current Science, vol. 97, no. 2, pp. 192–201, 2009.

39. D. L. Jeanmaire and R. P. Van Duyne, "Surface Raman spectroelectrochemistry Part I. Heterocyclic, aromatic, and aliphatic amines adsorbed on the anodized silver electrode," Journal of Electroanalytical Chemistry, vol. 84, no. 1, pp. 1–20, 1977.

40. M. G. Albrecht and J. A. Creighton, "Anomalously intense Raman spectra of pyridine at a silver electrode," Journal of the American Chemical Society, vol. 99, no. 15, pp. 5215–5217, 1977.

41. B. Pettinger, "Light scattering by adsorbates at Ag particles: quantum-

mechanical approach for energy transfer induced interfacial optical processes involving surface plasmons, multipoles, and electron-hole pairs," The Journal of Chemical Physics, vol. 85, no. 12, pp. 7442–7451, 1986.

42. A. M. Michaels, M. Nirmal, and L. E. Brus, "Surface enhanced Raman spectroscopy of individual rhodamine 6G molecules on large Ag nanocrystals," Journal of the American Chemical Society, vol. 121, no. 43, pp. 9932–9939, 1999.

43. P. Etchegoin, H. Liem, R. C. Maher et al., "A novel amplification mechanism for surface enhanced Raman scattering," Chemical Physics Letters, vol. 366, no. 1-2, pp. 115–121, 2002.

44. A. Otto, "On the electronic contribution to single molecule surface enhanced Raman spectroscopy,"Indian Journal of Physics, vol. 77B, pp. 63–73, 2003.

45. S. Shanmukh, L. Jones, J. Driskell, Y. Zhao, R. Dluhy, and R. A. Tripp, "Rapid and sensitive detection of respiratory virus molecular signatures using a silver nanorod array SERS substrate," Nano Letters, vol. 6, no. 11, pp. 2630–2636, 2006.

46. S. C. Pînzaru, L. M. Andronie, I. Domsa, O. Cozar, and S. Astilean, "Bridging biomolecules with nanoparticles: surface-enhanced Raman scattering from colon carcinoma and normal tissue," Journal of Raman Spectroscopy, vol. 39, no. 3, pp. 331–334, 2008.

47. S. P. Mulvaney, M. D. Musick, C. D. Keating, and M. J. Natan, "Glass-coated, analyte-tagged nanoparticles: a new tagging system based on detection with surface-enhanced Raman scattering,"Langmuir, vol. 19, no. 11, pp. 4784–4790, 2003.

48. W. E. Doering, M. E. Piotti, M. J. Natan, and R. G. Freeman, "SERS as a foundation for nanoscale, optically detected biological labels," Advanced Materials, vol. 19, no. 20, pp. 3100–3108, 2007.

49. L. Sun, C. Yu, and J. Irudayaraj, "Surface-enhanced Raman scattering based nonfluorescent probe for multiplex DNA detection," Analytical Chemistry, vol. 79, no. 11, pp. 3981–3988, 2007.

50. X. Qian, X. Zhou, and S. Nie, "Surface-enhanced raman nanoparticle beacons based on bioconjugated gold nanocrystals and long range plasmonic coupling," Journal of the American Chemical Society, vol. 130, no. 45, pp. 14934–14935, 2008.

51. K. K. Strelau, A. Brinker, C. Schnee, K. Weber, R. Möller, and J. Popp, "Detection of PCR products amplified from DNA of epizootic pathogens using magnetic nanoparticles and SERS," Journal of Raman

Spectroscopy, vol. 42, no. 3, pp. 243–250, 2011.

52. M. B. Wabuyele, F. Yan, and T. Vo-Dinh, "Plasmonics nanoprobes: detection of single-nucleotide polymorphisms in the breast cancer BRCA1 gene," Analytical and Bioanalytical Chemistry, vol. 398, no. 2, pp. 729–736, 2010.

53. J. Neng, M. H. Harpster, H. Zhang, J. O. Mecham, W. C. Wilson, and P. A. Johnson, "A versatile SERS-based immunoassay for immunoglobulin detection using antigen-coated gold nanoparticles and malachite green-conjugated protein A/G," Biosensors and Bioelectronics, vol. 26, no. 3, pp. 1009–1015, 2010.

54. Y. C. Cao, R. Jin, and C. A. Mirkin, "Nanoparticles with Raman spectroscopic fingerprints for DNA and RNA detection," Science, vol. 297, no. 5586, pp. 1536–1540, 2002.

55. Y. C. Cao, R. Jin, J. M. Nam, C. S. Thaxton, and C. A. Mirkin, "Raman dye-labeled nanoparticle probes for proteins," Journal of the American Chemical Society, vol. 125, no. 48, pp. 14676–14677, 2003.

56. J. D. Driskell and R. A. Tripp, "Label-free SERS detection of microRNA based on affinity for an unmodified silver nanorod array substrate," Chemical Communications, vol. 46, no. 19, pp. 3298–3300, 2010.

57. A. Barhoumi, D. Zhang, F. Tam, and N. J. Halas, "Surface-enhanced raman spectroscopy of DNA," Journal of the American Chemical Society, vol. 130, no. 16, pp. 5523–5529, 2008.

58. D. Graham, R. Stevenson, D. G. Thompson, L. Barrett, C. Dalton, and K. Faulds, "Combining functionalised nanoparticles and SERS for the detection of DNA relating to disease," Faraday Discussions, vol. 149, pp. 291–299, 2011.

59. C. V. Pagba, S. M. Lane, H. Cho, and S. Wachsmann-Hogiu, "Direct detection of aptamer-thrombin binding via surface-enhanced Raman spectroscopy," Journal of Biomedical Optics, vol. 15, no. 4, Article ID 047006, 2010.

60. B. Guven, N. Basaran-Akgul, E. Temur, U. Tamer, and I. H. Boyaci, "SERS-based sandwich immunoassay using antibody coated magnetic nanoparticles for Escherichia coli enumeration," Analyst, vol. 136, no. 4, pp. 740–748, 2011.

61. W. Cai, D. W. Shin, K. Chen et al., "Peptide-labeled near-infrared quantum dots for imaging tumor vasculature in living subjects," Nano Letters, vol. 6, no. 4, pp. 669–676, 2006.

62. J. Kneipp, H. Kneipp, M. McLaughlin, D. Brown, and K. Kneipp, "In

vivo molecular probing of cellular compartments with gold nanoparticles and nanoaggregates," Nano Letters, vol. 6, no. 10, pp. 2225–2231, 2006.

63. W. C. W. Chan and S. Nie, "Quantum dot bioconjugates for ultrasensitive nonisotopic detection," Science, vol. 281, no. 5385, pp. 2016–2018, 1998.

64. M. Bruchez Jr., M. Moronne, P. Gin, S. Weiss, and A. P. Alivisatos, "Semiconductor nanocrystals as fluorescent biological labels," Science, vol. 281, no. 5385, pp. 2013–2016, 1998.

65. H. Weller, "Colloidal semiconductor Q-particles: chemistry in the transition region between solid state and molecules," Angewandte Chemie - International Edition, vol. 32, no. 1, pp. 41–53, 1993.

66. D. Gerion, F. Chen, B. Kannan et al., "Room-temperature single-nucleotide polymorphism and multiallele DNA detection using fluorescent nanocrystals and microarrays," Analytical Chemistry, vol. 75, no. 18, pp. 4766–4772, 2003.

67. S. Pathak, S. K. Choi, N. Arnheim, and M. E. Thompson, "Hydroxylated quantum dots as luminescent probes for in situ hybridization," Journal of the American Chemical Society, vol. 123, no. 17, pp. 4103–4104, 2001.

68. I. L. Medintz, A. R. Clapp, H. Mattoussi, E. R. Goldman, B. Fisher, and J. M. Mauro, "Self-assembled nanoscale biosensors based on quantum dot FRET donors," Nature Materials, vol. 2, no. 9, pp. 630–638, 2003.

69. L. M. Devi and D. P. S. Negi, "Sensitive and selective detection of adenine using fluorescent ZnS nanoparticles," Nanotechnology, vol. 22, no. 24, Article ID 245502, 2011.

70. J. Gersten and A. Nitzan, "Spectroscopic properties of molecules interacting with small dielectric particles," The Journal of Chemical Physics, vol. 75, no. 3, pp. 1139–1152, 1981.

71. K. A. Kang, J. Wang, J. B. Jasinski, and S. Achilefu, "Fluorescence manipulation by gold nanoparticles: from complete quenching to extensive enhancement," Journal of Nanobiotechnology, vol. 9, article 16, 2011.

72. A. Quarta, R. D. Corato, L. Manna, A. Ragusa, and T. Pellegrino, "Fluorescent-magnetic hybrid nanostructures: preparation, properties, and applications in biology," IEEE Transactions on Nanobioscience, vol. 6, no. 4, pp. 298–308, 2007.

73. J. R. Lakowicz, J. Malicka, E. Matveeva, I. Gryczynski, and Z. Gryczynski, "Plasmonic technology: novel approach to ultrasensitive immunoassays," Clinical Chemistry, vol. 51, no. 10, pp. 1914–1922, 2005.

74. J. R. Lakowicz, "Plasmonics in biology and plasmon-controlled fluorescence," Plasmonics, vol. 1, no. 1, pp. 5–33, 2006.

75. I. L. Medintz and H. Mattoussi, "Quantum dot-based resonance energy transfer and its growing application in biology," Physical Chemistry Chemical Physics, vol. 11, no. 1, pp. 17–45, 2009.

76. P. C. Ray, G. K. Darbha, A. Ray, J. Walker, and W. Hardy, "Gold nanoparticle based FRET for DNA detection," Plasmonics, vol. 2, no. 4, pp. 173–183, 2007.

77. Y. Liu, Y. Wang, J. Jin, H. Wang, R. Yang, and W. Tan, "Fluorescent assay of DNA hybridization with label-free molecular switch: reducing background-signal and improving specificity by using carbon nanotubes," Chemical Communications, no. 6, pp. 665–667, 2009.

78. W. Bai, H. Zheng, Y. Long, X. Mao, M. Gao, and L. Zhang, "A carbon dots-based fluorescence turn-on method for DNA determination," Analytical Sciences, vol. 27, no. 3, pp. 243–246, 2011.

79. B. Tang, N. Zhang, Z. Chen et al., "Probing hydroxyl radicals and their imaging in living cells by use of FAM-DNA-Au nanoparticles," Chemistry, vol. 14, no. 2, pp. 522–528, 2008.

80. Z. S. Wu, J. H. Jiang, L. Fu, G. L. Shen, and R. Q. Yu, "Optical detection of DNA hybridization based on fluorescence quenching of tagged oligonucleotide probes by gold nanoparticles," Analytical Biochemistry, vol. 353, no. 1, pp. 22–29, 2006.

81. S. H. De Paoli Lacerda, J. J. Park, C. Meuse et al., "Interaction of gold nanoparticles with common human blood proteins," ACS Nano, vol. 4, no. 1, pp. 365–379, 2010.

82. S. Mayilo, M. A. Kloster, M. Wunderlich et al., "Long-range fluorescence quenching by gold nanoparticles in a sandwich immunoassay for cardiac troponin T," Nano Letters, vol. 9, no. 12, pp. 4558–4563, 2009.

83. X. He, J. Gao, S. S. Gambhir, and Z. Cheng, "Near-infrared fluorescent nanoprobes for cancer molecular imaging: status and challenges," Trends in Molecular Medicine, vol. 16, no. 12, pp. 574–583, 2010.

84. Y. T. Su, G. Y. Lan, W. Y. Chen, and H. T. Chang, "Detection of copper ions through recovery of the fluorescence of DNA-templated copper/silver nanoclusters in the presence of mercaptopropionic acid," Analytical Chemistry, vol. 82, no. 20, pp. 8566–8572, 2010.

85. Y. Cho, S. S. Lee, and J. H. Jung, "Recyclable fluorimetric and colorimetric mercury-specific sensor using porphyrin-functionalized Au@SiO$_2$ core/shell nanoparticles," Analyst, vol. 135, no. 7, pp. 1551–1555, 2010.

86. Y. Xiang, A. Tong, and Y. Lu, "Abasic site-containing DNAzyme and aptamer for label-free fluorescent detection of Pb^{2+} and adenosine with high sensitivity, selectivity, and tunable dynamic range," Journal of the American Chemical Society, vol. 131, no. 42, pp. 15352–15357, 2009.

87. R. Gill, M. Zayats, and I. Willner, "Semiconductor quantum dots for bioanalysis," Angewandte Chemie - International Edition, vol. 47, no. 40, pp. 7602–7625, 2008.

88. J. H. Lee, Y. M. Huh, Y. W. Jun et al., "Artificially engineered magnetic nanoparticles for ultra-sensitive molecular imaging," Nature Medicine, vol. 13, no. 1, pp. 95–99, 2007.

89. E. I. Altino☐lu and J. H. Adair, "Near infrared imaging with nanoparticles," Wiley Interdisciplinary Reviews: Nanomedicine and Nanobiotechnology, vol. 2, no. 5, pp. 461–477, 2010.

90. P. Vartholomeos, M. Fruchard, A. Ferreira, and C. Mavroidis, "MRI-guided nanorobotic systems for therapeutic and diagnostic applications," Annual Review of Biomedical Engineering, vol. 13, pp. 157–184, 2011.

91. E. V. Zagaynova, M. V. Shirmanova, M. Y. Kirillin et al., "Contrasting properties of gold nanoparticles for optical coherence tomography: phantom, in vivo studies and Monte Carlo simulation," Physics in Medicine and Biology, vol. 53, no. 18, pp. 4995–5009, 2008.

92. J. C. Kah, M. Olivo, T. H. Chow et al., "Control of optical contrast using gold nanoshells for optical coherence tomography imaging of mouse xenograft tumor model in vivo," Journal of Biomedical Optics, vol. 14, no. 5, Article ID 054015, 2009.

93. A. L. Oldenburg, M. N. Hansen, T. S. Ralston, A. Wei, and S. A. Boppart, "Imaging gold nanorods in excised human breast carcinoma by spectroscopic optical coherence tomography," Journal of Materials Chemistry, vol. 19, no. 35, pp. 6407–6411, 2009.

94. X. Yang, E. W. Stein, S. Ashkenazi, and L. V. Wang, "Nanoparticles for photoacoustic imaging," Wiley Interdisciplinary Reviews. Nanomedicine and nanobiotechnology, vol. 1, no. 4, pp. 360–368, 2009.

95. M. B. Mohamed, V. Volkov, S. Link, and M. A. El-Sayed, "The 'lightning' gold nanorods: fluorescence enhancement of over a million compared to the gold metal," Chemical Physics Letters, vol. 317, no. 6, pp. 517–523, 2000.

96. R. Damadian, "Tumor detection by nuclear magnetic resonance," Science, vol. 171, no. 3976, pp. 1151–1153, 1971.

97. D. S. Mathew and R. S. Juang, "An overview of the structure and

magnetism of spinel ferrite nanoparticles and their synthesis in microemulsions," Chemical Engineering Journal, vol. 129, no. 1–3, pp. 51–65, 2007.

98. S. Laurent, D. Forge, M. Port et al., "Magnetic iron oxide nanoparticles: synthesis, stabilization, vectorization, physicochemical characterizations and biological applications," Chemical Reviews, vol. 108, no. 6, pp. 2064–2110, 2008.

99. Y. W. Jun, Y. M. Huh, J. S. Choi et al., "Nanoscale size effect of magnetic nanocrystals and their utilisation for cancer diagnosis via magnetic resonance imaging," Journal of the American Chemical Society, vol. 127, no. 16, pp. 5732–5733, 2005.

100. R. T. Branca, Z. I. Cleveland, B. Fubara et al., "Molecular MRI for sensitive and specific detection of lung metastases," Proceedings of the National Academy of Sciences of the United States of America, vol. 107, no. 8, pp. 3693–3697, 2010.

101. D. Kim, M. K. Yu, T. S. Lee, J. J. Park, Y. Y. Jeong, and S. Jon, "Amphiphilic polymer-coated hybrid nanoparticles as CT/MRI dual contrast agents," Nanotechnology, vol. 22, no. 15, Article ID 155101, 2011.

102. D. Huang, E. A. Swanson, C. P. Lin et al., "Optical coherence tomography," Science, vol. 254, no. 5035, pp. 1178–1181, 1991.

103. J. M. Schmitt, "Optical Coherence Tomography (OCT): a review," IEEE Journal on Selected Topics in Quantum Electronics, vol. 5, no. 4, pp. 1205–1215, 1999.

104. H. G. Bezerra, M. A. Costa, G. Guagliumi, A. M. Rollins, and D. I. Simon, "Intracoronary optical coherence tomography: a comprehensive review. Clinical and research applications," JACC: Cardiovascular Interventions, vol. 2, no. 11, pp. 1035–1046, 2009.

105. A. M. Gobin, M. H. Lee, N. J. Halas, W. D. James, R. A. Drezek, and J. L. West, "Near-infrared resonant nanoshells for combined optical imaging and photothermal cancer therapy," Nano Letters, vol. 7, no. 7, pp. 1929–1934, 2007.

106. H. Y. Tseng, C. K. Lee, S. Y. Wu et al., "Au nanorings for enhancing absorption and backscattering monitored with optical coherence tomography," Nanotechnology, vol. 21, no. 29, Article ID 295102, 2010.

107. A. S. Paranjape, R. Kuranov, S. Baranov, et al., "Depth resolved photothermal OCT detection of macrophages in tissue using nanorose," Biomedical Optics Express, vol. 1, no. 1, pp. 2–16, 2010.

108. Y. Su, F. Zhang, K. Xu, J. Yao, and R. K. Wang, "A photoacoustic

tomography system for imaging of biological tissues," Journal of Physics D, vol. 38, no. 15, pp. 2640–2644, 2005.

109. K. S. Valluru, B. K. Chinni, and N. A. Rao, "Photoacoustic imaging: opening new frontiers in medical imaging," Journal of Clinical Imaging Science, vol. 1, article 24, 2011.

110. X. Yang, S. E. Skrabalak, Z. Y. Li, Y. Xia, and L. V. Wang, "Photoacoustic tomography of a rat cerebral cortex in vivo with Au nanocages as an optical contrast agent," Nano Letters, vol. 7, no. 12, pp. 3798–3802, 2007.

111. M. Eghtedari, A. Oraevsky, J. A. Copland, N. A. Kotov, A. Conjusteau, and M. Motamedi, "High sensitivity of in vivo detection of gold nanorods using a laser optoacoustic imaging system," Nano Letters, vol. 7, no. 7, pp. 1914–1918, 2007.

112. K. H. Song, C. Kim, K. Maslov, and L. V. Wang, "Noninvasive in vivo spectroscopic nanorod-contrast photoacoustic mapping of sentinel lymph nodes," European Journal of Radiology, vol. 70, no. 2, pp. 227–231, 2009.

113. M. B. Mohamed, V. Volkov, S. Link, and M. A. El-Sayed, "The 'lightning' gold nanorods: fluorescence enhancement of over a million compared to the gold metal," Chemical Physics Letters, vol. 317, no. 6, pp. 517–523, 2000.

114. H. Wang, T. B. Huff, D. A. Zweifel et al., "In vitro and in vivo two-photon luminescence imaging of single gold nanorods," Proceedings of the National Academy of Sciences of the United States of America, vol. 102, no. 44, pp. 15752–15756, 2005.

115. G. T. Boyd, Z. H. Yu, and Y. R. Shen, "Photoinduced luminescence from the noble metals and its enhancement on roughened surfaces," Physical Review B, vol. 33, no. 12, pp. 7923–7936, 1986.

116. N. J. Durr, T. Larson, D. K. Smith, B. A. Korgel, K. Sokolov, and A. Ben-Yakar, "Two-photon luminescence imaging of cancer cells using molecularly targeted gold nanorods," Nano Letters, vol. 7, no. 4, pp. 941–945, 2007.

117. K. Imura, T. Nagahara, and H. Okamoto, "Plasmon mode imaging of single gold nanorods," Journal of the American Chemical Society, vol. 126, no. 40, pp. 12730–12731, 2004.

118. K. T. Yong, I. Roy, H. Ding, E. J. Bergey, and P. N. Prasad, "Biocompatible near-infrared quantum dots as ultrasensitive probes for long-term in vivo imaging applications," Small, vol. 5, no. 17, pp. 1997–2004, 2009.

119. A. M. Smith, H. Duan, A. M. Mohs, and S. Nie, "Bioconjugated quantum dots for in vivo molecular and cellular imaging," Advanced Drug Delivery

Reviews, vol. 60, no. 11, pp. 1226–1240, 2008.

120. E. G. Soltesz, S. Kim, R. G. Laurence et al., "Intraoperative sentinel lymph node mapping of the lung using near-infrared fluorescent quantum dots," Annals of Thoracic Surgery, vol. 79, no. 1, pp. 269–277, 2005.

121. C. P. Parungo, Y. L. Colson, S. W. Kim et al., "Sentinel lymph node mapping of the pleural space," Chest, vol. 127, no. 5, pp. 1799–1804, 2005.

122. J. P. Zimmer, S. W. Kim, S. Ohnishi, E. Tanaka, J. V. Frangioni, and M. G. Bawendi, "Size series of small indium arsenide-zinc selenide core-shell nanocrystals and their application to in vivo imaging," Journal of the American Chemical Society, vol. 128, no. 8, pp. 2526–2527, 2006.

123. H. Kobayashi, Y. Hama, Y. Koyama et al., "Simultaneous multicolor imaging of five different lymphatic basins using quantum dots," Nano Letters, vol. 7, no. 6, pp. 1711–1716, 2007.

124. Y. Hama, Y. Koyama, Y. Urano, P. L. Choyke, and H. Kobayashi, "Simultaneous two-color spectral fluorescence lymphangiography with near infrared quantum dots to map two lymphatic flows from the breast and the upper extremity," Breast Cancer Research and Treatment, vol. 103, no. 1, pp. 23–28, 2007.

125. S. Kim, Y. T. Lim, E. G. Soltesz et al., "Near-infrared fluorescent type II quantum dots for sentinel lymph node mapping," Nature Biotechnology, vol. 22, no. 1, pp. 93–97, 2004.

126. Y. T. Lim, S. Kim, A. Nakayama, N. E. Stott, M. G. Bawendi, and J. V. Frangioni, "Selection of quantum dot wavelengths for biomedical assays and imaging," Molecular Imaging, vol. 2, no. 1, pp. 50–64, 2003.

127. J. D. Smith, G. W. Fisher, A. S. Waggoner, and P. G. Campbell, "The use of quantum dots for analysis of chick CAM vasculature," Microvascular Research, vol. 73, no. 2, pp. 75–83, 2007.

128. D. R. Larson, W. R. Zipfel, R. M. Williams et al., "Water-soluble quantum dots for multiphoton fluorescence imaging in vivo," Science, vol. 300, no. 5624, pp. 1434–1436, 2003.

129. X. Yu, L. Chen, K. Li et al., "Immunofluorescence detection with quantum dot bioconjugates for hepatoma in vivo," Journal of Biomedical Optics, vol. 12, no. 1, Article ID 014008, 2007.

130. H. Tada, H. Higuchi, T. M. Wanatabe, and N. Ohuchi, "In vivo real-time tracking of single quantum dots conjugated with monoclonal anti-HER2 antibody in tumors of mice," Cancer Research, vol. 67, no. 3, pp. 1138–1144, 2007.

131. M. E. Åkerman, W. C. W. Chan, P. Laakkonen, S. N. Bhatia, and E. Ruoslahti, "Nanocrystal targeting in vivo," Proceedings of the National Academy of Sciences of the United States of America, vol. 99, no. 20, pp. 12617–12621, 2002.

132. X. Gao, Y. Cui, R. M. Levenson, L. W. K. Chung, and S. Nie, "In vivo cancer targeting and imaging with semiconductor quantum dots," Nature Biotechnology, vol. 22, no. 8, pp. 969–976, 2004.

133. X. Huang, P. K. Jain, I. H. El-Sayed, and M. A. El-Sayed, "Determination of the minimum temperature required for selective photothermal destruction of cancer cells with the use of immunotargeted gold nanoparticles," Photochemistry and Photobiology, vol. 82, no. 2, pp. 412–417, 2006.

134. J. Chen, D. Wang, J. Xi et al., "Immuno gold nanocages with tailored optical properties for targeted photothermal destruction of cancer cells," Nano Letters, vol. 7, no. 5, pp. 1318–1322, 2007.

135. Y. Haba, C. Kojima, A. Harada, T. Ura, H. Horinaka, and K. Kono, "Preparation of poly(ethylene glycol)-modified poly(amido amine) dendrimers encapsulating gold nanoparticles and their heat-generating ability," Langmuir, vol. 23, no. 10, pp. 5243–5246, 2007.

136. D. K. Kirui, D. A. Rey, and C. A. Batt, "Gold hybrid nanoparticles for targeted phototherapy and cancer imaging," Nanotechnology, vol. 21, no. 10, Article ID 105105, 2010.

137. X. Huang, I. H. El-Sayed, W. Qian, and M. A. El-Sayed, "Cancer cell imaging and photothermal therapy in the near-infrared region by using gold nanorods," Journal of the American Chemical Society, vol. 128, no. 6, pp. 2115–2120, 2006.

138. X. Huang, I. H. El-Sayed, and M. A. El-Sayed, "Applications of gold nanorods for cancer imaging and photothermal therapy," Methods in Molecular Biology, vol. 624, pp. 343–357, 2010.

139. W. S. Kuo, C. N. Chang, Y. T. Chang et al., "Gold nanorods in photodynamic therapy, as hyperthermia agents, and in near-infrared optical imaging," Angewandte Chemie - International Edition, vol. 49, no. 15, pp. 2711–2715, 2010.

140. B. Van De Broek, N. Devoogdt, A. Dhollander et al., "Specific cell targeting with nanobody conjugated branched gold nanoparticles for photothermal therapy," ACS Nano, vol. 5, no. 6, pp. 4319–4328, 2011.

141. L. R. Hirsch, R. J. Stafford, J. A. Bankson et al., "Nanoshell-mediated near-infrared thermal therapy of tumors under magnetic resonance

guidance," Proceedings of the National Academy of Sciences of the United States of America, vol. 100, no. 23, pp. 13549–13554, 2003.

142. C. Loo, L. Hirsch, M. H. Lee et al., "Gold nanoshell bioconjugates for molecular imaging in living cells,"Optics Letters, vol. 30, no. 9, pp. 1012–1014, 2005.

143. C. Loo, A. Lowery, N. Halas, J. West, and R. Drezek, "Immunotargeted nanoshells for integrated cancer imaging and therapy," Nano Letters, vol. 5, no. 4, pp. 709–711, 2005.

144. X. Huang, P. K. Jain, I. H. El-Sayed, and M. A. El-Sayed, "Plasmonic photothermal therapy (PPTT) using gold nanoparticles," Lasers in Medical Science, vol. 23, no. 3, pp. 217–228, 2008.

145. S. B. Brown and S. H. Ibbotson, "Photodynamic therapy and cancer," BMJ, vol. 339, Article ID b2459, 2009.

146. J. Cadet, "The photodynamic therapy of cancer cells," Photochemistry and photobiology, vol. 87, no. 1, p. 1, 2011.

147. A. Lin and S. M. Hahn, "Photodynamic therapy: a light in the darkness?" Clinical Cancer Research, vol. 15, no. 13, pp. 4252–4253, 2009.

148. M. A. MacCormack, "Photodynamic therapy," Advances in Dermatology, vol. 22, pp. 219–258, 2006.

149. P. Zhang, W. Steelant, M. Kumar, and M. Scholfield, "Versatile photosensitizers for photodynamic therapy at infrared excitation," Journal of the American Chemical Society, vol. 129, no. 15, pp. 4526–4527, 2007.

150. B. Ungun, R. K. Prud›homme, S. J. Budijono et al., "Nanofabricated upconversion nanoparticles for photodynamic therapy," Optics Express, vol. 17, no. 1, pp. 80–86, 2009.

151. D. K. Chatterjee and Z. Yong, "Upconverting nanoparticles as nanotransducers for photodynamic therapy in cancer cells," Nanomedicine, vol. 3, no. 1, pp. 73–82, 2008.

152. H. S. Qian, H. C. Guo, P. C. L. Ho, R. Mahendran, and Y. Zhang, "Mesoporous-silica-coated up-conversion fluorescent nanoparticles for photodynamic therapy," Small, vol. 5, no. 20, pp. 2285–2290, 2009.

Chapter 8

NEURAL STEM CELLS: READY FOR THERAPEUTIC APPLICATIONS?

Simona Casarosa, Yuri Bozzi, and Luciano Conti

Center for Integrative Biology, Università degli Studi di Trento

ABSTRACT

Neural stem cells (NSCs) offer a unique and powerful tool for basic research and regenerative medicine. However, the challenges that scientists face in the comprehension of the biology and physiological function of these cells are still many. Deciphering NSCs fundamental biological aspects represents indeed a crucial step to control NSCs fate and functional integration following transplantation, and is essential for a safe and appropriate use of NSCs in injury/disease conditions. In this review, we focus on the biological properties of NSCs and discuss how these cells may be exploited to provide effective therapies for neurological disorders. We also review and discuss ongoing NSC-based clinical trials for these diseases.

REVIEW

Introduction

Conventional pharmacological treatments for most neurodegenerative conditions relieve some symptoms but rarely vary the course of the disease or halt its progression. Grafting of human fetal tissue has provided a proof of concept for cell therapy approaches to neurodegenerative diseases in a number of clinical studies, including treatment of Parkinson's and Huntington's disease patients [1]. Nonetheless, this does not represent a practical route for large-scale therapeutic applications due to limited availability and quality of human fetal tissue, as well as for ethical considerations.

To this regard, in the last years, great media consideration has brought neural stem cell (NSC) research into the spotlight. Most of this attention has been raised by the stimulating prospects of NSCs application for cell replacement therapies for neurological disorders, engendering hopes and expectations in the public and, particularly, in patients. Despite the evident benefits pledged by the NSC field and some encouraging preliminary studies in animal models, there still remains a gap between theory and practice. Indeed, while stem cell-based therapies are the current standard of care for blood tumors and are gaining agreement in the treatment of epidermal and corneal disorders, applications for diseases affecting the nervous system yet represent a pioneering field, being in the early phases of clinical scrutiny. What is still missing to effectively translate NSC research into clinical applications? Although important scientific progresses in the field have been achieved, we still lack a profound understanding of the basic biology of NSCs and how to manipulate these cells to provide reliable, safe and effective outcomes in cell-replacement approaches.

NSCs are immature cells present in the developing and adult Central Nervous System (CNS). Typically, NSCs are defined by three cardinal characteristics: self-renewal potential, neural tripotency (i.e., the capability to give rise to all of the major neural lineages: neurons, astrocytes and oligodendrocytes) and competence for *in vivo* regeneration (Figure 1; [2]). They have the potential to generate both neurons and glia of the developing brain and they also account for the limited regenerative potential in the adult brain. In the adult CNS, NSCs reside in defined regions ("neurogenic niches") that sustain their multipotency and regulate the balance between symmetrical self-renewal and fate-committed asymmetric divisions [3].

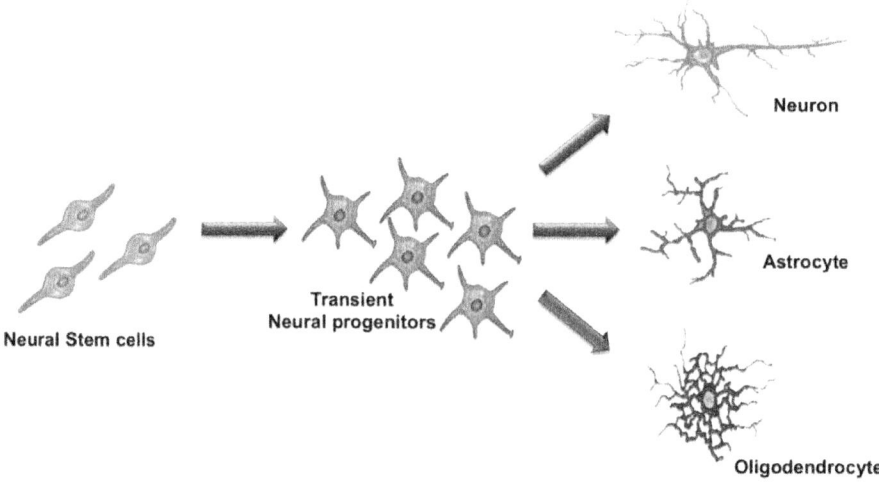

Figure 1. Cardinal neural stem cell properties.

Several studies have shown that NSCs can be extracted from neural tissue or generated from pluripotent cellular sources, genetically manipulated and differentiated *in vitro*[2]. In the past two decades, a variety of protocols for NSCs purification, generation and expansion in floating or adherent conditions have been described. However, beside these important progresses, the identification of the best sources for NSCs derivation and the optimization of approaches to stably proliferate them clonally *in vitro* still represents a major goal of NSC research.

Yet, for a realistic exploitation of NSCs for cell therapies, clinically-suitable NSC systems should hold specific key properties including (i) standardized production and scalability to good medical practice (GMP), (ii) karyotypic stability, (iii) ability to correctly integrate in the host tissue and (iv) differentiate into the required functional neural cells. In addition, NSCs should exhibit a reproducible, predictable and safe behavior following *in vivo* injection.

NSCs in Brain Organogenesis and Homeostasis

In the developing and adult CNS, different NSC populations dynamically appear following predetermined spatio-temporal developmental programs. Molecular and biological characteristics of NSCs greatly vary depending on the region and developmental stage considered [4].

Development of the vertebrate CNS starts with neural plate folding to originate the neural tube, consisting of radially elongated neuroepithelial cells (NEPs) [5]. NEPs develop definite identities and different fates depending on their positions along the rostrocaudal (R-C) and dorsoventral (D-V) axes of the neural tube. Patterning along the R-C axis leads to the initial distinction into prosencephalon, mesencephalon, rhombencephalon and spinal cord territories. NEPs are accountable for the first wave of neurogenesis in the neural tube. As development proceeds, NEPs convert themselves into another transitory NSC type, the so-called "radial glia" (RG) [6, 7]. This rapidly constitutes the main progenitor cell population in mid/late development and early postnatal life while disappearing at late postnatal and adult stages. Besides their ability to divide asymmetrically and to serve as progenitors of neurons and glia, RG cells constitute a scaffold on which neurons migrate in the developing brain. RG differentiation potential is less extensive compared to that of NEPs. Along with RG, another population of immature neural cells is constituted by Basal Progenitors (BPs) [8]. They are generated at early phases of development by NEPs and at later stages by RG. BPs mostly undergo one or two rounds of division, generating one or two pairs of neurons. Hence, BPs may be considered neurogenic transit-amplifying progenitors that specifically increase the production of neurons during restricted developmental time periods in definite brain areas (i.e. cerebral cortex).

At the end of neurogenesis (roughly at birth in mice), neurogenic RG cells are exhausted and residual RG cells are converted into a unique astrocyte-like subpopulation [9]. This population will make up the NSC pool of the adult brain, endorsed with neurogenesis and gliogenesis maintainance throughout adult life.

The concept that the adult brain retains the ability to self-renew some of its neurons has been broadly recognized in the last two decades and has represented a breakthrough in neurosciences. Pioneering studies from Altman and Das already reported the generation of new neurons in a variety of structures in the adult rat and cat including the olfactory bulb, hippocampus, and cerebral cortex [10]. However, their results were widely neglected until the early 1990s, when the formation of new neurons in adult rodent brain was clearly demonstrated [11, 12]. This led to the identification of the germinal zones of the adult brain. These are specialized niches located in the subventricular zone (SVZ) of the lateral ventricle wall and in the subgranular zone (SGZ) of the dentate gyrus of the hippocampus [3]. Whether NSCs reside in other regions of the adult mammalian brain is still disputed. Neuroblasts produced in the rodent SVZ migrate to the olfactory bulb following the rostral migratory stream (RMS), an anatomic structure well characterized in the rodent brain. The NSCs located in the SVZ, also called type B cells, generate actively dividing intermediate cells, named type C cells, which further divide giving rise to neuroblasts, referred to as type A cells that migrate away from the SVZ. These migrating neuroblasts are organized in chains that connect the SVZ to the olfactory bulb (constituting the RMS) where they gradually mature into functional GABAergic granule neurons. Fate-mapping studies actually reveal that type B cells are not developmentally restricted to neuronal lineages but can give rise also to glial progenies, suggesting they are authentic tripotent NSCs. The second germinal zone of the adult mammalian brain is the dentate gyrus of the hippocampus. Astrocyte-like NSCs, called type I progenitors, have been identified within the SGZ facing the dentate gyrus hilus. They share several properties with the type B cells of the adult SVZ, although they apparently exhibit a narrower developmental potential. Type I progenitors likely divide asymmetrically to produce immature proliferating progenitors, type II cells. These gradually differentiate into migrating neuroblasts that travel into the granule cell layer of the dentate gyrus, where they progressively mature into functional granule neurons. Differently from the type B cells of the SVZ, the progeny of type I progenitors does not migrate long distances, but remains localized in clusters closely connected to the parent cell. Additionally, hippocampal NSCs appear to be developmentally restricted to become granule neurons; currently, there is no evidence that type I progenitors can generate mature glial derivatives *in vivo*.

The discovery of NSCs and neurogenesis in the adult mammalian CNS has tremendously changed our view of the plasticity and function of the brain. This has prompted excitement for the possible exploitement of intrinsic neurogenic activity to cure brain diseases and rescue brain function after injury. Mobilization of endogenous NSCs has thus emerged as a potential therapeutic approach for neural repair. It is known that brain injury promotes the proliferation of adjacent NSCs, generating new astrocytes and neurons [13]. For example, focal ischemia transiently induces forebrain SVZ cell proliferation and neurogenesis. The NSCs in the SVZ and DG are also stimulated after traumatic brain injury or seizures [14], suggesting that adult neurogenesis may play a role in self-recovery mechanisms of the brain. However, the amount of spontaneously produced neuroblasts after brain injury is highly limited, and their survival and differentiation into mature neurons are far from obtaining regenerative effects.

It should be emphasized that, although our understanding of NSCs has increased dramatically over the past few years, there are still many major gaps regarding their *in vivo* control.

NSCs for Cell Replacement Approaches: Requirements & Available *In Vitro* Systems

A large number of studies have explored grafting behavior of several NSCs typologies (and their progeny) in a variety of preclinical studies and in some clinical investigations. Nevertheless, NSCs used for clinical applications should be safe, effective and accessible in large amount in GMP conditions. A variety of different sources for NSCs have been tested, including fetal- and adult CNS-derived NSCs, neural progenitors derived from pluripotent cells, and a range of non-neural stem cells, such as mesenchymal (MSCs) and bone marrow-derived (BMDSCs) stem cells. With these issues in mind, it should be remarked that up to now an ideal NSC system is not yet available to the clinic. Here, we will restrict our discussion to NSCs derived from neural tissue and from pluripotent stem cells. Advantages and disadvantages of each source and recent experimental evidence that highlight their potential use for clinical applications will be presented.

Fetal- and ADULT-DERIVED NSCs

The isolation of NSCs from their natural niches and their expansion in culture have been challenging issues, because the requirements to maintain these cells in their physiological state are yet poorly understood. In the early '90s, the identification of EGF and FGF-2 as key mitogens for NSCs led to set up culture conditions that support extended cell division of cells with NSCs properties [11, 12]. Since then, several studies reported that NSCs can be isolated from various regions of rodent (mouse and rat) and human brain at several developmental stages as well as from germinative areas of the adult brain. A widely used method is to culture NSCs as neurospheres (Figure 2; [15]). These are free-floating aggregates of neural progenitors, each, in theory, deriving from a single NSC. Their generation relies on neural tissue micro-dissection followed by exposure to defined mitogen-supplemented media. In such a procedure, primary cells are plated in low-attachment culture flasks in serum-free media supplemented with EGF and/or FGF-2. In these conditions, differentiating or differentiated cells are supposed to die, whereas NSCs respond to mitogens, divide and form floating aggregates (primary neurospheres) that can be dissociated and re-plated to generate secondary neurospheres. This procedure can be sequentially repeated several times to expand a NSC population.

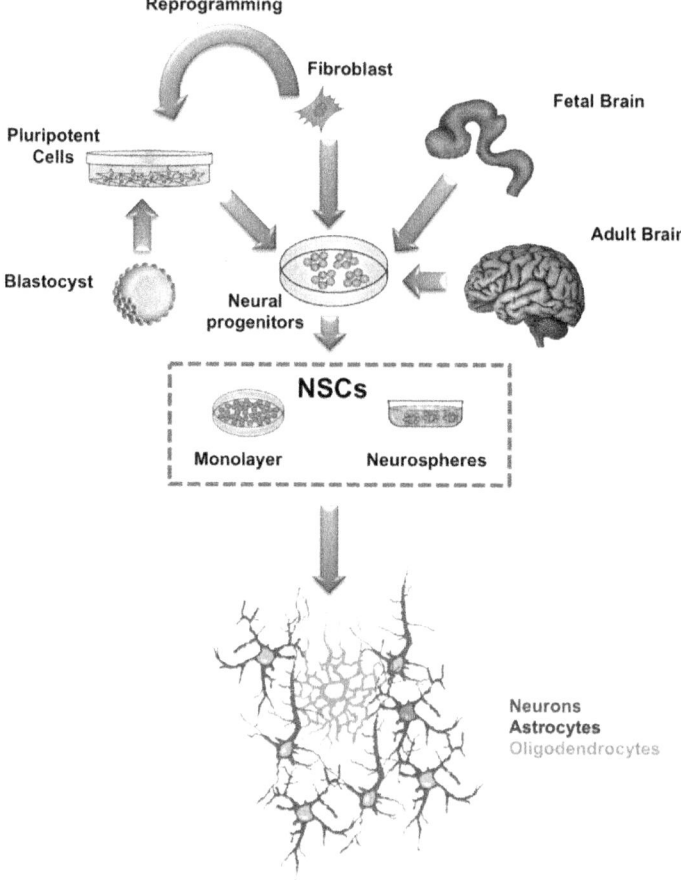

Figure 2. Sources and *in vitro* growth protocols for neural stem cell generation and expansion.

Complementary to neurosphere culture is adherent culture, in which cells are more easily monitored and have better access to growth factors (Figure 2). In the last decade, several groups reported the generation and expansion of adherent NSC lines from neural tissue of rodent and human origin. According to this procedure, NSCs can be competently expanded as adherent clonal homogeneous NSC lines by exposure to specific mitogens, such as EGF and/or FGF2 [16, 17]. In these conditions, cells divide symmetrically, retaining their tripotential differentiation capacity. Adherent culture regimens have been shown to allow for cultures with less differentiated cells compared to the neurosphere assay, where cell–cell contacts and non-uniform mitogens exposure is thought to stimulate differentiation programs [2].

Although several studies have attempted to provide comparisons between fetal- and adult-derived NSCs, systematic side-by-side analyses are still few and do not allow to draw any solid conclusions. In fact, results might be hampered by culture conditions, especially for human NSCs. Nonetheless, major differences between fetal- and adult-derived human NSCs have been reported in terms of both biological and molecular properties. Fetal-derived human NSCs generally exhibit a shorter doubling time, a more extensive expansion potential *in vitro* and better integrative potential following grafting in animal models [18–21]. Noteworthy, substantial differences have been also described when comparing human NSCs derived from different brain areas of the same fetus [22].

NSCs from Pluripotent Stem Cells

Neuralization protocols applied to mouse and human pluripotent stem cells, including embryonic stem cells (ESCs) and induced pluripotent stem cells (iPSCs), allow for the generation of NSCs populations. ESCs are derived from the inner cell mass (ICM) of blastocyst stage mammalian embryos [23, 24]. They are characterized by an intrinsic capacity for self-renewal and the ability to generate all cell types derived from the three embryonic germ layers (pluripotency). In the last years, the advent of iPSC technology has completely revolutioned the "pluripotency" field, avoiding the requirement of embryos as source of rodent and, most importantly, of human pluripotent stem cells. Moreover, the use of iPSC opens new possibilities for studies of human development and disorders, further increasing the potential biomedical applications of this type of cells [25]. iPSCs are the product of a reprogramming procedure that allows the conversion of somatic cells directly into pluripotent cells [26, 27]. This technology is straightforward, robust and since its discovery has been implemented in terms of efficiency and reproducibility. iPSCs closely resemble ESCs with respect to expression of pluripotency markers, self-renewal potential, and multilineage differentiation potential. Both murine and human pluripotent stem cells can be exposed to neuralizing *in vitro* protocols to generate a large amount of NSCs or progenitor cells. In these conditions, pluripotent stem cells undergo progressive lineage restrictions similar to those observed during normal fetal development, leading to the generation of a range of distinct neural precursor populations [2].

Generally, two main procedures to generate NSCs from pluripotent stem cells have been developed. The first strategy relies on the formation of embryoid bodies (EBs), three-dimensional (3D) aggregates. EBs recapitulate many aspects of cell differentiation occurring during early mammalian embryogenesis and give rise to cells of the three germ layers, including neural cells. EBs dissociated and plated in adhesion on coated plastic surfaces in defined media will produce rosette-like neural cells corresponding to the NEPs of the developing brain [28, 29]. This NSC population can be subsequently enriched, although no efficient methods for their extensive expansion have been reported. EB-independent procedures based on adherent monolayer protocols have been also described [30, 31].

Electrophysiology studies have shown that pluripotent-derived NSCs efficiently generate fully mature neurons *in vitro*, as well as functionally integrated neurons after transplantation in the mammalian CNS [32, 33]. Nevertheless, major limitations to therapeutic applications of pluripotent-derived NSCs are represented by safety concerns and caveats about their clinical-grade production. Indeed, grafted pluripotent cells can form teratomas, implying that in a clinical setting residual undifferentiated pluripotent stem cells should be excluded from the cell preparation before grafting. Protocols for avoiding teratocarcinoma formation *in vivo* after transplantation of ESC/iPSCs-derived cells are under scrutiny. In this view, recent studies have reported the direct conversion of adult somatic cells into NSCs, thus opening a new path to generate NSCs without contamination of undifferentiated pluripotent stem cells [34].

The ability to generate patient-specific iPSCs and NSCs clearly provides enormous prospective for future personalized medicine, although too little is yet known about these cells to make any firm prediction.

Possible Therapeutic Actions of Grafted NSCs in Different Neurodegenerative Conditions

Although the capacity of NSCs to divide and appropriately differentiate *in vitro* has attracted much attention for clinical translation, it does not assure that these cells functionally incorporate into the recipient tissue and produce efficient restoration of compromised functions after grafting. In order to generate therapeutic benefits in specific neurological diseases, grafted cells have to accomplish a certain grade of morphological, anatomical and functional integration into the impaired host CNS tissue.

Neurodegenerative disorders embody a heterogeneous collection of chronic and progressive diseases characterized by distinct aetiologies, anatomical impairments and symptoms [35]. Some of these disorders, such as Huntington's disease (HD), are acquired in an entirely genetic manner. Alzheimer's disease (AD), amyotrophic lateral sclerosis (ALS), and Parkinson's disease (PD) mainly occur sporadically, although familiar forms caused by inheritance of gene mutations are known. On the other hand, the CNS can also be affected by other non-degenerative conditions, such as spinal cord injury and stroke, with no genetic heritable components.

By virtue of this extreme heterogeneity, different specific requirements should be envisaged when considering cell replacement as a possible therapeutic strategy. We can distinguish between (i) "neuronal" CNS degenerative disorders caused by a prominent loss of specific neuronal populations and (ii) "non neuronal" CNS degenerative conditions characterized by loss of non neuronal elements.

In the case of neuronal degeneration, the success of cell replacement strictly depends on the complexity and accuracy of the pattern of connectivity that needs to be restored. In PD, affected dopaminergic neurons in the substantia nigra (SN) exert a modulatory action on striatal target circuits mostly through the release of dopamine. This system is defined as "paracrine" and even a partial pattern repair may lead to a significant functional recovery in such conditions. Indeed, in PD the donor cells can be transplanted directly into the target region to circumvent the problem of long-distance neuritic growth in the adult CNS. Despite the ectopic location, if grafted cells are able to re-establish a regulated and efficient release of dopamine, they can lead to a clinically relevant functional recovery. However, cell-based treatment strategies are extremely difficult for other diseases such as HD, ALS, trauma, stroke, and AD, which are characterized by the need of a complex pattern repair.

Differently, "non neuronal" CNS degenerative syndromes such as multiple sclerosis (MS) characterized by severe inflammation, oligodendroglial degeneration and axonal demyelination, represent a good target for cell replacement, due to their limited requirements for pattern repair [36]. In MS, grafted cells should produce oligodendrocytes able to restore axonal myelination in order to lead to a functional rescue.

It should also be emphasized that although pattern repair is critical to obtain permanent efficacy in the brain, cells transplanted into the brain may also be beneficial via the release of molecules that may either stimulate the regenerative potential of local cells (where present) or increase the survival of the remaining host elements, thus slowing disease progression (Figure 3; [37]). Also, immunomodulatory activity of the grafted cells could be of benefit in diseases such as MS where a prominent disease-associated inflammation contributes to the establishment and progression of the disease (Figure 3).

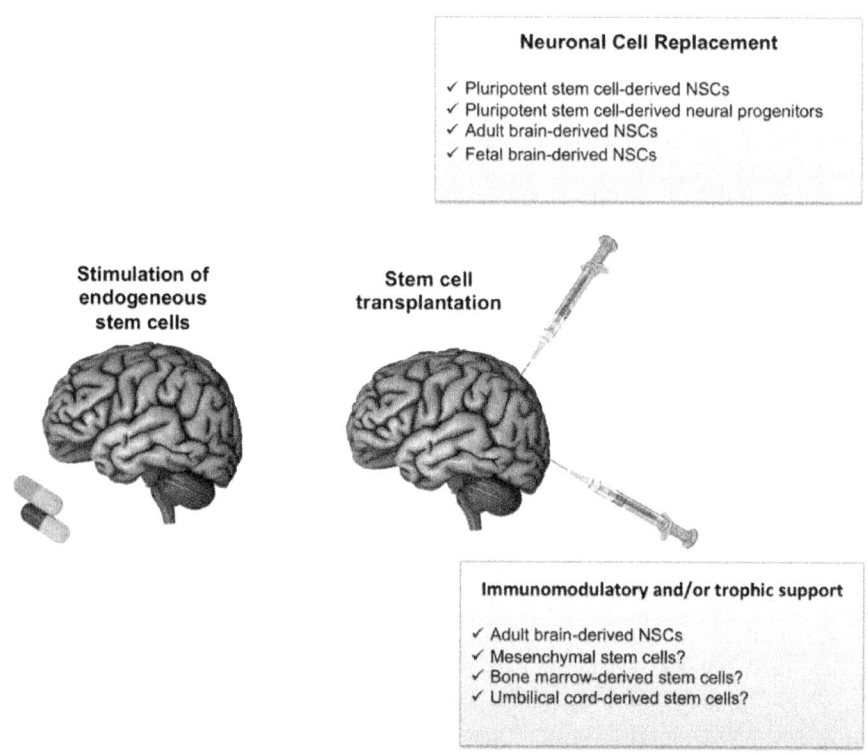

Figure 3. Therapeutical strategies for neural stem cells exploitation in CNS diseases.

Current Clinical Trials Involving NSCs for Neurodegenerative Diseases

A large number of studies have explored the grafting behavior of several NSCs typologies (and their progenies) in preclinical studies. Moreover, a few attempts are already being made to translate these discoveries into the clinical setting. Currently, the registry and results database of publicly and privately supported clinical studies with human participants conducted around the world (http://

www.clinicaltrial.gov/) reports 880 international clinical trials employing the use of stem cells for treatment of patients affected by several CNS disorders (query terms: "stem cells AND nervous system") among which 89 are testing NSC injection approaches (query terms: "neural stem cell AND injection AND nervous system disease"). If restricting the search to non-tumor diseases, the database returns 51 studies (query terms: "neural stem cell AND injection AND nervous system disease NOT tumor") with only 27 of these currently open. Interestingly, if we analyze these results more carefully, only 5 studies are actually open to explore the potential clinical relevance of NSCs while the remaining are testing the regenerative potential of mesenchymal stem cells whose actual uselfness in brain diseases is far from being a solid preclinical reality.

To date, few Phase I and II clinical studies employing NSCs (a dozen in highly debilitating CNS disorders; details reported in Table 1) have been performed, with the main objective to demonstrate safety and practicability and to explore the potential effectiveness of the treatments. These clinical trials are testing only few NSCs products, mainly consisting of fetal-tissue-derived allogenic NSCs. Among them, HuCNS-SC, a StemCells Inc. (Palo Alto, CA) product candidate, is a purified population of allogenic NSCs derived from human fetal (16–20 weeks) brain tissue, sorted using the CD133 marker and expanded in culture as neurospheres. These cells are routinely stored in a frozen state and reanimated before transplantation [38, 39]. In May 2006 a clinical trial (the first using human NSCs) was approved at Oregon Health and Science University (OHSU, Portland, OR, USA) for the use of these cells for lysosomal storage diseases (*ClinicalTrials.gov identifier, NCT00337636*), rare (one in 100,000 children) fatal autosomal recessive progressive neurological disorders caused by mutations in enzymes that ultimately lead to accumulation of neurotoxic lipofuscins. Patients generally lose their vision, develop seizures and dementia, and die before their teens. In this application NSCs serve as "Trojan Horses" to deliver enzymes to other cells within the brain, a concept established in rodent models of enzyme deficiencies. Preclinical studies indicated that intracerebral grafts into animal model of infantile neuronal ceroid lipofuscinosis (NCL, Batten's disease) integrate into the host brain, produce and release the defective enzyme resulting in protection of affected neurons. In this open-label, dose-escalating phase I trial, single donor HuCNS-SC were injected into 6 patients with either infantile or late infantile NCL. Enrollment in the trial was limited to patients in the advanced stages of the disease (patients with significant neurological and cognitive impairment, whose developmental age was demonstrated to be less than two-thirds of their chronological age). Two dose levels were administered with the first 3 patients receiving a target dose of approximately 500 million cells and the other 3 patients receiving a target dose of approximately 1 billion cells. The cells were grafted directly into each patient's brain (injections were performed into eight areas of each child's brain) and patients received immunosuppression for 12 months after transplantation. The grafting procedure and combination with prolonged immune suppression were both well tolerated thus showing a favorable safety profile with transplanted cells, neurosurgical procedure, and immunosuppression regimen. In addition, transplanted cells showed long-term survival. Five patients (one died for disease) completed the 12-month Phase I study and were subsequently enrolled in a subsequent four-year, long-term observational study, with three of the five surviving to the end of the four-year study [40]. Three of the six patients transplanted with HuCNS-SC cells have now survived more than five years post-transplant, and it is noteworthy that each have stable quality-of-life measures, considering that they suffer a progressive neurodegenerative disorder. Assessment of the patients' cognitive and neurological function revealed stable scores in some tests, but the clinical outcomes were generally consistent with the expected course of impairment associated with this neurodegenerative disease with no safety concerns attributed to the grafts. Examination of the brains from three patients who expired due to causes related to the underlying disease showed evidence of engraftment, migration and long-term survival of the HuCNS-SC cells following transplantation and the planned cessation of immunosuppression. Grafted cells provided widespread global replacement enzyme and bystander neuroprotection. Although these results look very promising, more detailed analyses and a longer patients' follow up will be fundamental to draw more solid conclusions. In 2009, the Company completed the Phase I safety study and in October 2010 embarked on a Phase Ib safety and efficacy trial (*clinicaltrials. gov identifier no. NCT01238315*) in 6 children with less advanced Batten's disease and therefore most likely to benefit from a timely neural stem cell transplant. However, the study was discontinued because of the failure to enroll patients meeting the study criteria (out of 22 initial prospects, none of them met the entry criteria).

Table 1. Clinical trials involving NSCs for neurodegenerative diseases

NCT number	Title	Recruitment	Conditions	Interventions	Sponsor/Collaborators	Phases	Enrollment	Start date	Completion date
NCT00337636	Study of HuCNS-SC Cells in Patients With Infantile or Late Infantile Neuronal Ceroid Lipofuscinosis (NCL)	Completed	Neuronal Ceroid Lipofuscinosis	Biological: HuCNS-SC	StemCells, Inc.	Phase 1	6	May 2006	September 2009
NCT01005004	Study of Human Central Nervous System (CNS) Stem Cells Transplantation in Pelizaeus-Merzbacher Disease (PMD) Subjects	Completed	Pelizaeus-Merzbacher Disease	Biological: HuCNS-SC cells implantation	StemCells, Inc.	Phase 1	4	November 2009	December 2012
NCT01151124	Pilot Investigation of Stem Cells in Stroke	Active, not recruiting	Stroke	Biological: CTX0E03 neural stem cells implantation	ReNeuron Limited	Phase 1	12	June 2010	March 2015
NCT01217008	Safety Study of GRNOPC1 in Spinal Cord Injury	Completed	Spinal Cord Injury	Biological: hES-derived GRNOPC1 implantation	Asterias Biotherapeutics, Inc.	Phase 1	5	October 2010	July 2013
NCT01238315	Safety and Efficacy Study of HuCNS-SC in Subjects With Neuronal Ceroid Lipofuscinosis	Withdrawn	Neuronal Ceroid Lipofuscinosis	Biological: HuCNS-SC cells implantation	StemCells, Inc.	Phase 1	0	November 2010	April 2011

NCT01321333	Study of Human Central Nervous System Stem Cells (HuCNS-SC) in Patients With Thoracic Spinal Cord Injury	Active, not recruiting	Thoracic Spinal Cord Injury|Spinal Cord Injury|Spinal Cord Injury Thoracic|Spinal Cord Trauma	Biological: HuCNS-SC cells implantation	StemCells, Inc.	Phase 1-2	12	March 2011	December 2015
NCT01348451	Human Spinal Cord Derived Neural Stem Cell Transplantation for the Treatment of Amyotrophic Lateral Sclerosis	Active, not recruiting	Amyotrophic Lateral Sclerosis	Biological: human neural spinal cord stem cells implantation	Neuralstem Inc.	Phase 1	18	January 2009	August 2014
NCT01640067	Human Neural Stem Cell Transplantation in Amyotrophic Lateral Sclerosis (ALS)	Recruiting	Amyotrophic Lateral Sclerosis	Biological: Human Neural Stem Cells implantation	Azienda Ospedaliera Santa Maria, Terni, Italy|Azienda Ospedaliero Universitaria Maggiore della Carita|Università di Padova Italy	Phase 1	18	December 2011	September 2016
NCT01725880	Long-Term Follow-Up of Transplanted Human Central Nervous System Stem Cells (HuCNS-SC) in Spinal Cord Trauma Subjects	Enrolling by invitation	Spinal Cord Injury	Observation	StemCells, Inc.	Phase 1-2	12	November 2012	March 2019
NCT01730716	Dose Escalation and Safety Study of Human Spinal Cord Derived Neural Stem Cell Transplantation for the Treatment of Amyotrophic Lateral Sclerosis	Enrolling by invitation	Amyotrophic Lateral Sclerosis	Biological: Human spinal cord stem cell implantation	Neuralstem Inc.	Phase 2	18	May 2013	April 2014

NCT01772810	Safety Study of Human Spinal Cord-derived Neural Stem Cell Transplantation for the Treatment of Chronic SCI	Not yet recruiting	Spinal Cord Injury (SCI)	Biological: Human spinal cord stem cells implantation	Neuralstem Inc.	Phase 1	8	May 2014	March 2016
NCT02117635	Pilot Investigation of Stem Cells in Stroke Phase II Efficacy	Not yet recruiting	Ischaemic Stroke|Cerebral Infarction|Hemiparesis|Arm Paralysis	Biological: CTX DP	ReNeuron Limited	Phase 2	41	June 2014	June 2017

A second open-label phase I clinical trial (*ClinicalTrials.gov NCT01005004*) has been sponsored by StemCell Inc. at the University of California, San Francisco (UCSF) using HuCNS-SC brain transplantation for Pelizaeus-Merzbacher disease (PMD). PMD is a X-linked congenital leukodystrophy characterized by defective myelination [41]. Preclinical studies in animal models showed that intracerebral injection of HuCNS-SC produced oligodendrocytes leading to remyelination of neurons affected by the mutated gene for PMD [42]. This phase I clinical trial has enrolled four young children with an early severe form of PMD. Each patient received a total brain dose of 300 million cells through two injections into the frontal white matter area of each hemisphere. Immunosuppression was administered for 9 months following transplantation. The patients have been followed for twelve months after transplantation; during this period they underwent regular neurological assessments and MRI analyses. Data regarding this clinical trial have been published in October 2012, indicating a good safety profile for the HuCNS-SC cells and the transplantation procedure [43]. Clinical assessment revealed small gains in motor and cognitive function in three of the four patients; the fourth patient remained clinically stable. Moreover, MRI inspections suggest myelination in the region of the transplantation, which progressed over time and persisted after the withdrawal of immunosuppression at nine months. Upon completion of the Phase I trial, all four patients have been enrolled into a 4-year long-term follow up study.

StemCells Inc. has also sponsored a Phase I/II clinical trial (*clinicaltrials.gov identifier no. NCT01321333*) at the University Hospital Balgrist (Zurich, Switzerland) for assaying safety and preliminary efficacy of intramedullary spinal cord transplantation of human HuCNS-SC neurospheres in subjects with thoracic (T2-T11) spinal cord trauma. The study began in March 2011 and includes 12 subjects who suffered spinal cord injury (SCI) in the 3 to 12 months prior to cell transplantation. Each subject received a total fixed dose of approximately 20 million cells injected directly into the thoracic spinal cord near the injury. The first patients have been transplanted with no safety concerns arising concerning the surgery and the cellular transplant [44]. This trial is estimated to end in March 2016. In November 2012 the consequential long-term follow up of the 12 patients subjected to HuCNS-SC transplantation has started and it will last until March 2018 (*clinicaltrials.gov identifier no. NCT01725880*).

Recently, also Neuralstem Inc. has received FDA approval to initiate a Phase I safety trial using proprietary NSCs for chronic spinal cord injury (*clinicaltrials.gov identifier no. NCT01772810*). The cell line used in this trial (NSI-566RSC) is derived from the cervical and upper thoracic regions of the spinal cord from a single 8-week human fetus after an elective abortion [45]. The cells are serially expanded as monolayer in serum-free media supplemented with FGF-2 as the sole mitogen to maintain proliferation and prevent differentiation. This trial will enroll up to eight SCI patients with thoracic lesions (T2-T12) one to two years post-injury with no motor or sensory function in the relevant segments at and below injury (complete paralysis). All patients in the trial will receive six injections close to injury site with the first four patients receiving 100,000 cells per injection and the second four patients 200,000 cells per injection. The patients will also receive immunosuppressive therapy, which will last for three months, as tolerated. Late this year, Neuralstem Inc. will also start an acute spinal cord injury trial in Seoul in collaboration with their South Korean partner CJ CheilJedang.

NSI-566 NSCs have also been used in clinical trials for the treatment of ALS (Lou Gehrig's disease). In 2009 Neuralstem Inc. sponsored the first NSC-based phase I safety trial in ALS (*clinicaltrials.gov identifier no. NCT01348451*) at the Emory University School of Medicine (Atlanta, GA, USA). The trial enrolled 18 patients divided in 5 groups characterized by slightly different inclusion criteria and receiving different procedures in terms of site and number of injections (number of cells/injection: 100,000). The first group (cohort A) included patients with advanced ALS who are unable to walk, and received transplant injections in one or both sides of the lower spinal cord. The trial then progressed to patients who were still ambulatory. The first three of these patients (Cohort B) received five unilateral injections while the next three patients (Cohort C) received ten bilateral injections in the same lumbar region. In 2011 Neuralstem received approval to move into the cervical (upper back) stage of the trial. In this stage, two groups of three patients each were included. The first three of these patients (Cohort D) were ambulatory with some arm dysfunction and received grafts on one side of their neck. The last group (Cohort E) included ALS patients still ambulatory, who received both injections on one side of their neck and injections on both sides of their lower spine. The last patient in this Phase I trial was treated in August 2012 and the trial was concluded in February 2013. The clinical assessments on the first 12 grafted patients demonstrated no evidence of acceleration of disease progression with the planned 18 months post-transplantation follow up [46]. In April 2013, following conclusion of its Phase I FDA-approved trial, Neuralstem received approval to start a Phase II dose escalation and safety trial (*clinicaltrials.gov identifier no. NCT01730716*) to be performed in three centers: Emory University Hospital in Atlanta, Georgia, site of Phase I; ALS Clinic at the University of Michigan Health System, in Ann Arbor, Michigan, and Massachusetts General Hospital in Boston. This Phase II trial is designed to treat up to 15 ambulatory patients, in five different dosing cohorts, advancing up to a maximum of 40 injections, and 400,000 cells per injection based on safety. The first 12 patients will receive injections in the cervical region of the spinal cord only, where the stem cells could help preserve breathing function. The final three patients will receive both cervical and lumbar injections.

Another ALS Phase I clinical trial using allogenic fetal brain-derived human NSCs has been approved to the Azienda Ospedaliera Santa Maria (Terni, Italy) in June 2012. This study includes a total of 18 ALS patients that will be treated with intraspinal implanted allogeneic foetal-derived neurospheres (*clinicaltrials.gov identifier no. NCT01640067*).

Fetal NSCs are also being used for treatment of disabled ischemic stroke patients by the company ReNeuron Limited (UK). For this clinical application, ReNeuron uses its proprietary allogenic foetal-derived brain human NSCs (CTX0E03), which were derived from human fetal brain tissue following genetic modification with a conditional immortalizing gene, *c*-mycER [47]. This transgene generates a fusion protein that stimulates cell proliferation in the presence of a synthetic drug, 4-hydroxy-tamoxifen (4-OHT). The cell line is clonal, expands rapidly in culture and has a normal human karyotype (46 XY). In the absence of growth factors and 4-OHT, the cells undergo growth arrest and differentiate into neurons and astrocytes. The Phase I clinical trial (*clinicaltrials.gov identifier no. NCT01151124*) started on June 2012 at the Glasgow Southern General Hospital (Glasgow, Scotland). The study is designed to test the safety of CTX0E03 NSCs product by direct single dose injection into the damaged brains of 12 male patients 60 years of age or over who remain moderately to severely disabled 6 months to 5 years following an ischemic stroke. The trial will consider four ascending doses of CTX0E03 cells (4 dosage groups of three patients at each dose level receiving 2 million, 5 million, 10 million or 20 million cells). Clinical outcomes will be measured over 24 months and patients will be invited to participate in a long-term follow-up trial for a further 8 years.

To date, only one study has exploited the use of human ES-derived neural cells for the treatment of CNS injury in a clinical setting. In 2010, Geron Corporation started a Phase I Safety clinical trial with human ES cell-derived oligodendrocyte progenitors (GRNOPC1 cells) [48–50] in patients with neurologically complete, subacute spinal cord injury (*ClinicalTrials.gov identifier: NCT01217008*). The study was designed to test the safety of the Geron GRNOPC1 cell product by direct single dose (2 million cells) injection into the damaged spinal cord of five patients with neurologically complete spinal cord injuries (between 7 and 14 days post injury). The enthusiasm about this study was abated only a year later, when Geron suddenly and surprisingly stopped the trial and decided to give up on the stem cell division. In 2013

Asterias Biotherapeutics, Inc., a subsidiary of BioTime, purchased Geron's stem cell division and announced that it will resume the spinal cord trial.

Is There a Rationale for Stem Cell-Based Treatments for Neurodevelopmental Disorders?

An increasing number of recent studies indicate that neurological and neuropsychiatric disorders including epilepsy, autism and schizophrenia may arise from altered brain development. According to this view, neurodevelopmental disorders may emerge from altered neurogenetic processes that lead to misplacement or loss of neurons and their connections in the postnatal brain. It is therefore not surprising that, in recent years, much attention has been paid to the possible use of NSCs to understand (and possibly treat) these pathologies. Currently, NSCs transplantation strategies have been essentially tested in rodent models of temporal lobe epilepsy (TLE). In both humans and rodents, TLE is accompanied by massive cell loss in the hippocampus and limbic areas [51]. In the past ten years, NSCs transplantation (both during the latent and chronic phases) has been widely tested as a tool to counteract cell loss and ameliorate TLE symptoms in rodents. NSCs are attractive as donor cells for grafting in TLE for a number of reasons: 1) they can be expanded in culture for extended periods; 2) they migrate extensively into the hippocampal layers; 3) they can differentiate into inhibitory GABAergic neurons as well as astrocytes secreting anticonvulsant factors (such as the glial cell-line derived neurotrophic factor GDNF); 4) they produce neurotrophic factors that can stimulate hippocampal neurogenesis from endogenous pools of NSCs [52]. Different strategies have been used to test the protective effects of transplanted NSCs, including transplantation of embryonic or adult hippocampal NSCs in both the KA and pilocarpine models of TLE [52]. Antiepileptic and neuroprotective effects have also been reported after transplantation of different types of stem cells. Grafting of human umbilical cord stem cells [53] or genetically-engineered bone marrow mesenchymal stem cells [54] have been shown to ameliorate seizures in the pilocarpine rat model of TLE.

Currently, research performed on animal models does not provide a rationale for the use of stem cells in neurodevelopmental disorders, with the only exception of epilepsy. Indeed, massive cell loss is detected in the epileptic brain, thus offering a rationale for cell replacement therapies. On the contrary, current research does not support the idea that stem cell transplantation may work in the case of other neurodevelopmental disorders such as autism, since no massive cell loss is observed in the brain of autistic patients. Accordingly, a Phase I trial to test the effect of stem cells transplantation in 20 patients with temporal lobe epilepsy is currently ongoing (*clinicaltrials.gov identifier no.NCT00916266*). The study, authorized on July 2008 to the Instituto do Cerebro de Brasilia (Brazil), has the aim to evaluate autologous bone marrow derived stem cells (BMDSCs) transplantation as a safe and potentially beneficial treatment for patients with temporal lobe refractory epilepsy. As primary outcomes were considered evaluation of seizure frequency, hippocampal volume and cognitive performance. Although the study should have been completed in June 2012, no results are currently available. Beyond the lack of solid results, it should be emphasized that the rationale for using BMDSCs transplantation in TLE patients is rather inappropriate, since these cells can not replace neural cell lost in these patients (see also below, "Clinical testing of non-neural cells for CNS diseases").

More surprisingly, one completed and five ongoing trials have been approved to test the efficacy of stem cell transplantation in patients affected by autism spectrum disorders (query terms: "stem cells AND autism") (details are reported in Table 2). None of these studies is actually injecting NSCs or pluripotent cells derivatives: they are focussed on using autologous BMSC (*clinicaltrials.gov identifier no. NCT01740869, NCT01974973*), adipose-derived MSC (*clinicaltrials.gov identifier no. NCT01502488*) and human cord blood mononuclear cells (*clinicaltrials.gov identifier no.NCT01343511, NCT01638819, NCT01836562*). Considering the young age and the overall number of patients enrolled (more than 350), the poor rationale and the total absence of published reports from these studies, great attention should be payed to these trials.

Table 2. Clinical trials for neurodevelopmental disorderss

NCT number	Title	Recruitment	Conditions	Interventions	Sponsor/Collaborators	Phases	Enrollment	Start date	Completion date
NCT01343511	Safety and Efficacy of Stem Cell Therapy in Patients With Autism	Completed	Autism	Biological: human cord blood mononuclear cells implantation|Biological: human cord blood mononuclear cells and human umbilical cord mesenchymal stem cells Injection	Shenzhen Beike Bio-Technology Co., Ltd.|Shandong Jiaotong Hospital|Association for the Handicapped Of Jinan	Phase 1-2	37	March 2009	May 2011
NCT01502488	Adipose Derived Stem Cell Therapy for Autism	Not yet recruiting	Autism	Procedure: Fat Harvesting and Stem Cell Injection	Ageless Regenerative Institute|Instituto de Medicina Regenerativa, S.A. de C.V.	Phase 1-2	10	October 2014	January 2017
NCT01638819	Autologous Cord Blood Stem Cells for Autism	Recruiting	Autism	Biological: Autologous Cord Blood Stem Cells Injection|Biological: Placebo	Sutter Health	Phase 2	30	August 2012	August 2014
NCT01740869	Autologous Bone Marrow Stem Cells for Children With Autism Spectrum Disorders	Recruiting	Autism|Autism Spectrum	Biological: Stem cells Injection	Hospital Universitario Dr. Jose E. Gonzalez	Phase 1-2	30	November 2012	Not specified

NCT01836562	A Clinical Trial to Study the Safety and Efficacy of Bone Marrow Derived Autologous Cells for the Treatment of Autism	Recruiting	Autism	Biological: Autologous Cord Blood Stem Cells Injection	Chaitanya Hospital, Pune	Phase 1-2	100	March 2011	April 2014
NCT01974973	Stem Cell Therapy in Autism Spectrum Disorders	Recruiting	Autism Spectrum Disorders	Procedure: Autologous bone marrow mononuclear cell transplantation|Procedure: Autologous bone marrow mononuclear cell transplantation in patients with autism	Neurogen Brain and Spine Institute	Phase 1	150	August 2009	October 2014

Retinal Dystrophies and Stem Cell Treatments

Another field in which a strong interest has been raised for cell treatments based on NSCs or cell derivatives of human pluripotent stem cells is represented by retinopathies. Many studies are undergoing in order to characterize the appropriate source of cells for these therapeutic approaches. The retina is subject to different degenerative diseases, which can be both age related (such as AMD, Age-related Macular Degeneration) and inherited (LCA, Leber's Congenital Amaurosis, and RP, Retinitis pigmentosa). These are all characterized by degeneration of photoreceptors and/or retinal pigmented epithelium (RPE cells). A number of different therapeutical approaches have been meaningfully explored to cure human photoreceptor degenerations. These include environmental modifications (nutrient supplementation and avoidance of light), drugs (i.e. neurotrophic factors), and gene therapy to provide the healthy version of the mutated gene, retinal prostheses [55].

A large body of transplantation studies has been carried out in order to assess the molecular features characterizing cells with the ability to integrate and generate functional photoreceptors in degenerating retinae. These studies defined that, in rodents, integration of donor cells is best obtained by using post-mitotic photoreceptor precursors [56, 57]. They also showed that, for donor cells to integrate in retinal tissue, they should have specific molecular characteristics and should be derived from retinae that have not reached complete maturation. Despite the promise, the low numbers of integrating cells impaired a real functional recovery in the transplanted eyes, even if some restoration of vision was observed [58]. Moreover, a possible transition to the clinic would be blocked by the obvious ethical concerns in the use of human retinal progenitor cells. For these reasons, many reseachers turned their attention to obtaining postmitotic photoreceptor precursors *in vitro*, by differentiation of pluripotent cells such as ESCs and iPSCs. Integration capabilities of the *in vitro*differentiated cells have also been tested by subretinal injections in mice [59–61]. All these studies assessed terminal differentiation and integration of pluripotent cells-derived photoreceptors and, when possible, functionality, although showing alternative results.

To date, the clinicaltrials.gov registry reports 50 international clinical trials employing different stem cells types (query terms: "stem cells AND retinopathy"), mostly autologous BMSCs, for treatment of patients affected by retinal dystrophies. Among these, an age related macular degneration (AMD) Phase I/II trial sponsored by Stem Cell Inc. is ongoing in three US centers: (Byers Eye Institute at Stanford, Stanford Hospital and Clinics, Palo Alto, California; New York Eye and Ear Infirmary, New York; Retina Foundation of the Southwest Recruiting Dallas, Texas) (*clinicaltrials.gov identifier no. NCT01632527*). This study, involving 16 patients affected by geographic atrophy secondary to AMD, is designed to investigate the safety and preliminary efficacy of unilateral subretinal transplantation of HuCNS-SC cells by means of a single transplant procedure. Immunosuppressive agents will be administered orally to all subjects for a period of three months after surgery. Participants will be monitored for complications as well as structural evidence of successful engraftment and changes to vision.

Interestingly, the field of retinopathies represents the arena where the largest number of human ESC-derived products is being tested in the clinics. Currently, seven trials are open employing terminally differentiated human ESC-derived hRPE (details are reported in Table 3). Six of these studies are sponsored by Advanced Cell Technology (ACT; USA) and CHA Bio & Diostech (South Korea) for the use of the ACT's MA09-hRPE cells for treatment of Stargardt's Macular Dystrophy patients (SMD; *clinicaltrials.gov identifier no. NCT01469832, NCT01345006* and *NCT01625559*), dry AMD patients (*clinicaltrials.gov identifier no. NCT01674829, NCT01344993*) or Myopic Macular Degeneration patients (MMD; *clinicaltrials.gov identifier no.NCT02122159*). In these studies, the number of grafted cells varies between 50,000 and 200,000 cells in order to define the optimal dosage. The preliminary results of two of these trials have been published, showing safety and some promising efficacy in vision restoration. For this reason, MA09-hRPE cells from ACT have been recently granted by the FDA the orphan drug designation for use in the treatment of Stargardt's Macular Dystrophy (SMD). This represents the first time that orphan drug status has been granted for the use of an embryonic stem cell derived therapy in treating an unmet medical need. As a result, ACT is entitled to obtain several benefits aimed at fostering clinical exploitation of this cell product, including tax credits, access to grant funding for clinical trials, accelerated FDA approval and allowance for marketing exclusivity after drug approval for a period of as long as seven years.

Table 3. Clinical trials involving stem cells for retinal dystrophies

NCT number	Title	Recruitment	Conditions	Interventions	Sponsor/Collaborators	Phases	Enrollment	Start date	Completion date
NCT01344993	Safety and Tolerability of Sub-retinal Transplantation of hESC Derived RPE (MA09-hRPE) Cells in Patients With Advanced Dry Age Related Macular Degeneration	Recruiting	Dry Age Related Macular Degeneration	Biological: MA09-hRPE implantation	Advanced Cell Technology	Phase 1-2	16	April 2011	December 2014
NCT01345006	Sub-retinal Transplantation of hESC Derived RPE (MA09-hRPE) Cells in Patients With Stargardt's Macular Dystrophy	Recruiting	Stargardt's Macular Dystrophy	Biological: MA09-hRPE implantation	Advanced Cell Technology	Phase 1-2	16	April 2011	December 2014
NCT01469832	Safety and Tolerability of Sub-retinal Transplantation of Human Embryonic Stem Cell Derived Retinal Pigmented Epithelial (hESC-RPE) Cells in Patients With Stargardt's Macular Dystrophy (SMD)	Recruiting	Stargardt's Macular Dystrophy|Fundus Flavimaculatus|Juvenile Macular Dystrophy	Biological: MA09-hRPE implantation	Advanced Cell Technology	Phase 1-2	16	November 2011	December 2014
NCT01625559	Safety and Tolerability of MA09-hRPE Cells in Patients With Stargardt's Macular Dystrophy (SMD)	Recruiting	Stargardt's Macular Dystrophy	Biological: MA09-hRPE implantation	CHA Bio & Diostech	Phase 1	3	September 2012	October 2014

NCT	Title	Status	Condition	Intervention	Sponsor	Phase	Enrollment	Start Date	Completion Date
NCT01632527	Study of Human Central Nervous System Stem Cells (HuCNS-SC) in Age-Related Macular Degeneration (AMD)	Recruiting	Age Related Macular Degeneration|Macular Degeneration|AMD	Drug: HuCNS-SC cells implantation	StemCells, Inc.	Phase 1-2	16	June 2012	June 2014
NCT01674829	A Phase I/IIa, Open-Label, Single-Center, Prospective Study to Determine the Safety and Tolerability of Sub-retinal Transplantation of Human Embryonic Stem Cell Derived Retinal Pigmented Epithelial (MA09-hRPE) Cells in Patients With Advanced Dry Age-related Macular Degeneration (AMD)	Recruiting	Dry Age Related Macular Degeneration	Biological: MA09-hRPE implantation	CHA Bio & Diostech	Phase 1-2	12	September 2012	April 2016
NCT01691261	A Study Of Implantation Of Human Embryonic Stem Cell Derived Retinal Pigment Epithelium In Subjects With Acute Wet Age Related Macular Degeneration And Recent Rapid Vision Decline	Not yet recruiting	Age Related Macular Degeneration	Biological: PF-05206388 implantation	Pfizer|University College, London	Phase 1	10	April 2014	July 2016
NCT02122159	Research With Retinal Cells Derived From Stem Cells for Myopic Macular Degeneration	Not yet recruiting	Myopic Macular Degeneration	Biological: MA09-hRPE implantation	University of California, Los Angeles|Advanced Cell Technology, Inc.	Phase 1-2	Not specified	Not specified	Not specified

The remaining study is represented by a non-randomized safety/efficacy study phase I open label trial of RPE replacement aiming at evaluating the safety and feasibility/efficacy of treating subjects with wet AMD in whom there is rapidly progressing vision loss (*clinicaltrials.gov identifier no. NCT01691261*). The study has been sponsored by Pfizer and will be performed at University College, London (UK). It will include 10 patients that will receive Pf-05206388 RPE cells immobilized on a polyester membrane as a monolayer, derived from human ESCs. The implanted membrane is approximately 6 mm × 3 mm and is intended to be life-long.

Clinical Testing of Non-Neural Cells for CNS Diseases

While in this review we have mainly discussed ongoing clinical studies based on human NSCs or neural derivatives of pluripotent stem cells, most of the currently open Phase I/II clinical trials for treatment of brain diseases are actually testing therapeutic potential of cells of non-neural origin. Mesenchymal stem cells (MSCs) derived from bone marrow (BMD-MSCs) or adipose tissue and bone marrow derived stem cells (BMDSCs) have been proposed by some scientists and many clinicians as reliable extra-neural sources of multipotential stem cells for brain repair [62, 63]. From a clinical point of view, these non-neural cell sources provide the advantage of easy accessibility and of autologous approaches, thus minimizing immune reactions. Nonetheless, although these cells have been widely used in clinical settings for non neural diseases and their safety has been demonstrated when injected in several body districts (different from the CNS), particularly with autologous transplants, their sustained therapeutic benefit for brain disorders has not been consistently obtained, neither in preclinical nor in clinical studies. Several optimistic reports have been published regarding MSCs conversion into NSCs or neurons but actually the proof of functional neurons generation from MSCs is still missing. Indeed, these reports based their conclusions on morphological inspection or on the expression of few neuronal markers rather than demonstrating that these neuronal-like cells exhibit all the morphological, antigenic and functional key properties of neurons (i.e. presence of mature synaptic structures, electrical excitability, controlled neurotransmitter release) or showing an effect in disease models [64, 65]. Overall, the currently available evidences indicate that the number of BMDSCs or MSCs able to differentiate into neurons, if any, is extremely low and irrelevant when thinking of their potential clinical exploitation. However, several laboratories continue to declare that MSCs can be converted into neurons, thus encouraging many unsubstantiated clinical trials testing these cells for brain repair. To this regard, the clinical outcome published in 2010 regarding the first open-label pilot clinical trial with autologous BMD-

MSCs transplanted into the striatum of patients with advanced Parkinson's Disease is emblematic [66]. Seven PD patients aged 22 to 62 years received a single-dose (1 million per kg body weight) unilateral transplantation of autologous cells into the sublateral ventricular zone by stereotaxic surgery. Patients were followed up for a period that ranged from 10 to 36 months and the results indicate that the protocol seems to be safe with no adverse events occurring during the observation period. Nonetheless, the number of patients recruited and the uncontrolled type of trial did not permit demonstration of effectiveness of the treatment. Indeed, the clinical improvements were only marginal and possibly due to a placebo effect. Regardless of the capability of MSCs to differentiate into neurons, many trials have been pushed by the belief that MSCs might provide benefit to patients affected from brain diseases by virtue of neuroprotective/immunomodulatory properties, thus supporting diseased cells and controlling or adjusting inflammation within the patient's CNS. As such, many preclinical studies have been performed with MSCs but at the present time it is yet unproven that they might exert substantial and enduring effects in any neurological condition. Finally, long-term follow up should be performed in order to be confident about the safety of injecting MSCs (or any other non-neural cell type) in the brain. Indeed, it is still unknown (i) what eventually happens to these cells in the brain, (ii) if they survive long term in the lesioned region and (iii) if their "ectopic nature" might induce adverse effects in the future.

Beside the use of MSCs and BD-MSCs, hematopoietic stem cells (HSCs) represent an interesting source of non neural cells, exploitable for treating some CNS disorders. Indeed, HSCs have shown very promising clinical results for gene therapy treatments of congenital leukodystrophies [67, 68].

CONCLUSIONS

NSCs have become one of the most intensively studied cell types in biology. Our knowledge of their identities and properties has been radically revolutionized by the possibility to isolate and expand them *in vitro*. One can anticipate that a rigorous assessment of the functional qualities of NSCs combined with emerging knowledge of CNS microenvironment following injury will allow us to exploit the advantages offered by these cells in the clinical setting. It should be emphasized that the NSCs potential to regenerate the brain relies also on the competence of other cells to contribute to repairing processes. NSCs and pluripotent stem cells neural derivatives are now under scrutiny in early Phase I/II trials for CNS disorders. Also new systems based on direct conversion of non neural cells into NSCs and into specific neuronal populations are bringing new possible strategies of intervention. While it is yet too early to predict the outcome of these trials, initial results indicate no safety concerns. Although the field is moving forward every year and new trials are continuously being planned and started, to date none was successful, thus reducing the hope that stem cells may be a valid CNS therapy in the next few years.

Before envisaging any therapeutic application of such cells for CNS disorders, we need to confront several, and still unsolved, problems: (i) the ideal stem cell source for transplantation in each specific disease context; (ii) the appropriate number of cells to transplant; (iii) a clinically-applicable transplantation strategy; (iv) the right disease stage for cell transplantation; and, finally, (v) the most appropriate *in vivo* and/or *in vitro* manipulations to obtain the proper cells to be transplanted. With time, progress will be made, but it is only by following a correct preclinical research and well-established methods of translation from the laboratory to the clinic that this can happen. The growing interest and participation in stem cell therapies of big pharmaceutical companies and their collaborating partners worldwide will represent an important step in order to increase the number of new well-defined clinical trials in the next few years. It is mandatory that scientists and clinicians should work side by side in order to pay attention not to abandon these principles in the hurry to move to clinic and to responsibly communicate to the public and patients [69]. The real risk is that the entire field of stem cell research and therapies might become a fertile ground for commercially driven illusion sellers that peddle the poisoning concept of stem cell research as an alchemic science.

ACKNOWLEDGEMENTS

Data reported in this review were identified by searches (as of April 20, 2014). Abstracts and reports from meetings were not included, and only papers published in English were reviewed. Data about ongoing clinical trials were obtained from http://www.clinicaltrial.gov/ (as of April 20, 2014). Due to the large amount of bibliographic material available on this subject, we apologize to those authors whose studies have not been cited in this review. This work was funded by grants from the University of Trento (CIBIO start-up to L.C., Y.B. and S.C.).

AUTHORS' CONTRIBUTION

SC, YB and LC organized, composed sections, and edited the final manuscript. LC prepared the figures and tables, which were edited by all the authors. All authors read and approved the final manuscript.

REFERENCES

1. Lindvall O, Barker RA, Brustle O, Isacson O, Svendsen CN: Clinical translation of stem cells in neurodegenerative disorders. Cell Stem Cell. 2012, 10: 151-155. 10.1016/j.stem.2012.01.009.
2. Conti L, Cattaneo E: Neural stem cell systems: physiological players or *in vitro* entities?. Nat Rev Neurosci. 2010, 11: 176-187.
3. Fuentealba LC, Obernier K, Alvarez-Buylla A: Adult neural stem cells bridge their niche. Cell Stem Cell. 2012, 10: 698-708. 10.1016/j.stem.2012.05.012.
4. Gage FH, Temple S: Neural stem cells: generating and regenerating the brain. Neuron. 2013, 80: 588-601. 10.1016/j.neuron.2013.10.037.
5. Grabel L: Developmental origin of neural stem cells: the glial cell that could. Stem Cell Rev. 2012, 8: 577-585. 10.1007/s12015-012-9349-8.
6. Malatesta P, Hartfuss E, Gotz M: Isolation of radial glial cells by fluorescent-activated cell sorting reveals a neuronal lineage. Development. 2000, 127: 5253-5263.
7. Pollard SM, Conti L: Investigating radial glia *in vitro*. Prog Neurobiol. 2007, 83: 53-67. 10.1016/j.pneurobio.2007.02.008.
8. Miyata T, Kawaguchi A, Saito K, Kawano M, Muto T, Ogawa M: Asymmetric production of surface-dividing and non-surface-dividing cortical progenitor cells. Development. 2004, 131: 3133-3145. 10.1242/dev.01173.
9. Alvarez-Buylla A, Lim DA: For the long run: maintaining germinal niches in the adult brain. Neuron. 2004, 41: 683-686. 10.1016/S0896-6273(04)00111-4.
10. Altman J: Are new neurons formed in the brains of adult mammals?. Science. 1962, 135: 1127-1128. 10.1126/science.135.3509.1127.
11. Reynolds BA, Tetzlaff W, Weiss S: A multipotent EGF-responsive striatal embryonic progenitor cell produces neurons and astrocytes. J Neurosci. 1992, 12: 4565-4574.
12. Reynolds BA, Weiss S: Generation of neurons and astrocytes from isolated cells of the adult mammalian central nervous system. Science. 1992, 255: 1707-1710. 10.1126/science.1553558.
13. Christie KJ, Turnley AM: Regulation of endogenous neural stem/progenitor cells for neural repair-factors that promote neurogenesis and gliogenesis in the normal and damaged brain. Front Cell Neurosci. 2012, 6: 70.

14. Parent JM, Kron MM: Neurogenesis and Epilepsy. Jasper's Basic Mechanisms of the Epilepsies. Edited by: Noebels JL, Avoli M, Rogawski MA, Olsen RW, Delgado-Escueta AV. 2012, Bethesda (MD), USA: Published by Oxford University Press, 4
15. Reynolds BA, Rietze RL: Neural stem cells and neurospheres–reevaluating the relationship. Nat Methods. 2005, 2: 333-336. 10.1038/nmeth758.
16. Conti L, Pollard SM, Gorba T, Reitano E, Toselli M, Biella G, Sun Y, Sanzone S, Ying QL, Cattaneo E, Smith A: Niche-independent symmetrical self-renewal of a mammalian tissue stem cell. PLoS Biol. 2005, 3: e283-10.1371/journal.pbio.0030283.
17. Koch P, Opitz T, Steinbeck JA, Ladewig J, Brustle O: A rosette-type, self-renewing human ES cell-derived neural stem cell with potential for *in vitro* instruction and synaptic integration. Proc Natl Acad Sci U S A. 2009, 106 (9): 3225-3230. 10.1073/pnas.0808387106.
18. Palmer TD, Schwartz PH, Taupin P, Kaspar B, Stein SA, Gage FH: Cell culture. Progenitor cells from human brain after death. Nature. 2001, 411: 42-43. 10.1038/35075141.
19. Rossi F, Cattaneo E: Opinion: neural stem cell therapy for neurological diseases: dreams and reality. Nat Rev Neurosci. 2002, 3: 401-409. 10.1038/nrn809.
20. Maisel M, Herr A, Milosevic J, Hermann A, Habisch HJ, Schwarz S, Kirsch M, Antoniadis G, Brenner R, Hallmeyer-Elgner S, Lerche H, Schwarz J, Storch A: Transcription profiling of adult and fetal human neuroprogenitors identifies divergent paths to maintain the neuroprogenitor cell state. Stem Cells. 2007, 25: 1231-1240. 10.1634/stemcells.2006-0617.
21. Marei HE, Ahmed AE, Michetti F, Pescatori M, Pallini R, Casalbore P, Cenciarelli C, Elhadidy M: Gene expression profile of adult human olfactory bulb and embryonic neural stem cell suggests distinct signaling pathways and epigenetic control. PLoS One. 2012, 7: e33542-10.1371/journal.pone.0033542.
22. Fan Y, Marcy G, Lee ES, Rozen S, Mattar CN, Waddington SN, Goh EL, Choolani M, Chan JK: Regionally-specified second trimester fetal neural stem cells reveals differential neurogenic programming. PLoS One. 2014, 9: e105985-10.1371/journal.pone.0105985.
23. Evans MJ, Kaufman MH: Establishment in culture of pluripotential cells from mouse embryos. Nature. 1981, 292: 154-156. 10.1038/292154a0.

24. Thomson JA, Itskovitz-Eldor J, Shapiro SS, Waknitz MA, Swiergiel JJ, Marshall VS, Jones JM: Embryonic stem cell lines derived from human blastocysts. Science. 1998, 282: 1145-1147.
25. Yamanaka S: Induced pluripotent stem cells: past, present, and future. Cell Stem Cell. 2012, 10: 678-684. 10.1016/j.stem.2012.05.005.
26. Takahashi K, Tanabe K, Ohnuki M, Narita M, Ichisaka T, Tomoda K, Yamanaka S: Induction of pluripotent stem cells from adult human fibroblasts by defined factors. Cell. 2007, 131: 861-872. 10.1016/j.cell.2007.11.019.
27. Takahashi K, Yamanaka S: Induction of pluripotent stem cells from mouse embryonic and adult fibroblast cultures by defined factors. Cell. 2006, 126: 663-676. 10.1016/j.cell.2006.07.024.
28. Okabe S, Forsberg-Nilsson K, Spiro AC, Segal M, McKay RD: Development of neuronal precursor cells and functional postmitotic neurons from embryonic stem cells *in vitro*. Mech Dev. 1996, 59: 89-102. 10.1016/0925-4773(96)00572-2.
29. Zhang SC, Wernig M, Duncan ID, Brustle O, Thomson JA: *In vitro* differentiation of transplantable neural precursors from human embryonic stem cells. Nat Biotechnol. 2001, 19: 1129-1133. 10.1038/nbt1201-1129.
30. Chambers SM, Fasano CA, Papapetrou EP, Tomishima M, Sadelain M, Studer L: Highly efficient neural conversion of human ES and iPS cells by dual inhibition of SMAD signaling. Nat Biotechnol. 2009, 27: 275-280. 10.1038/nbt.1529.
31. Ying QL, Stavridis M, Griffiths D, Li M, Smith A: Conversion of embryonic stem cells into neuroectodermal precursors in adherent monoculture. Nat Biotechnol. 2003, 21: 183-186. 10.1038/nbt780.
32. Emborg ME, Liu Y, Xi J, Zhang X, Yin Y, Lu J, Joers V, Swanson C, Holden JE, Zhang SC: Induced pluripotent stem cell-derived neural cells survive and mature in the nonhuman primate brain. Cell Rep. 2013, 3: 646-650. 10.1016/j.celrep.2013.02.016.
33. Kriks S, Shim JW, Piao J, Ganat YM, Wakeman DR, Xie Z, Carrillo-Reid L, Auyeung G, Antonacci C, Buch A, Yang L, Beal MF, Surmeier DJ, Kordower JH, Tabar V, Studer L: Dopamine neurons derived from human ES cells efficiently engraft in animal models of Parkinson's disease. Nature. 2011, 480: 547-551.
34. Thier M, Wörsdörfer P, Lakes YB, Gorris R, Herms S, Opitz T, Seiferling D, Quandel T, Hoffmann P, Nöthen MM, Brüstle O, Edenhofer F: Direct conversion of fibroblasts into stably expandable neural stem cells. Cell Stem Cell. 2012, 10 (4): 473-479. 10.1016/j.stem.2012.03.003.

35. Lindvall O, Kokaia Z: Stem cells in human neurodegenerative disorders–time for clinical translation?. J Clin Invest. 2010, 120: 29-40. 10.1172/JCI40543.
36. Giusto E, Donega M, Cossetti C, Pluchino S: Neuro-immune interactions of neural stem cell transplants: from animal disease models to human trials. Exp Neurol. 2013, 260C: 19-32.
37. Martino G, Pluchino S, Bonfanti L, Schwartz M: Brain regeneration in physiology and pathology: the immune signature driving therapeutic plasticity of neural stem cells. Physiol Rev. 2011, 91: 1281-1304. 10.1152/physrev.00032.2010.
38. Tamaki S, Eckert K, He D, Sutton R, Doshe M, Jain G, Tushinski R, Reitsma M, Harris B, Tsukamoto A, Gage F, Weissman I, Uchida N: Engraftment of sorted/expanded human central nervous system stem cells from fetal brain. J Neurosci Res. 2002, 69: 976-986. 10.1002/jnr.10412.
39. Uchida N, Buck DW, He D, Reitsma MJ, Masek M, Phan TV, Tsukamoto AS, Gage FH, Weissman IL: Direct isolation of human central nervous system stem cells. Proc Natl Acad Sci U S A. 2000, 97: 14720-14725. 10.1073/pnas.97.26.14720.
40. Selden NR, Al-Uzri A, Huhn SL, Koch TK, Sikora DM, Nguyen-Driver MD, Guillaume DJ, Koh JL, Gultekin SH, Anderson JC, Vogel H, Sutcliffe TL, Jacobs Y, Steiner RD: Central nervous system stem cell transplantation for children with neuronal ceroid lipofuscinosis. J Neurosurg Pediatr. 2013, 11: 643-652. 10.3171/2013.3.PEDS12397.
41. Garbern JY: Pelizaeus-Merzbacher disease: genetic and cellular pathogenesis. Cell Mol Life Sci. 2007, 64: 50-65. 10.1007/s00018-006-6182-8.
42. Uchida N, Chen K, Dohse M, Hansen KD, Dean J, Buser JR, Riddle A, Beardsley DJ, Wan Y, Gong X, Nguyen T, Cummings BJ, Anderson AJ, Tamaki SJ, Tsukamoto A, Weissman IL, Matsumoto SG, Sherman LS, Kroenke CD, Back SA: Human neural stem cells induce functional myelination in mice with severe dysmyelination. Sci Transl Med. 2012, 4: 155ra136.
43. Gupta N, Henry RG, Strober J, Kang SM, Lim DA, Bucci M, Caverzasi E, Gaetano L, Mandelli ML, Ryan T, Perry R, Farrell J, Jeremy RJ, Ulman M, Huhn SL, Barkovich AJ, Rowitch DH: Neural stem cell engraftment and myelination in the human brain. Sci Transl Med. 2012, 4: 155ra137.
44. Tsukamoto A, Uchida N, Capela A, Gorba T, Huhn S: Clinical translation of human neural stem cells. Stem Cell Res Ther. 2013, 4: 102-10.1186/scrt313.

45. Guo X, Johe K, Molnar P, Davis H, Hickman J: Characterization of a human fetal spinal cord stem cell line, NSI-566RSC, and its induction to functional motoneurons. J Tissue Eng Regen Med. 2010, 4: 181-193. 10.1002/term.223.

46. Glass JD, Boulis NM, Johe K, Rutkove SB, Federici T, Polak M, Kelly C, Feldman EL: Lumbar intraspinal injection of neural stem cells in patients with amyotrophic lateral sclerosis: results of a phase I trial in 12 patients. Stem Cells. 2012, 30: 1144-1151. 10.1002/stem.1079.

47. Pollock K, Stroemer P, Patel S, Stevanato L, Hope A, Miljan E, Dong Z, Hodges H, Price J, Sinden JD: A conditionally immortal clonal stem cell line from human cortical neuroepithelium for the treatment of ischemic stroke. Exp Neurol. 2006, 199: 143-155. 10.1016/j.expneurol.2005.12.011.

48. Lebkowski J: GRNOPC1: the world's first embryonic stem cell-derived therapy. Interview with Jane Lebkowski. Regen Med. 2011, 6: 11-13. 10.2217/rme.11.77.

49. Okamura RM, Lebkowski J, Au M, Priest CA, Denham J, Majumdar AS: Immunological properties of human embryonic stem cell-derived oligodendrocyte progenitor cells. J Neuroimmunol. 2007, 192: 134-144. 10.1016/j.jneuroim.2007.09.030.

50. Zhang YW, Denham J, Thies RS: Oligodendrocyte progenitor cells derived from human embryonic stem cells express neurotrophic factors. Stem Cell Dev. 2006, 15: 943-952. 10.1089/scd.2006.15.943.

51. Majores M, Schoch S, Lie A, Becker AJ: Molecular neuropathology of temporal lobe epilepsy: complementary approaches in animal models and human disease tissue. Epilepsia. 2007, 48 (Suppl 2): 4-12.

52. Shetty AK: Neural Stem Cell Therapy for Temporal Lobe Epilepsy. Jasper's Basic Mechanisms of the Epilepsies. Edited by: Noebels JL, Avoli M, Rogawski MA, Olsen RW, Delgado-Escueta AV. 2012, Bethesda (MD), USA: Published by Oxford University Press, 4

53. Costa-Ferro ZS, de Borba CF, de Freitas Souza BS, Leal MM, da Silva AA, de Bellis Kuhn TI, Forte A, Sekiya EJ, Soares MB, Dos Santos RR: Antiepileptic and neuroprotective effects of human umbilical cord blood mononuclear cells in a pilocarpine-induced epilepsy model. Cytotechnology. 2014, 66: 193-199. 10.1007/s10616-013-9557-3.

54. Long Q, Qiu B, Wang K, Yang J, Jia C, Xin W, Wang P, Han R, Fei Z, Liu W: Genetically engineered bone marrow mesenchymal stem cells improve functional outcome in a rat model of epilepsy. Brain Res. 2013, 1532: 1-13.

55. Stone EM: Progress toward effective treatments for human photoreceptor degenerations. Curr Opin Genet Dev. 2009, 19: 283-289. 10.1016/j.gde.2009.03.006.
56. Lakowski J, Han YT, Pearson RA, Gonzalez-Cordero A, West EL, Gualdoni S, Barber AC, Hubank M, Ali RR, Sowden JC: Effective transplantation of photoreceptor precursor cells selected via cell surface antigen expression. Stem Cells. 2011, 29: 1391-1404.
57. MacLaren RE, Pearson RA, MacNeil A, Douglas RH, Salt TE, Akimoto M, Swaroop A, Sowden JC, Ali RR: Retinal repair by transplantation of photoreceptor precursors. Nature. 2006, 444: 203-207. 10.1038/nature05161.
58. Pearson RA, Barber AC, Rizzi M, Hippert C, Xue T, West EL, Duran Y, Smith AJ, Chuang JZ, Azam SA, Xue T, West EL, Duran Y, Smith AJ, Chuang JZ, Azam SA, Luhmann UF, Benucci A, Sung CH, Bainbridge JW, Carandini M, Yau KW, Sowden JC, Ali RR: Restoration of vision after transplantation of photoreceptors. Nature. 2012, 485: 99-103. 10.1038/nature10997.
59. Hambright D, Park KY, Brooks M, McKay R, Swaroop A, Nasonkin IO: Long-term survival and differentiation of retinal neurons derived from human embryonic stem cell lines in un-immunosuppressed mouse retina. Mol Vis. 2012, 18: 920-936.
60. Tucker BA, Park IH, Qi SD, Klassen HJ, Jiang C, Yao J, Redenti S, Daley GQ, Young MJ: Transplantation of adult mouse iPS cell-derived photoreceptor precursors restores retinal structure and function in degenerative mice. PLoS One. 2011, 6: e18992-10.1371/journal.pone.0018992.
61. West EL, Gonzalez-Cordero A, Hippert C, Osakada F, Martinez-Barbera JP, Pearson RA, Sowden JC, Takahashi M, Ali RR: Defining the integration capacity of embryonic stem cell-derived photoreceptor precursors. Stem Cells. 2012, 30: 1424-1435. 10.1002/stem.1123.
62. Sadan O, Melamed E, Offen D: Bone-marrow-derived mesenchymal stem cell therapy for neurodegenerative diseases. Expert Opin Biol Ther. 2009, 9: 1487-1497. 10.1517/14712590903321439.
63. Wislet-Gendebien S, Laudet E, Neirinckx V, Rogister B: Adult bone marrow: which stem cells for cellular therapy protocols in neurodegenerative disorders?. J Biomed Biotechnol. 2012, 2012: 601560.
64. Sanchez-Ramos JR: Neural cells derived from adult bone marrow and umbilical cord blood. J Neurosci Res. 2002, 69: 880-893. 10.1002/jnr.10337.

65. Wislet-Gendebien S, Wautier F, Leprince P, Rogister B: Astrocytic and neuronal fate of mesenchymal stem cells expressing nestin. Brain Res Bull. 2005, 68: 95-102. 10.1016/j.brainresbull.2005.08.016.

66. Venkataramana NK, Kumar SK, Balaraju S, Radhakrishnan RC, Bansal A, Dixit A, Rao DK, Das M, Jan M, Gupta PK, Totey SM: Open-labeled study of unilateral autologous bone-marrow-derived mesenchymal stem cell transplantation in Parkinson's disease. Transl Res. 2010, 155: 62-70. 10.1016/j.trsl.2009.07.006.

67. Biffi A, Montini E, Lorioli L, Cesani M, Fumagalli F, Plati T, Baldoli C, Martino S, Calabria A, Canale S, Benedicenti F, Vallanti G, Biasco L, Leo S, Kabbara N, Zanetti G, Rizzo WB, Mehta NA, Cicalese MP, Casiraghi M, Boelens JJ, Del Carro U, Dow DJ, Schmidt M, Assanelli A, Neduva V, Di Serio C, Stupka E, Gardner J, von Kalle C: Lentiviral hematopoietic stem cell gene therapy benefits metachromatic leukodystrophy. Science. 2013, 341: 1233158-10.1126/science.1233158.

68. Cartier N, Hacein-Bey-Abina S, Bartholomae CC, Veres G, Schmidt M, Kutschera I, Vidaud M, Abel U, Dal-Cortivo L, Caccavelli L, Mahlaoui N, Kiermer V, Mittelstaedt D, Bellesme C, Lahlou N, Lefrère F, Blanche S, Audit M, Payen E, Leboulch P, l'Homme B, Bougnères P, Von Kalle C, Fischer A, Cavazzana-Calvo M, Aubourg P: Hematopoietic stem cell gene therapy with a lentiviral vector in X-linked adrenoleukodystrophy. Science. 2009, 326: 818-823. 10.1126/science.1171242.

69. Cattaneo E, Bonfanti L: Therapeutic potential of neural stem cells: greater in people's perception than in their brains?. Front Neurosci. 2014, 8: 79.

Chapter 9

STRATEGIES OF MUCOSAL IMMUNOTHERAPY FOR ALLERGIC DISEASES

Yi-Ling Ye[1], Ya-Hui Chuang[2], and Bor-Luen Chiang[3]

[1]Department of Biotechnology, National Formosa University, Yunlin, Taiwan

[2]Department of Clinical Laboratory Sciences and Medical Biotechnology, National Taiwan University, Taipei, Taiwan

[3]Graduate Institute of Clinical Medicine, National Taiwan University, Taipei, Taiwan

ABSTRACT

Incidences of allergic disease have recently increased worldwide. Allergen-specific immunotherapy (SIT) has long been a controversial treatment for allergic diseases. Although beneficial effects on clinically relevant outcomes have been demonstrated in clinical trials by subcutaneous immunotherapy (SCIT), there remains a risk of severe and sometimes fatal anaphylaxis. Mucosal immunotherapy is one advantageous choice because of its non-injection routes of administration and lower side-effect profile. This study reviews recent progress in mucosal immunotherapy for allergic diseases. Administration routes, antigen quality and quantity, and adjuvants used are major considerations in this field. Also, direct uses of unique probiotics, or specific cytokines, have been discussed. Furthermore, some researchers have reported new therapeutic ideas that combine two or more strategies. The most important strategy for development of mucosal therapies for allergic diseases is the improvement of antigen formulation, which includes continuous searching for efficient adjuvants, collecting more information about dominant T-cell epitopes of allergens, and having the proper combination of each. In clinics, when compared to other mucosal routes, sublingual immunotherapy (SLIT) is a preferred choice for therapeutic administration, although local and systemic side effects have been reported. Additionally, not every allergen has the same beneficial effect. Further studies are needed to determine the benefits of mucosal immunotherapy for different allergic diseases after comparison of the different administration routes in children and adults. Data collected from large, well-designed, double-blind, placebo-controlled, and randomized trials, with post-treatment follow-up, can provide robust substantiation of current evidence.

INTRODUCTION

Allergic diseases are a global health problem and result from a complex interaction between genetic and environmental factors. Type 2 T helper (Th2) immune responses play a critical role in the development of allergic diseases.[1] The cellular response to allergens, occurring in the skin, leads to atopic dermatitis. The disruption of the skin barrier initiates the subsequent atopic development toward allergic airway disease. Allergic airway disease encompasses a variety of symptoms and conditions that affect the mucosal lining of the airways, from the nose (allergic rhinitis) to the lungs (asthma).[2,3,4] Atopic dermatitis and allergic rhinitis in children have been reported to be significant risk factors for subsequent development of asthma.[5,6,7]

The treatment of allergic diseases is based on allergen avoidance, pharmacological treatment and immunotherapy. In the pharfmacological therapy of atopic dermatitis, only symptomatic anti-inflammatory and anti-allergic treatments, local or systemic, exist. However, no prophylactic or long-term treatment regimens are available at present to prevent, attenuate or cure sensitizations in atopic dermatitis patients.[8] Current available pharmacological agents for the treatment of allergic rhinitis include intranasal corticosteroids, H1 antihistamines, decongestants, cromolyn sodium, leukotriene antagonists and anticholinergics.[9] Medications to treat asthma can be classified as controllers or relievers. In controller treatment of asthma, corticosteroids and long-acting β2-agonists in fixed-combination inhalers are currently the most effective therapy.[10,11] However, long-term side effects of corticosteroid inhalation, such as osteoporosis and stunting of growth, need more consideration.[12] Inhalation of corticosteroids does not seem to modify the course of the disease significantly and is not curative because asthma symptoms and inflammation rapidly recur when treatment is discontinued. Also, a small percentage of patients do not respond to the inhaled corticosteroids.[13,14] Immunotherapy is the only controller treatment currently available with the potential to change the natural history of allergic disease and delay the allergic march observed in many atopic individuals.[15] According to current guidelines for asthma treatment (GINA), the appropriate immunotherapy requires the identification and use of a single well-defined clinically relevant antigen. Antigen-specific immunotherapy (SIT), which often uses the subcutaneous route (subcutaneous immunotherapy; SCIT), is the first choice for induction of hyporesponsiveness to the respective allergens.[16] Specific immunotherapy should be considered only after strict environmental avoidance and pharmacological intervention, including inhaled glucocorticosteroids, have failed to control asthmatic symptoms.[17] SCIT involves the injection of increasing amounts of the allergen under the skin. The long-term time course for these injections may reduce the efficacy of SCIT treatment, owing to the side effects[18] from accompanying potent Th2 adjuvants.[19, 20] The possibility of local or systemic adverse effects (such as anaphylaxis) must be considered. The review of SCIT trials[21] found that immunotherapy could reduce asthma symptoms, the need for medications and the risk of severe asthma attacks after future exposure to the allergen. Immunotherapy was also found to be possibly as effective as inhaled steroids. Overall, there was a significant improvement in asthma symptom scores (standardized mean difference: −0.59; 95% confidence interval: −0.83 to −0.35]). Furthermore, it would have been necessary to treat three patients (95% confidence interval: 3–5) with immunotherapy to avoid worsening of asthma symptoms in one patient and to treat four patients (95% confidence interval: 3–6) with immunotherapy to avoid increased medication in one patient. Immunotherapy was found to reduce allergen-specific bronchial hyper-reactivity, with some reduction in non-specific bronchial hyper-reactivity as well. However, if 16 patients were treated with immunotherapy, one would be expected to develop a local adverse reaction. Also, if nine patients were treated with immunotherapy, one would be expected to develop a systemic reaction, of any severity. A review by the Mayo Clinic in Rochester confirmed the safety and efficacy of allergen immunotherapy for allergic rhinitis and conjunctivitis, allergic forms of asthma and insect stings based on numerous well-designed scientific studies.[22] Additionally, national and international guidelines confirm the clinical efficacy of injection immunotherapy in rhinitis and asthma, as well as the safety, provided that recommendations are followed.[23] Thus far, SIT is not indicated for atopic dermatitis without accompanying allergic rhinitis or asthma, and only with the caution that it might induce exacerbations manifesting in atopic dermatitis or relapses of latent atopic dermatitis. In the study by Werfel et al.,[24] adult patients with severe forms of atopic dermatitis benefited from SIT with house dust mite (HDM) allergen extract lasting 12 months.

Improved strategies and targets for immunomodulation of allergic diseases should consider the following: (i) fewer side effects; (ii) antigen-specific modulation for long-term effects; and (iii) non-injection routes. Mucosal immunotherapy is an ideal choice based on these considerations. The mucosa-associated lymphoid tissues are the largest mammalian lymphoid organ system. Unique characteristics of the mucosal immune system, including the large production of secretary IgA antibodies and routine maintenance of immune tolerance, contribute to the efficacy of mucosal immunotherapy.[25] Akbari et al. found that pulmonary dendritic cells (DCs) collected from antigen-exposed mice produced IL-10 and lead to the development of IL-10 secretion by CD4⁺ T regulatory 1-like cells.[26] In another study, mucosal DCs derived from mesenteric lymph nodes produced transforming growth factor-β (TGF-β), which induced the development of Th3 cells.[27] However, environmental factors, level of antigen exposure and DC subtype each contribute to the results obtained in these studies. The characteristics of mucosal DCs critical for Th-type development still require further study.[28] Hufnagl et al. indicated that mucosal application of peptides is superior to systemic application for preventing both local and systemic polyallergic Th2 immune responses, which suggests that mucosal tolerance induction is an attractive strategy for the primary and secondary prevention of allergic lung pathology.[29]

MUCOSAL IMMUNOTHERAPEUTIC STRATEGIES FOR ALLERGIC DISEASES

In animal experiments, the successful application of mucosal immunotherapy for allergic diseases depends on antigen dose or formulation, mucosal adjuvants and Th-type immune manipulation. These studies have also led to the development of combination therapies. Here, we clarify these strategies and summarize the effect of treatment in the experimental model or clinic in Tables 1 and 2 (combination strategies). Although the outcome of mucosal immunotherapy in human trials still needs to be clarified, non-injection immunotherapy is an attractive therapy for allergic diseases.

Table 1. Summary of mucosal therapy for allergic diseases.

Strategy	*Animal model*	*Human studies*
Antigen dose and formulation	Recombinant form/i.n.;[43] major T epitopes/i.h.,[45] i.n.[46] and oral;[48] monomeric allergoid/oral (intragastric administration)[47]	Monomeric allergoid (Lais, Lofarma S.p.A., Milan) for mite-sensitized patients with rhinitis and intermittent asthma/ SLIT[48, 49]
Adjuvants for mucosal immunotherapy	Cholera toxin B/i.t.[52]	No recent study for allergic diseases
	CpG ODNs/i.t.[53] and i.n.[61]	CpG ODNs for atopic asthma (RDPC*)/i.h.[64]

	Chitin/oral[70] and i.n.;[71] Chitosan/i.n.[77]	Phase I/IIa study on chitin microparticles for allergic rhinitis subjects (ClinicalTrials.gov, identifier: NCT00443495)
Probiotics	*Lactobacillus* spp., *Bifidobacterium* spp./oral[81, 82]	Clinical trial results are still diverse for each study/oral[87]

Abbreviations: i.h., inhalation; i.n., intranasal; i.t., intratracheal; ODN, oligodeoxynucleotide; RDPC*, randomized, double-blind, placebo-controlled clinical trials; SLIT, sublingual immunotherapy.

Table 2. Combination effect of mucosal therapy for allergic diseases.

Combination effect	*Animal model*
Antigen–adjuvant	HDM–CTB/i.n.,[88] OVA–CTB/oral,[91] i.n.[92] or rBet–CTB/i.n.[92]
	Thiolated ODN with maleimide-activated OVA/i.t.[95]
	Der f-CS nanovaccine/i.n.;[98] chitosan-pDer p2[100] or pDer p1 nanoparticles/oral[101]
	Mucoadhesive chitosan-formulated OVA/sublingual[99]
Antigen–probiotics	Co-administration/oral[102] and i.n.[104]
	Recombinant Der p1 111–139[105] and Bet v 1[107] producing probiotics/i.n.
Induction of Th1, Tr or anti-Th2 immune response	IL-13 peptide-based virus-like particle vaccine/i.n.[108]
	Chitosan/IFN-γ pDNA nanoparticles (CIN)/i.n.[109]
	Coadministration of live lactococci producing IL-12 and BLG/i.n.[110]

Abbreviations: BLG, bovine β-lactoglobulin; CS, chitosan; CIN, chitosan/IFN-γ pDNA nanoparticle; CTB, cholera toxin B; HDM, house dust mite; Der p, *Dermatophagoides pteronyssinus*; i.h., inhalation; i.n., intranasal; i.t., intratracheal; ODN, oligodeoxynucleotide; OVA, ovalbumin.

Antigen dose and Formulation

Mechanisms of mucosal tolerance induced by high- and low-dose antigens are different.[30, 31] High doses of oral or mucosal antigen lead to T cell receptor activation without costimulation and the simultaneous presence of inhibitory ligands leads to anergy[32] or deletion.[16, 33] Low-dose tolerance is induced by regulatory cells, such as Th3,[18, 19] T regulatory 1 cells[34, 35] and CD4$^+$CD25$^+$regulatory cells.[36, 37] Also, CD8$^+$ T cells, *via* the production of

TGF-β,[38, 39] and γ/δT cells[40] have been identified as acting as regulatory cells during oral tolerance induction. However, the characterization and function of, and the interactions between, different types of regulatory T cells still require further study.[41]

The use of allergen extracts brings forth the possibility of *de novo* sensitization against natural allergen components delivered in allergen extract preparations. Alternatively, using the allergen in recombinant form,[42, 43, 44] or only the major T-cell epitopes,[12, 45] has enhanced treatment efficacy and safety. Additionally, peptide immunotherapy using peptides against multiple immunodominant allergen-specific T-cell epitopes is a safe and promising strategy for allergy control.[46] Allergens can be further modified through the production of allergoids, which are allergen extracts that have been polymerized into larger aggregates by a chemical reaction. According to the theoretical concept, this chemical modification is hypothesized to result in reduced allergenicity and maintained immunogenicity in mouse models and in clinics.[47, 48, 49]

Takagi *et al.* found that mice orally fed with transgenic rice seeds co-expressing the Cryj I and Cryj II peptide-defining T-cell epitopes before challenge with cedar pollen inhibited the development of serum allergen-specific IgE and IgG antibodies and Th cell proliferative responses. The serum levels of IL-4, IL-5, IL-13 and histamine were significantly decreased, and the development of pollen-induced clinical symptoms was inhibited in this mouse model. These results indicate the potential of transgenic rice seeds in the production and mucosal delivery of allergen-specific T-cell epitope peptides for the induction of oral tolerance to pollen allergens.[50]

Adjuvants for Mucosal Immunotherapy

Cholera Toxin B (CTB)

CTB is produced by *Vibrio cholera*. Despite being a transmucosal carrier-delivery system for induction of peripheral immunological tolerance, CTB has also been used as a non-toxic mucosal immunomodulatory adjuvant through its binding ability to the asialo-GM-1 receptor on B cells, T cells and DCs.[51] Smits *et al.* found that intratracheal administration of CTB can suppress allergic inflammation through the induction of airway luminal IgA secretions in a TGF-b-dependent manner, which is necessary for its preventive and curative effect.[52]

CpG Oligodeoxynucleotides (CpG ODNs)

CpG ODNs contain unmethylated CpG motifs, which confer the immunostimulatory properties of bacterial DNA through the ability to induce immune responses.[53,54] CpG ODNs can enhance Th1 immune responses,[55,56] suppress Th2 responses[57,58] and induce regulatory T cells.[59,60] These findings suggest that CpG ODNs can be a therapeutic approach for the treatment of Th2-mediated allergic asthma. The immunomodulatory effects of CpG ODNs on the development of HDM *Dermatophagoides farinae* (Der f)-induced airway inflammation and remodeling in mice have been reported.[53] Simultaneous intratracheal instillation of CpG ODNs with Der f at the first allergen exposure showed significant inhibition of inflammation in a dose-dependent manner of CpG. For intranasal therapy, Ramaprakash *et al.* found that intranasal CpG therapy attenuated experimental fungal asthma in both a TLR9-dependent and an independent manner.[61] A clinical study also showed CpG ODNs to have promising experimental and clinical results in allergic rhinitis.[62,63] However, a subsequent study of CpG ODNs (delivered by nebulization) showed fewer benefits in asthma.[56] Although CpG ODNs could increase the expression of IFN-γ and IFN-γ-inducible genes, they did not sufficiently inhibit allergen-induced responses in asthmatic subjects.[56,64] However, the ability of CpG ODNs to promote Th1 responses has already led to the design of phase I clinical trials with allergy patients.[65,66]

Chitin/Chitosan

Chitin is a key structural component of helminths, arthropods and fungi.[67,68] The immune response to chitin is still considered controversial.[69,70,71,72,73] Oral[70] and intratracheal[71] administration of chitin has been shown to downmodulate allergic airway inflammation in murine models. Controversially, Reese *et al.*[73] found that intranasal administration of chitin resulted in eosinophil and basophil accumulation in helminth-infected mice. The sensitizing role for chitin may, through alternatively activated macrophages, mediate eosinophil recruitment *via* leukotriene B4 production. Several factors, such as the administration route or particle size, may account for the Th1 *vs.* Th2 response to chitin. There are still many controversial and unsolved issues in this field to be discussed.[74]

Chitosan is formed naturally through the action of chitin deacetylases or by the deacetylation of chitin oligosaccharides.[75] It is a natural biodegradable mucoadhesive polysaccharide derived from crustacean shells. This slowly degrading polymer has been shown to increase transcellular and paracellular transport of macromolecules across intestinal epithelial monolayers.[75,

[76] Chen et al. found that soluble chitosan delivered intranasally with water during allergen sensitization[77] could attenuate airway inflammation in the Der f-induced murine allergy model. Furthermore, a phase I/IIa study on chitin microparticles delivered by the nasal route to subjects suffering from allergic rhinitis has entered clinical trials.

Probiotics

Using unique strains of probiotics can improve immunomodulatory effects of mucosal therapy.[78] Probiotics are dietary supplements that contain beneficial bacteria such as *Lactobacillus GG* (LGG) and are effective in preventing early atopy in children through the modulation of intestinal microbiota.[79, 80] In animal models of asthma, orally administered probiotics can strain-dependently decrease allergen-specific IgE production and modulate systemic cytokine production.[78] Certain probiotics (LGG or *Bifidobacterium lactis* and *Lactobacillus reuteri*) have been shown to decrease airway hyper-responsiveness and inflammation by inducing regulatory mechanisms.[81, 82] However, definitive conclusions are lacking because of the variety of experimental protocols used. Before using probiotics for asthma prevention, further studies using molecular methods to test for microbiota[83] and large-scale analyses are required.

In clinics, the implementation of probiotics for primary prevention early in infancy is increasingly being discussed as the optimal time point for intervention. A recent meta-analysis of several clinical trials suggests that pre- and post-natal probiotic interventions are effective in preventing the development of pediatric dermatitis,[84] although the effects on allergy development are less clear. Additionally, a double-blind, placebo-controlled study was conducted to examine the effectiveness of LGG and *L. gasseri* TMC0356 in alleviating Japanese cedar pollinosis, a seasonal allergic rhinitis caused by Japanese cedar pollen. Fermented milk prepared with these two bacteria, or placebo yoghurt, was administered to 40 subjects with a clinical history of Japanese cedar pollinosis for 10 weeks.[85] The allergic rhinitis alleviating effects of LGG and *L. gasseri* (TMC0356) might be due at least partly to their specific downregulation of the human Th2 immune response. In the clinical trials, randomized, placebo-controlled, double-blind studies of *Lactobacillus plantarum* No. 14 administration to female students with seasonal allergic diseases found *L. plantarum* No. 14 to strongly induce the gene expression of Th1-type cytokines. This study highlights the clinical effects of *L. plantarum* No. 14 on seasonal allergic diseases,[86] but a Cochrane systematic review concluded that, when the results for the different probiotic strains used in clinical trials are pooled, probiotics are not effective for the treatment of atopic dermatitis.[87] Also, synbiotics (90% short-chain galacto-oligosaccharides, 10% long-chain fructo-oligosaccharides: Immunofortis and *Bifidobacteriu breve* M-16V) had no effect on bronchial inflammation and the late asthmatic response but did significantly reduce systemic production of Th2-cytokines after allergen challenge and improved peak expiratory flow for patients with asthma and HDM allergy.[88]

The other randomized, double-blind, placebo-controlled, allergy-prevention trial used a combination of LGG, *L. rhamnosus* LC705, *B. breve* Bb99 and *Propionibacterium freudenreichii* ssp. *Shermanii* prenatally and during the 6 months after birth. Probiotics might also enhance IgA responses in the gut and regulate inflammatory cytokines, both of which are immunomodulatory effects that could prevent progression of atopy and potential development of disease.[89] To date, the evidence suggesting that probiotics can be used to treat or prevent allergic diseases in children remains controversial. Data from the recent randomized, double-blinded, placebo-controlled clinical trials using probiotics for the treatment of allergic diseases in children have been collected but are insufficient to strongly recommend probiotics as a standard treatment or preventative measure for pediatric allergic disease. Additional studies with standardized designs, bacterial strains, dosages and durations should be performed for different allergic diseases of children.[90]

Combination Effect

An important issue in mucosal immunotherapy is how to improve efficacy. Combining the proper adjuvant with the specific allergens will contribute to more efficient antigen-SIT. We have focused on examples of the combination strategy for the treatment of allergic diseases in mouse models, which are summarized in Table 2. The efficacy of the combination strategy in human studies is still unclear.

CTB–Ag

Many studies have shown that the mucosal administration of relevant antigens (Ag) together with, and preferably linked to, the non-toxic B subunit of cholera toxin (CTB) by either oral[91] or intranasal[92] administration represents a highly effective way to maximize oral tolerance induction for immunotherapeutic purposes and is superior to the administration of Ag alone. Sun et al.[51] found that using *N*-suc-cinimidyl [3-(2-pyridyl)dithio]-propionate as a conjugator to link different antigens can achieve immunotolerance through a single oral administration of low-dose antigen. In a study of the HDM allergen, Lee and Mo found that immune tolerance could be induced through intranasal application of a HDM and CTB conjugate in the murine allergic rhinitis model and that the effect can last for 4 weeks.[92] Interestingly, Wiedermann et al. found that the tolerogenic or immunogenic properties of CTB strongly depend on the nature of the coupled allergen.[93] The clinical trials for mucosal respiratory or gastrointestinal allergies have yet to be performed.[94]

CpG ODN-Conjugated Ag (CpG ODN–Ag)

Shirota et al.[95] found antigen-conjugated CpG ODN (mixing thiolated CpG ODN with maleimide-activated ovalbumin (OVA)) to be a novel antigen-specific immunomodulator that could regulate murine airway eosinophilia and Th2 cells. Interestingly, the CpG ODN–Ag conjugate was 100-fold more effective than the unconjugated mixture at inducing Th1 differentiation *in vitro* in an IL-12-dependent manner. Mucosal or intratracheal administration of CpG ODN with allergens or CpG ODN–Ag has also been applied to the different animal models of allergic disease.[96] A variety of clinical trials are currently ongoing to determine the efficacy of CpG ODNs as a therapeutic tool for atopic diseases. In the review by Gupta and Agrawal, therapeutic applications of CpG ODNs in allergy and asthma are discussed. CpG ODNs may be used alone or as an adjuvant for immunotherapy to treat these disorders.[97]

Chitosan–Ag

Liu et al.[98] tested immunotherapeutic efficacy of intranasal administration of Der f entrapped in chitosan microparticles in sensitized mice. Mice treated with the intranasal Der f–chitosan nanovaccine prior to challenge displayed alleviated airway hyper-reactivity, lung inflammation and mucus production, and had fewer eosinophilic cells in the bronchoalveolar lavage fluid (BALF). The IL-4 cytokine levels in BALF and from splenocytes were reduced, but IgA and IFN-g in serum were increased. Liu et al. also observed that IL-10 levels were increased among splenocytes and in BALF, which contributed to the increase in regulatory T cells in the spleen. These results illustrate how intranasal administration of the Der f–chitosan nano-vaccine plays a role in immunological protection against murine allergic asthma by inducing regulatory T cells and Th1-type reactions.

Interestingly, Saint-Lu et al.[99] tested two types of chitosan microparticles, differing in size and surface charge, for the *in vitro* capacity to improve antigen uptake and presentation by murine bone marrow-derived dendritic cells or purified oral antigen-presenting cells (CD11b$^+$ CD11c$^-$ cells in buccal floor and lingual tissues). Also, OVA-sensitized BALB/c mice were treated sublingually with soluble or chitosan-formulated OVA twice a week for 2 months. Saint et al. found that only a mucoadhesive, especially one that is positively charged, and a micro particulate form of chitosan enhances OVA uptake, processing and presentation by murine bone marrow-derived dendritic cells, and oral antigen-presenting cells. Sublingual administration of such chitosan-formulated OVA particles enhances tolerance induction in mice with established asthma. Mucoadhesive chitosan microparticles represent a promising formulation for use in sublingual allergy vaccines. In other studies, chitosan nanoparticles, containing plasmid DNA encoding the HDM allergen*Dermatophagoides pteronyssinus* 2 (Der p2)[100] or Der p1,[101] induced IFN-γ in serum and prevented subsequent sensitization of Th2 cell-regulated allergen-specific IgE responses following oral vaccination in mice. The data on Der p2 also indicate that oral administration of chitosan–Der p2 DNA nanoparticles results in the expression of Der p2 by epithelial cells in both the stomach and small intestine. Levels of IFN-γ from chitosan–DNA nanoparticle-treated mice were higher than those in the phosphate-buffered saline-treated group, the group receiving chitosan nanoparticles without the Der p2 plasmid, and those given the naked Der p2 plasmid.[100]

Co-Administration with Ag and Probiotics/Recombinant Probiotics

Recent experimental studies have shown a reduction in IgG1 or IgE when the specific lactic acid bacteria (LAB) strain *Lactobacillus casei* strain Shirota was orally administered[102] or injected[103] together with the particular allergen. In the murine model of birch pollen allergy, Repa *et al.* demonstrated that intranasal co-application of *Lactococcus lactis* and *L. plantarum* strains with the recombinant Bet v 1 protein, before and after sensitization with the allergen, resulted in a shift from Th2 to Th1 responses characterized by a marked reduction in the IgE/IgG2a ratio and increased IFN-γ production.[104] Successful immunomodulation was further demonstrated by the suppression of allergen-induced basophil degranulation. These results indicate that combined mucosal application of LAB with a specific allergen could be another prophylactic and therapeutic approach to allergy treatment. Furthermore, recombinant *L. plantarum* expressing the HDM antigen Der p1 also could reduce Th2 cytokines in sensitized mice.[105] Similarly, therapeutic effects from administration of recombinant *L. lactis* and *L. plantarum* strains expressing Bet v 1 have also been reported. Intranasal or intragastric pretreatment with the Bet v 1-producing LAB (*L. lactis* and *L. plantarum*) strains led to significantly reduced allergen-specific IgE and increased IgG2a levels, indicating a shift to non-allergic Th1 responses.[106, 107] However, in sensitized mice, mucosal application of these recombinant strains did not sufficiently reduce allergic immune responses, which indicates that LAB-inducing immunosuppressive cytokines, such as TGF-β or IL-10, rather than Th1-like cytokines, may be more beneficial in therapeutic settings.

Mucosal Induction of Th1, Regulatory T or Anti-Th2 Immune Response

Ma et al.[108] showed that intranasal vaccination with the IL-13 peptide-based virus-like particle vaccine could induce more effective suppression than subcutaneous immunization, characterized by OVA-driven Th2 patterns of antibody responses, airway IL-13 and eosinophil accumulation. Another example is the study that used chitosan/IFN-γ pDNA nanoparticles to generate *in situ* production of IFN-γ and *in vivo* effects.[109] Mucosal chitosan/IFN-γ pDNA nanoparticle therapy was found to have both prophylactic and therapeutic effects in the OVA animal model by reducing allergen-induced airway inflammation and airway hyper-responsiveness. This effect was dependent on signal transducer and activator of transcription 4 (STAT4) signaling. Importantly, chitosan alone could not efficiently alleviate airway inflammation in this study. Similarly, intranasal co-administration of live lactococci producing IL-12 and bovine β-lactoglobulin, a major cows› milk allergen, could also improve the efficiency of tolerance induction by intranasal administration of bovine β-lactoglobulin.[110] Interestingly, IL-10-inducing adjuvants such as 1alpha, 25-dihydoxyvitamin D3 plus dexamethasone and *L. plantarum*[111] can both significantly enhance sublingual immunotherapy (SLIT) efficacy in a murine asthma model.

ROUTES OF MUCOSAL THERAPY

Many routes for mucosal immunotherapy have been proposed and investigated, including oral (straight swallow), nasal or trachea, and sublingual. Here, we have summarized each route and discussed the efficiency of the immunotherapy. The dose, side effects and technical limitations for clinical use are taken into consideration.

Oral

Oral delivery is attractive because of the ease of administration. This type of administration has direct access to the gastrointestinal tract, which has an abundant mucosal immune system. Oral administration offers improved convenience and leads to compliance with patients, thereby reducing overall healthcare costs.[112] Many animal studies have reported that immune therapy for allergic diseases could be given by oral administration of OVA or purified allergens.[113, 114, 115] In addition to the adjustment of antigen dose and the frequency of dosing[116] for the mucosal adjuvants, including CTB, CpG ODNs and chitosan as previously mentioned, oral tolerance is enhanced by IL-4,[19] IL-10,[19,116, 117] anti-IL-12,[118] TGF-â,[119] Flt-3 ligand[120] and anti-CD40 ligand.[121] Antigen absorption following oral administration is less dangerous in regards to the airways or skin, which suggests that the mouth is likely to be a tolerogenic site. Interestingly, in the study of Allam *et al.*, the administration of the Phlp5 allergen (a major grass pollen allergen) was found to induce oral mucosal Langerhans cells to bind to Phlp5 in a dose- and time-dependent manner and to lead to an increased production of the tolerogenic cytokines IL-10 and TGF-β and an enhanced migratory capacity, but decelerated maturation, of oral Langerhans cells.[122] However, in clinical trials, oral immunotherapy requires doses thousands of times higher than those for conventional SCIT owing to gastrointestinal degradation after ingestion.[123] Moreover, gastrointestinal side effects seemed to increase with increasing oral administration doses of pollen birch extract in the birch pollinosis study.[124] For these reasons, it is no longer considered a feasible option for immunotherapy in clinics.[125]

Nasal or Tracheal Administration

Nasal or tracheal administration is a promising route for immunotherapy because doses are lower than for the oral route. Through the intratracheal administration route, Haneda *et al.* found that TGF-β secreted by T cells plays an important role in the down-modulation of immune responses to high doses of antigens, which might otherwise induce deleterious inflammation in the airway mucosal tissues.[126] Honey *et al.* addressed the mechanisms underlying peripheral T-cell tolerance following intranasal or inhalation administration of a high dose of immunogenic peptide (p1 111–139) derived from the HDM allergen Der p1[127, 128] and found this treatment to involve a downregulation of the Th cell response. Data from clinical trials for intranasal immunotherapy have been reported,[129, 130] but the administration of this therapy required great skill and sometimes the immunotherapy itself was found to provoke allergic responses in patients.[131] The clinical trial results for tracheal administration have suggested that clinical efficacy is unproven and that the risk/benefit ratio is unfavorable.[132, 133]

SLIT

The mechanisms behind SLIT include the production of blocking IgG4 antibodies,[134, 135] the presence of high numbers of tolerogenic DC subsets, the induction of regulatory T cells,[136, 137] and the programming of the immune system toward a regulatory state of unresponsiveness to specific allergens. SLIT may also increase IL-10, which has a clear role in suppressing allergic immune responses.[138] The study of O'Hehir *et al.* found that TGF-β mediates the immunological suppression seen early in clinically effective sublingual HDM immunotherapy and the increase in regulatory T cells with suppressor function.[139] In a mouse model of rhinitis,[140] SLIT can reduce allergic symptoms. Brimnes *et al.* established a mouse model using a clinically relevant allergen to produce hallmarks of allergic rhinitis.[141] Using this model, SLIT was demonstrated to reduce allergic symptoms in a time and dose-dependent manner.

Comparisons of skin biopsies from the subcutaneous injection sites[142] and the oral mucosa[143] of SLIT-treated allergic subjects confirmed the negligible presence of inflammatory cells.[144] In addition, the high doses administered with oral immunotherapy resulted in significant local reactions, including gastrointestinal bleeding, which were possibly able to interfere with antigen absorption and thus with immunization. However, it was evident that the sublingual mucosa could tolerate higher allergen levels than the mucosa in the nose or skin.[145]

Clinically, use of the sublingual route is supported by numerous controlled trials showing its efficacy in asthma and rhinitis in adults and children.[143, 146] Additionally, no severe adverse events occurred during the trial, and the most common adverse events were mild asthma attack and local rash. Cao et al. evaluated the safety and efficacy of SLIT with Der f drops in Der f allergic asthma and/or rhinitis patients.[147] After 25 weeks of treatment, the SLIT and placebo groups did not show significant difference in the production of Der f-specific IgE antibody, while specific IgG4 increased significantly in SLIT patients after 25 weeks of treatment compared to those in a control group. The peak expiratory flow rates and rhinitis symptoms in the SLIT group improved, and the medical score of asthma significantly decreased. Furthermore, no severe adverse events occurred in the trial, and the most common adverse events were mild asthma and local rash. However, SLIT is not always a safe alternative to subcutaneous therapy.[148] The most frequent side effect of oral sublingual therapy is itching after antigen intake. Shortness of breath, wheezing and severe asthma attacks have also been reported,[149] yet SLIT is now accepted by the World Health Organization as a valid alternative to subcutaneous therapy in children.[63] The magnitude of clinical efficacy is reported to range between 20% and 50%, owing to the reduction of symptoms and medical scores, and is greater than the effects of placebo therapy.[43] Optimal therapeutic doses are still unknown, but range from three to five times or as high as 375 times the doses used in subcutaneous immunotherapy. The review written by Larenas-Linnemann discusses the shortcomings of SLIT in terms of efficacy, dosing, timing of treatment and patient selection, which all need to be taken into serious consideration.[150]

Evidence for beneficial effects from SLIT has been confirmed in children with allergic rhinitis[151, 152, 153] or asthma[154] caused by pollen exposure. SLIT was also found to prevent the progression from allergic rhinitis to asthma.[155] However, for HDM-induced asthma, therapeutic effects for patients cannot be determined without data from randomized, large population-based, high-quality studies.[146, 156, 157, 158,159, 160, 161]

As discussed in a systematic review of SLIT for allergic rhinitis,[162] the therapeutic manipulation of SLIT should be interpreted with caution. The use of different allergens, optimal doses, duration of treatment and the application in children or adults should be further examined for SLIT.[155] Interestingly, in the study of Marogna et al., the clinical effects of a monomeric allergoid were assessed across three different maintenance doses in mite-sensitized patients with rhinitis and intermittent asthma. In this clinical trial, SLIT with monomeric allergoids produced clinically significant results across a wide range of doses. The absence of significant side effects, even at high doses, was probably due to the low level of allergenicity.[163]

For immunotherapy of atopic dermatitis, SCIT is not indicated for use because of the likelihood that it could induce exacerbations of manifest atopic dermatitis or relapses of latent atopic dermatitis. In children treated either with SLIT or placebo, Pajno et al. reported a benefit from SLIT exclusively in children with atopic dermatitis sensitized against HDMs and in those with mild-to-moderate variants of atopic dermatitis, adjudged by the SCORAD index (SCORing Atopic Dermatitis).[164] However, two patients in the SLIT group were excluded from the study owing to intense generalized flush reactions occurring within 1 h after sublingual allergen administration.

CONCLUSION

The putative value of SIT is not only in its use as a causal therapy for already manifest sensitizations, but also in its use as a preventive measure to avoid the development of further sensitizations and to counteract the atopic march early in life. This is of particular importance in light of recent developments that provide clear evidence for a genetically determined skin barrier dysfunction that predisposes a subgroup of patients with atopic dermatitis to the manifestation of numerous sensitizations and concomitant asthma.[165] Although SIT is an effective treatment for many allergic diseases, certain drawbacks, such as the long duration of treatment and the risk of anaphylactic reactions, need to be taken into account. Many gene-based strategies for immunotherapy aimed at reversing or preventing abnormal immune regulation and restoring Th1-predominated responses or regulatory T cell function have been developed,[166, 167, 168, 169, 170] but therapeutic efficacy depends highly on delivery efficiency and target selection.[171] Mucosal immunotherapy is a better strategy for treating allergic disease because of its non-injection routes and low side-effect profile. SLIT is a better choice for prophylactic and therapeutic approaches to allergy treatment, but the outcomes for allergic disease from the use of different sensitizing allergens will still need further definition. Also, a challenge remains in evaluating whether results from experimental animal studies will hold true in humans.

Continuous improvements have been made in allergen preparation, such as the introduction of highly purified allergoids[172] and recombinant allergens,[173] the targeting of dominant T-cell epitopes of allergen,[12, 45] and the refinement of treatment schedules,[174] as well as in the concomitant use of adjuvants.[175] Additional types of mucosal adjuvant candidates have been explored, including living parasites,[176] IL-10, TGF-β-inducing compounds[111, 177] and natural compounds derived from plants and herbs.[178, 179]

The collection of data from large, well-designed, double-blind, placebo-controlled, randomized trials with post-treatment follow-up, will provide robust substantiation of current evidence. SCIT has demonstrated long-term clinical effects and the potential to preventing the development of asthma in children with allergic rhinoconjunctivitis for up to 7 years after treatment termination.[180] The role of SCIT in adult asthma treatment is still limited. Mucosal immunotherapy studies in adults and children with allergic diseases that use different types of allergens and different routes of administration and evaluate the side effects from each route will improve our knowledge on this issue.[181]

We anticipate that continued growth in the understanding of imunotherapeutic strategies for allergic diseases will offer therapies with lower doses, greater safety and more effective application in the future.

REFERENCES

1. Bloomfield SF, Stanwell-Smith R, Crevel RW, Pickup J. Too clean, or not too clean: the hygiene hypothesis and home hygiene. *Clin Exp Allergy* 2006; 36: 402–425.

2. Togias A. Mechanisms of nose-lung interaction. *Allergy* 1999; 54 (Suppl 57): 94–105.

3. Passalacqua G, Ciprandi G, Canonica GW. United airways disease: therapeutic aspects. *Thorax* 2000; 55 (Suppl 2): S26–S27.

4. Passalacqua G, Ciprandi G, Canonica GW. The nose–lung interaction in allergic rhinitis and asthma: united airways disease. *Curr Opin Allergy Clin Immunol* 2001; 1: 7–13.

5. Bousquet J, Khaltaev N, Cruz AA, Denburg J, Fokkens WJ, Togias A et al. Allergic Rhinitis and its Impact on Asthma (ARIA) 2008 update (in collaboration with the World Health Organization, GA²LEN and AllerGen).*Allergy* 2008; 63 (Suppl 86): 8–160.

6. Linneberg A, Henrik Nielsen N, Frolund L, Madsen F, Dirksen A, Jorgensen T. The link between allergic rhinitis and allergic asthma: a prospective population-based study. The Copenhagen Allergy Study. *Allergy* 2002; 57: 1048–1052.

7. Zheng T, Yu J, Oh MH, Zhu Z. The atopic march: progression from atopic dermatitis to allergic rhinitis and asthma. *Allergy Asthma Immunol Res* 2011;3: 67–73.

8. Akdis CA, Akdis M, Bieber T, Bindslev-Jensen C, Boguniewicz M, Eigenmann P*et al*. Diagnosis and treatment of atopic dermatitis in children and adults: European Academy of Allergology and Clinical Immunology/American Academy of Allergy, Asthma and Immunology/ PRACTALL Consensus Report. *J Allergy Clin Immunol* 2006; 118: 152–169.

9. Greiner AN, Meltzer EO. Overview of the treatment of allergic rhinitis and nonallergic rhinopathy. *Proc Am Thorac Soc* 2011; 8: 121–131.

10. Chung KF, Caramori G, Adcock IM. Inhaled corticosteroids as combination therapy with beta-adrenergic agonists in airways disease: present and future.*Eur J Clin Pharmacol* 2009; 65: 853–871.

11. Papi A. Treatment strategies in mild asthma. *Curr Opin Pulm Med* 2009; 15: 29034.

12. Pedersen S. Do inhaled corticosteroids inhibit growth in children? *Am J Respir Crit Care Med* 2001; 164: 521–535.

13. Heaney LG, Robinson DS. Severe asthma treatment: need for characterising patients. *Lancet* 2005; 365: 974–976.

14. Wenzel S. Severe asthma in adults. *Am J Respir Crit Care Med* 2005; 172: 149–160.

15. Nagai H, Teramachi H, Tuchiya T. Recent advances in the development of anti-allergic drugs. *Allergol Int* 2006; 55: 35–42.

16. Chen Y, Inobe J, Marks R, Gonnella P, Kuchroo VK, Weiner HL. Peripheral deletion of antigen-reactive T cells in oral tolerance. *Nature* 1995; 376: 177–180.

17. Bousquet J, Lockey R, Malling HJ, Alvarez-Cuesta E, Canonica GW, Chapman MD *et al*. Allergen immunotherapy: therapeutic vaccines for allergic diseases. World Health Organization. American academy of Allergy, Asthma and Immunology. *Ann Allergy Asthma Immunol* 1998; 81: 401–405.

18. Marth T, Zeitz Z, Ludviksson B, Strober W, Kelsall B. Murine model of oral tolerance. Induction of Fas-mediated apoptosis by blockade of interleukin-12.*Ann NY Acad Sci* 1998; 859: 290–294.

19. Inobe J, Slavin AJ, Komagata Y, Chen Y, Liu L, Weiner HL. IL-4 is a differentiation factor for transforming growth factor-beta secreting Th3 cells and oral administration of IL-4 enhances oral tolerance in experimental allergic encephalomyelitis. *Eur J Immunol* 1998; 28: 2780–2790.

20. Weiner HL. Induction and mechanism of action of transforming growth factor-beta-secreting Th3 regulatory cells. *Immunol Rev* 2001; 182: 207–214.

21. Abramson MJ, Puy RM, Weiner JM. Injection allergen immunotherapy for asthma. *Cochrane Database Syst Rev* 2010; (8): CD001186.

22. Rank MA, Li JT. Allergen immunotherapy. *Mayo Clin Proc* 2007; 82: 1119–1123.

23. Passalacqua G, Durham SR. Allergic rhinitis and its impact on asthma update: allergen immunotherapy. *J Allergy Clin Immunol* 2007; 119: 881–891.

24. Werfel T, Breuer K, Rueff F, Przybilla B, Worm M, Grewe M et al. Usefulness of specific immunotherapy in patients with atopic dermatitis and allergic sensitization to house dust mites: a multi-centre, randomized, dose-response study. *Allergy* 2006; 61: 202–205.

25. Holmgren J, Czerkinsky C. Mucosal immunity and vaccines. *Nat Med* 2005;11: S45–S53.

26. Akbari O, DeKruyff RH, Umetsu DT. Pulmonary dendritic cells producing IL-10 mediate tolerance induced by respiratory exposure to antigen. *Nat Immunol* 2001; 2: 725–731.

27. Schroder NW. The role of innate immunity in the pathogenesis of asthma. *Curr Opin Allergy Clin Immunol* 2009; 9: 38–43.

28. Holgate ST. Pathogenesis of asthma. *Clin Exp Allergy* 2008; 38: 872–897.

29. Hufnagl K, Focke M, Gruber F, Hufnagl P, Loupal G, Scheiner O et al. Airway inflammation induced after allergic poly-sensitization can be prevented by mucosal but not by systemic administration of polypeptides. *Clin Exp Allergy* 2008; 38: 1192–1202.

30. Mowat AM. Anatomical basis of tolerance and immunity to intestinal antigens. *Nat Rev Immunol* 2003; 3: 331–341.

31. Wu HY, Weiner HL. Oral tolerance. *Immunol Res* 2003; 28: 265–284.

32. Mayer L, Shao L. Therapeutic potential of oral tolerance. *Nat Rev Immunol* 2004; 4: 407–419.

33. Appleman LJ, Boussiotis VA. T cell anergy and costimulation. *Immunol Rev* 2003; 192: 161–180.

34. Groux H, O'Garra A, Bigler M, Rouleau M, Antonenko S, de Vries JE et al. A CD4⁺ T-cell subset inhibits antigen-specific T-cell responses and prevents colitis. *Nature* 1997; 389: 737–742.
35. Battaglia M, Blazar BR, Roncarolo MG. The puzzling world of murine T regulatory cells. *Microbes Infect* 2002; 4: 559–566.
36. Groux H. Type 1 T-regulatory cells: their role in the control of immune responses. *Transplantation* 2003; 75: 8S–12S.
37. Sakaguchi S, Sakaguchi N, Shimizu J, Yamazaki S, Sakihama T, Itoh M et al. Immunologic tolerance maintained by CD25⁺ CD4⁺ regulatory T cells: their common role in controlling autoimmunity, tumor immunity, and transplantation tolerance. *Immunol Rev* 2001; 182: 18–32.
38. Thorstenson KM, Khoruts A. Generation of anergic and potentially immunoregulatory CD25⁺CD4 T cells *in vivo* after induction of peripheral tolerance with intravenous or oral antigen. *J Immunol* 2001; 167: 188–195.
39. Miller A, Lider O, Roberts AB, Sporn MB, Weiner HL, Suppressor T cells generated by oral tolerization to myelin basic protein suppress both in vitro and in vivo immune responses by the release of transforming growth factor beta after antigen-specific triggering. *Proc Natl Acad Sci USA* 1992; 89: 421–425.
40. Horwitz DA, Zheng SG, Gray JD. The role of the combination of IL-2 and TGF-beta or IL-10 in the generation and function of CD4⁺ CD25⁺ and CD8⁺regulatory T cell subsets. *J Leukoc Biol* 2003; 74: 471–478.
41. Wan YY. Regulatory T cells: immune suppression and beyond. *Cell Mol Immunol* 2010; 7: 204–210.
42. Valenta R, Lidholm J, Niederberger V, Hayek B, Kraft D, Gronlund H. The recombinant allergen-based concept of component-resolved diagnostics and immunotherapy (CRD and CRIT). *Clin Exp Allergy* 1999; 29: 896–904.
43. Wiedermann U. Prophylaxis and therapy of allergy by mucosal tolerance induction with recombinant allergens or allergen constructs. *Curr Drug Targets Inflamm Allergy* 2005; 4: 577–583.
44. Niederberger V, Horak F, Vrtala S, Spitzauer S, Krauth MT, Valent P et al. Vaccination with genetically engineered allergens prevents progression of allergic disease. *Proc Natl Acad Sci USA* 2004; 101 (Suppl 2): 14677–14682.
45. Hoyne GF, O'Hehir RE, Wraith DC, Thomas WR, Lamb JR. Inhibition of T cell and antibody responses to house dust mite allergen by inhalation

of the dominant T cell epitope in naive and sensitized mice. *J Exp Med* 1993; 178: 1783–1788.

46. Hufnagl K, Winkler B, Focke M, Valenta R, Scheiner O, Renz H et al. Intranasal tolerance induction with polypeptides derived from 3 noncross-reactive major aeroallergens prevents allergic polysensitization in mice. *J Allergy Clin Immunol* 2005; 116: 370–376.

47. Petrarca C, Lazzarin F, Pannellini T, Iezzi M, Braga M, Mistrello G et al. Monomeric allergoid intragastric administration induces local and systemic tolerogenic response involving IL-10-producing $CD4^+CD25^+$ T regulatory cells in mice. *Int J Immunopathol Pharmacol* 2010; 23: 1021–1031.

48. Gammeri E, Arena A, D'Anneo R, La Grutta S. Safety and tolerability of ultra-rush (20 minutes) sublingual immunotherapy in patients with allergic rhinitis and/or asthma. *Allergol Immunopathol (Madr)* 2005; 33: 221–223.

49. Gammeri E, Arena A, D'Anneo R, La Grutta S. Safety and tolerability of ultra-Rush (20 minutes) sublingual immunotherapy in patients with allergic rhinitis and/or asthma. *Allergol Immunopathol (Madr)* 2005; 33: 142–144.

50. Takagi H, Hiroi T, Yang L, Tada Y, Yuki Y, Takamura K et al. A rice-based edible vaccine expressing multiple T cell epitopes induces oral tolerance for inhibition of Th2-mediated IgE responses. *Proc Natl Acad Sci USA* 2005; 102: 17525–17530.

51. Sun JB, Holmgren J, Czerkinsky C. Cholera toxin B subunit: an efficient transmucosal carrier-delivery system for induction of peripheral immunological tolerance. *Proc Natl Acad Sci USA* 1994; 91: 10795–10799.

52. Smits HH, Gloudemans AK, van Nimwegen M, Willart MA, Soullie T, Muskens F et al. Cholera toxin B suppresses allergic inflammation through induction of secretory IgA. *Mucosal Immunol* 2009; 2: 331–339.

53. Hirose I, Tanaka H, Takahashi G, Wakahara K, Tamari M, Sakamoto T et al. Immunomodulatory effects of CpG oligodeoxynucleotides on house dust mite-induced airway inflammation in mice. *Int Arch Allergy Immunol* 2008; 147: 6–16.

54. Krieg AM. CpG motifs in bacterial DNA and their immune effects. *Annu Rev Immunol* 2002; 20: 709–760.

55. Klinman DM, Yi AK, Beaucage SL, Conover J, Krieg AM. CpG motifs present in bacteria DNA rapidly induce lymphocytes to secrete interleukin

6, interleukin 12, and interferon gamma. *Proc Natl Acad Sci USA* 1996; 93: 2879–2883.

56. Fonseca DE, Kline JN. Use of CpG oligonucleotides in treatment of asthma and allergic disease. *Adv Drug Deliv Rev* 2009; 61: 256–262.

57. Shirota H, Sano K, Kikuchi T, Tamura G, Shirato K. Regulation of T-helper type 2 cell and airway eosinophilia by transmucosal coadministration of antigen and oligodeoxynucleotides containing CpG motifs. *Am J Respir Cell Mol Biol* 2000; 22: 176–182.

58. Hayashi T, Beck L, Rossetto C, Gong X, Takikawa O, Takabayashi K *et al*. Inhibition of experimental asthma by indoleamine 2,3-dioxygenase. *J Clin Invest* 2004; 114: 270–279.

59. Moseman EA, Liang X, Dawson AJ, Panoskaltsis-Mortari A, Krieg AM, Liu YJ *et al*. Human plasmacytoid dendritic cells activated by CpG oligodeoxynucleotides induce the generation of $CD4^+CD25^+$ regulatory T cells.*J Immunol* 2004; 173: 4433–4442.

60. Oldenhove G, de Heusch M, Urbain-Vansanten G, Urbain J, Maliszewski C, Leo O *et al*. $CD4^+$ $CD25^+$ regulatory T cells control T helper cell type 1 responses to foreign antigens induced by mature dendritic cells *in vivo*. *J Exp Med* 2003;198: 259–266.

61. Ramaprakash H, Hogaboam CM. Intranasal CpG therapy attenuated experimental fungal asthma in a TLR9-dependent and -independent manner.*Int Arch Allergy Immunol* 2009; 152: 98–112.

62. Tulic MK, Fiset PO, Christodoulopoulos P, Vaillancourt P, Desrosiers M, Lavigne F *et al*. Amb a 1-immunostimulatory oligodeoxynucleotide conjugate immunotherapy decreases the nasal inflammatory response. *J Allergy Clin Immunol* 2004; 113: 235–241.

63. Allergen immunotherapy: therapeutic vaccines for allergic diseases. Geneva: January 27–29 1997. *Allergy* 1998; 53 (Suppl 44) : 1–42.

64. Gauvreau GM, Hessel EM, Boulet LP, Coffman RL, O'Byrne PM. Immunostimulatory sequences regulate interferon-inducible genes but not allergic airway responses. *Am J Respir Crit Care Med* 2006; 174: 15–20.

65. Davis HL, Suparto II, Weeratna RR, Jumintarto , Iskandriati DD, Chamzah SS*et al*. CpG DNA overcomes hyporesponsiveness to hepatitis B vaccine in orangutans. *Vaccine* 2000; 18: 1920–1924.

66. Vollmer J. Progress in drug development of immunostimulatory CpG oligodeoxynucleotide ligands for TLR9. *Expert Opin Biol Ther* 2005; 5: 673–682.

67. Synowiecki J, Al-Khateeb NA. Production, properties, and some new applications of chitin and its derivatives. *Crit Rev Food Sci Nutr* 2003; 43: 145–171.
68. Biagini G, Muzzarelli RA, Giardino R, Castaldini C. *Advances in Chitin and Chitosan*. New York: Elsevier Applied Science, 1992.
69. Burton OT, Zaccone P. The potential role of chitin in allergic reactions. *Trends Immunol* 2007; 28: 419–422.
70. Shibata Y, Foster LA, Bradfield JF, Myrvik QN. Oral administration of chitin down-regulates serum IgE levels and lung eosinophilia in the allergic mouse. *J Immunol* 2000; 164: 1314–1321.
71. Strong P, Clark H, Reid K. Intranasal application of chitin microparticles down-regulates symptoms of allergic hypersensitivity to *Dermatophagoides pteronyssinus* and *Aspergillus fumigatus* in murine models of allergy. *Clin Exp Allergy* 2002; 32: 1794–1800.
72. Zhu Z, Zheng T, Homer RJ, Kim YK, Chen NY, Cohn L et al. Acidic mammalian chitinase in asthmatic Th2 inflammation and IL-13 pathway activation. *Science* 2004; 304: 1678–1682.
73. Reese TA, Liang HE, Tager AM, Luster AD, van Rooijen N, Voehringer D et al. Chitin induces accumulation in tissue of innate immune cells associated with allergy. *Nature* 2007; 447: 92–96.
74. Lee CG. Chitin, chitinases and chitinase-like proteins in allergic inflammation and tissue remodeling. *Yonsei Med J* 2009; 50: 22–30.
75. Tsigos I, Bouriotis V. Purification and characterization of chitin deacetylase from Colletotrichum lindemuthianum. *J Biol Chem* 1995; 270: 26286–26291.
76. Angelova N, Hunkeler D. Effect of preparation conditions on properties and permeability of chitosan–sodium hexametaphosphate capsules. *J Biomater Sci Polym Ed* 2001; 12: 1317–1337.
77. Chen CL, Wang YM, Liu CF, Wang JY. The effect of water-soluble chitosan on macrophage activation and the attenuation of mite allergen-induced airway inflammation. *Biomaterials* 2008; 29: 2173–2182.
78. Borchers AT, Selmi C, Meyers FJ, Keen CL, Gershwin ME. Probiotics and immunity. *J Gastroenterol* 2009; 44: 26–46.
79. Reinecker HC, Steffen M, Witthoeft T, Pflueger I, Schreiber S, MacDermott RP et al. Enhanced secretion of tumour necrosis factor-alpha, IL-6, and IL-1 beta by isolated lamina propria mononuclear cells from patients with ulcerative colitis and Crohn's disease. *Clin Exp Immunol* 1993; 94: 174–181.

80. Kalliomaki M, Salminen S, Arvilommi H, Kero P, Koskinen P, Isolauri E. Probiotics in primary prevention of atopic disease: a randomised placebo-controlled trial. *Lancet* 2001; 357: 1076–1079.
81. Feleszko W, Jaworska J, Rha RD, Steinhausen S, Avagyan A, Jaudszus A *et al*. Probiotic-induced suppression of allergic sensitization and airway inflammation is associated with an increase of T regulatory-dependent mechanisms in a murine model of asthma. *Clin Exp Allergy* 2007; 37: 498–505.
82. Forsythe P, Inman MD, Bienenstock J. Oral treatment with live *Lactobacillus reuteri* inhibits the allergic airway response in mice. *Am J Respir Crit Care Med* 2007; 175: 561–569.
83. Penders J, Stobberingh EE, van den Brandt PA, Thijs C. The role of the intestinal microbiota in the development of atopic disorders. *Allergy* 2007;62: 1223–1236.
84. Lee J, Seto D, Bielory L. Meta-analysis of clinical trials of probiotics for prevention and treatment of pediatric atopic dermatitis. *J Allergy Clin Immunol* 2008; 121: 116–121.
85. Kawase M, He F, Kubota A, Hiramatsu M, Saito H, Ishii T *et al*. Effect of fermented milk prepared with two probiotic strains on Japanese cedar pollinosis in a double-blind placebo-controlled clinical study. *Int J Food Microbiol* 2009; 128: 429–434.
86. Nagata Y, Yoshida M, Kitazawa H, Araki E, Gomyo T. Improvements in seasonal allergic disease with *Lactobacillus plantarum* No. 14. *Biosci Biotechnol Biochem* 2010; 74: 1869–1877.
87. Tang ML, Lahtinen SJ, Boyle RJ. Probiotics and prebiotics: clinical effects in allergic disease. *Curr Opin Pediatr* 2010; 22: 626–634.
88. van de Pol MA, Lutter R, Smids BS, Weersink EJ, van der Zee JS. Synbiotics reduce allergen-induced T-helper 2 response and improve peak expiratory flow in allergic asthmatics. *Allergy* 2011; 66: 39–47.
89. Kukkonen K, Kuitunen M, Haahtela T, Korpela R, Poussa T, Savilahti E. High intestinal IgA associates with reduced risk of IgE-associated allergic diseases.*Pediatr Allergy Immunol* 2010; 21 (Pt 1): 67–73.
90. Pan SJ, Kuo CH, Lam KP, Chu YT, Wang WL, Hung CH. Probiotics and allergy in children—an update review. *Pediatr Allergy Immunol* 2010; 21: e659–e666.
91. Rask C, Holmgren J, Fredriksson M, Lindblad M, Nordstrom I, Sun JB *et al*. Prolonged oral treatment with low doses of allergen conjugated to cholera toxin B subunit suppresses immunoglobulin E antibody responses in sensitized mice. *Clin Exp Allergy* 2000; 30: 1024–1032.

92. Lee CH, Mo JH. Recent advances in immunotherapy of allergic rhinitis. *Curr Allergy Asthma Rep* 2008; 8: 269–271.

93. Wiedermann U, Jahn-Schmid B, Lindblad M, Rask C, Holmgren J, Kraft D *et al*. Suppressive versus stimulatory effects of allergen/cholera toxoid (CTB) conjugates depending on the nature of the allergen in a murine model of type I allergy. *Int Immunol* 1999; 11: 1131–1138.

94. Sun JB, Czerkinsky C, Holmgren J. Mucosally induced immunological tolerance, regulatory T cells and the adjuvant effect by cholera toxin B subunit. *Scand J Immunol* 2010; 71: 1–11.

95. Shirota H, Sano K, Kikuchi T, Tamura G, Shirato K. Regulation of murine airway eosinophilia and Th2 cells by antigen-conjugated CpG oligodeoxynucleotides as a novel antigen-specific immunomodulator. *J Immunol* 2000; 164: 5575–5582.

96. Hussain I, Kline JN. DNA, the immune system, and atopic disease. *J Investig Dermatol Symp Proc* 2004; 9: 23–28.

97. Gupta GK, Agrawal DK. CpG oligodeoxynucleotides as TLR9 agonists: therapeutic application in allergy and asthma. *BioDrugs* 2010; 24: 225–235.

98. Liu Z, Guo H, Wu Y, Yu H, Yang H, Li J. Local nasal immunotherapy: efficacy of *Dermatophagoides farinae*–chitosan vaccine in murine asthma. *Int Arch Allergy Immunol* 2009; 150: 221–228.

99. Saint-Lu N, Tourdot S, Razafindratsita A, Mascarell L, Berjont N, Chabre H *et al*. Targeting the allergen to oral dendritic cells with mucoadhesive chitosan particles enhances tolerance induction. *Allergy* 2009; 64: 1003–1013.

100. Li GP, Liu ZG, Liao B, Zhong NS. Induction of Th1-type immune response by chitosan nanoparticles containing plasmid DNA encoding house dust mite allergen Der p 2 for oral vaccination in mice. *Cell Mol Immunol* 2009; 6: 45–50.

101. Chew JL, Wolfowicz CB, Mao HQ, Leong KW, Chua KY. Chitosan nanoparticles containing plasmid DNA encoding house dust mite allergen, Der p 1 for oral vaccination in mice. *Vaccine* 2003; 21: 2720–2729.

102. Matsuzaki T, Yamazaki R, Hashimoto S, Yokokura T. The effect of oral feeding of *Lactobacillus casei* strain Shirota on immunoglobulin E production in mice. *J Dairy Sci* 1998; 81: 48–53.

103. Shida K, Takahashi R, Iwadate E, Takamizawa K, Yasui H, Sato T *et al*.*Lactobacillus casei* strain Shirota suppresses serum immunoglobulin E and immunoglobulin G1 responses and systemic anaphylaxis in a food allergy model. *Clin Exp Allergy* 2002; 32: 563–570.

104. Repa A, Grangette C, Daniel C, Hochreiter R, Hoffmann-Sommergruber K, Thalhamer J *et al*. Mucosal co-application of lactic acid bacteria and allergen induces counter-regulatory immune responses in a murine model of birch pollen allergy. *Vaccine* 2003; 22: 87–95.

105. Kruisselbrink A, Heijne Den Bak-Glashouwer MJ, Havenith CE, Thole JE, Janssen R. Recombinant *Lactobacillus plantarum* inhibits house dust mite-specific T-cell responses. *Clin Exp Immunol* 2001; 126: 2–8.

106. Daniel C, Repa A, Mercenier A, Wiedermann U, Wells J. The European LABDEL project and its relevance to the prevention and treatment of allergies. *Allergy* 2007; 62: 1237–1242.

107. Daniel C, Repa A, Wild C, Pollak A, Pot B, Breiteneder H *et al*. Modulation of allergic immune responses by mucosal application of recombinant lactic acid bacteria producing the major birch pollen allergen Bet v 1. *Allergy* 2006; 61: 812–819.

108. Ma Y, Ma AG, Peng Z. A potential immunotherapy approach: mucosal immunization with an IL-13 peptide-based virus-like particle vaccine in a mouse asthma model. *Vaccine* 2007; 25: 8091–8099.

109. Kumar M, Kong X, Behera AK, Hellermann GR, Lockey RF, Mohapatra SS. Chitosan IFN-gamma-pDNA nanoparticle (CIN) therapy for allergic asthma. *Genet Vaccines Ther* 2003; 1: 3.

110. Cortes-Perez NG, Ah-Leung S, Bermudez-Humaran LG, Corthier G, Wal JM, Langella P *et al*. Intranasal coadministration of live lactococci producing interleukin-12 and a major cow's milk allergen inhibits allergic reaction in mice. *Clin Vaccine Immunol* 2007; 14: 226–233.

111. van Overtvelt L, Lombardi V, Razafindratsita A, Saint-Lu N, Horiot S, Moussu H *et al*. IL-10-inducing adjuvants enhance sublingual immunotherapy efficacy in a murine asthma model. *Int Arch Allergy Immunol* 2008; 145: 152–162.

112. Tighe H, Corr M, Roman M, Raz E. Gene vaccination: plasmid DNA is more than just a blueprint. *Immunol Today* 1998; 19: 89–97.

113. Shin JH, Kang JM, Kim SW, Cho JH, Park YJ. Effect of oral tolerance in a mouse model of allergic rhinitis. *Otolaryngol Head Neck Surg* 2010; 142: 370–375.

114. Xie QM, Wu X, Wu HM, Deng YM, Zhang SJ, Zhu JP *et al*. Oral administration of allergen extracts from *Dermatophagoides farinae* desensitizes specific allergen-induced inflammation and airway hyperresponsiveness in rats. *Int Immunopharmacol* 2008; 8: 1639–1645.

115. Faria AM, Weiner HL. Oral tolerance. *Immunol Rev* 2005; 206: 232–259.
116. Faria AM, Maron R, Ficker SM, Slavin AJ, Spahn T, Weiner HL. Oral tolerance induced by continuous feeding: enhanced up-regulation of transforming growth factor-beta/interleukin-10 and suppression of experimental autoimmune encephalomyelitis. *J Autoimmun* 2003; 20: 135–145.
117. Slavin AJ, Maron R, Weiner HL. Mucosal administration of IL-10 enhances oral tolerance in autoimmune encephalomyelitis and diabetes. *Int Immunol* 2001;13: 825–833.
118. Marth T, Strober W, Kelsall BL. High dose oral tolerance in ovalbumin TCR-transgenic mice: systemic neutralization of IL-12 augments TGF-beta secretion and T cell apoptosis. *J Immunol* 1996; 157: 2348–2357.
119. Thorbecke GJ, Schwarcz R, Leu J, Huang C, Simmons WJ. Modulation by cytokines of induction of oral tolerance to type II collagen. *Arthritis Rheum* 1999; 42: 110–118.
120. Edwan JH, Perry G, Talmadge JE, Agrawal DK. Flt-3 ligand reverses late allergic response and airway hyper-responsiveness in a mouse model of allergic inflammation. *J Immunol* 2004; 172: 5016–5023.
121. Hanninen A, Martinez NR, Davey GM, Heath WR, Harrison LC. Transient blockade of CD40 ligand dissociates pathogenic from protective mucosal immunity. *J Clin Invest* 2002; 109: 261–267.
122. Allam JP, Wurtzen PA, Reinartz M, Winter J, Vrtala S, Chen KW et al. Phl p 5 resorption in human oral mucosa leads to dose-dependent and time-dependent allergen binding by oral mucosal Langerhans cells, attenuates their maturation, and enhances their migratory and TGF-beta1 and IL-10-producing properties. *J Allergy Clin Immunol* 2010; 126: 638–645.
123. Taudorf E, Laursen LC, Djurup R, Kappelgaard E, Pedersen CT, Soborg M et al. Oral administration of grass pollen to hay fever patients. An efficacy study in oral hyposensitization. *Allergy* 1985; 40: 321–335.
124. Taudorf E, Laursen LC, Lanner A, Bjorksten B, Dreborg S, Soborg M et al. Oral immunotherapy in birch pollen hay fever. *J Allergy Clin Immunol* 1987;80: 153–161.
125. Canonica GW, Passalacqua G. Noninjection routes for immunotherapy. *J Allergy Clin Immunol* 2003; 111: 437–448.
126. Haneda K, Sano K, Tamura G, Shirota H, Ohkawara Y, Sato T et al. Transforming growth factor-beta secreted from $CD4^+$ T cells ameliorates antigen-induced eosinophilic inflammation. A novel high-dose tolerance in the trachea. *Am J Respir Cell Mol Biol* 1999; 21: 268–274.

127. Hoyne GF, Askonas BA, Hetzel C, Thomas WR, Lamb JR. Regulation of house dust mite responses by intranasally administered peptide: transient activation of CD4⁺ T cells precedes the development of tolerance *in vivo*. *Int Immunol* 1996; 8: 335–342.

128. Hoyne GF, O'Hehir RE, Wraith DC, Thomas WR, Lamb JR. Inhibition of T cell and antibody responses to house dust mite allergen by inhalation of the dominant T cell epitope in naive and sensitized mice. *J Exp Med* 1993; 178: 1783–1788.

129. Marcucci F, Sensi LG, Caffarelli C, Cavagni G, Bernardini R, Tiri A *et al.* Low-dose local nasal immunotherapy in children with perennial allergic rhinitis due to Dermatophagoides. *Allergy* 2002; 57: 23–28.

130. Pocobelli D, del Bono A, Venuti L, Falagiani P, Venuti A. Nasal immunotherapy at constant dosage: a double-blind, placebo-controlled study in grass-allergic rhinoconjunctivitis. *J Investig Allergol Clin Immunol* 2001;11: 79–88.

131. Bjorksten B. Local immunotherapy is not documented for clinical use. *Allergy* 1994; 49: 299–301.

132. Crimi E, Voltolini S, Troise C, Gianiorio P, Crimi P, Brusasco V *et al.* Local immunotherapy with *Dermatophagoides* extract in asthma. *J Allergy Clin Immunol* 1991; 87: 721–728.

133. Tari MG, Mancino M, Monti G. Immunotherapy by inhalation of allergen in powder in house dust allergic asthma—a double-blind study. *J Investig Allergol Clin Immunol* 1992; 2: 59–67.

134. Markert UR. Local immunotherapy in allergy: prospects for the future. *Chem Immunol Allergy* 2003; 82: 127–135.

135. Ozdemir C. An immunological overview of allergen specific immunotherapy—subcutaneous and sublingual routes. *Ther Adv Respir Dis* 2009; 3: 253–262.

136. O'Hehir RE, Sandrini A, Anderson GP, Rolland JM. Sublingual allergen immunotherapy: immunological mechanisms and prospects for refined vaccine preparation. *Curr Med Chem* 2007; 14: 2235–2244.

137. Bohle B, Kinaciyan T, Gerstmayr M, Radakovics A, Jahn-Schmid B, Ebner C. Sublingual immunotherapy induces IL-10-producing T regulatory cells, allergen-specific T-cell tolerance, and immune deviation. *J Allergy Clin Immunol* 2007; 120: 707–713.

138. Akdis CA, Barlan IB, Bahceciler N, Akdis M. Immunological mechanisms of sublingual immunotherapy. *Allergy* 2006; 61 (Suppl 81): 11–14.

139. O'Hehir RE, Gardner LM, de Leon MP, Hales BJ, Biondo M, Douglass JA et al. House dust mite sublingual immunotherapy: the role for transforming growth factor-beta and functional regulatory T cells. *Am J Respir Crit Care Med* 2009;180: 936–947.

140. Scadding G, Durham S. Mechanisms of sublingual immunotherapy. *J Asthma* 2009; 46: 322–334.

141. Brimnes J, Kildsgaard J, Jacobi H, Lund K. Sublingual immunotherapy reduces allergic symptoms in a mouse model of rhinitis. *Clin Exp Allergy* 2007; 37: 488–497.

142. Eberlein-Konig B, Jung C, Rakoski J, Ring J. Immunohistochemical investigation of the cellular infiltrates at the sites of allergoid-induced late-phase cutaneous reactions associated with pollen allergen-specific immunotherapy. *Clin Exp Allergy* 1999; 29: 1641–1647.

143. Wilson DR, Lima MT, Durham SR. Sublingual immunotherapy for allergic rhinitis: systematic review and meta-analysis. *Allergy* 2005; 60: 4–12.

144. Marcucci F, Sensi L, Incorvaia C, Di Cara G, Moingeon P, Frati F. Oral reactions to sublingual immunotherapy: a bioptic study. *Allergy* 2007; 62: 1475–1477.

145. Marcucci F, Sensi L, Di Cara G, Gidaro G, Incorvaia C, Frati F. Sublingual reactivity to rBET V1 and rPHL P1 in patients with oral allergy syndrome. *Int J Immunopathol Pharmacol* 2006; 19: 141–148.

146. Stelmach I, Kaczmarek-Wozniak J, Majak P, Olszowiec-Chlebna M, Jerzynska J. Efficacy and safety of high-doses sublingual immunotherapy in ultra-rush scheme in children allergic to grass pollen. *Clin Exp Allergy* 2009; 39: 401–408.

147. Cao LF, Lu Q, Gu HL, Chen YP, Zhang Y, Lu M et al. Clinical evaluation for sublingual immunotherapy of allergic asthma and atopic rhinitis with Dermatophagoides Farinae Drops. *Zhonghua Er Ke Za Zhi* 2007; 45: 736–741. Chinese.

148. Agostinis F, Foglia C, Landi M, Cottini M, Lombardi C, Canonica GW et al. The safety of sublingual immunotherapy with one or multiple pollen allergens in children. *Allergy* 2008; 63: 1637–1639.

149. Cochard MM, Eigenmann PA. Sublingual immunotherapy is not always a safe alternative to subcutaneous immunotherapy. *J Allergy Clin Immunol* 2009;124: 378–379.

150. Larenas-Linnemann D. Sublingual immunotherapy in children: complete and updated review supporting evidence of effect. *Curr Opin Allergy*

Clin Immunol 2009; 9: 168–176.

151. Esch RE, Bush RK, Peden D, Lockey RF. Sublingual-oral administration of standardized allergenic extracts: phase 1 safety and dosing results. *Ann Allergy Asthma Immunol* 2008; 100: 475–481.

152. Nuhoglu Y, Ozumut SS, Ozdemir C, Ozdemir M, Nuhoglu C, Erguven M. Sublingual immunotherapy to house dust mite in pediatric patients with allergic rhinitis and asthma: a retrospective analysis of clinical course over a 3-year follow-up period. *J Investig Allergol Clin Immunol* 2007; 17: 375–378.

153. Wahn U, Tabar A, Kuna P, Halken S, Montagut A, de Beaumont O *et al*. Efficacy and safety of 5-grass-pollen sublingual immunotherapy tablets in pediatric allergic rhinoconjunctivitis. *J Allergy Clin Immunol* 2009; 123: 160–166.

154. Bufe A, Eberle P, Franke-Beckmann E, Funck J, Kimmig M, Klimek L *et al*. Safety and efficacy in children of an SQ-standardized grass allergen tablet for sublingual immunotherapy. *J Allergy Clin Immunol* 2009; 123: 167–173.

155. Fiocchi A, Fox AT. Preventing progression of allergic rhinitis: the role of specific immunotherapy. *Arch Dis Child Educ Pract Ed* 2011; 96: 91–100.

156. Broide DH. Immunomodulation of allergic disease. *Annu Rev Med* 2009; 60: 279–291.

157. Eifan AO, Akkoc T, Yildiz A, Keles S, Ozdemir C, Bahceciler NN *et al*. Clinical efficacy and immunological mechanisms of sublingual and subcutaneous immunotherapy in asthmatic/rhinitis children sensitized to house dust mite: an open randomized controlled trial. *Clin Exp Allergy* 2010; 40: 922–932.

158. Pham-Thi N, Scheinmann P, Fadel R, Combebias A, Andre C. Assessment of sublingual immunotherapy efficacy in children with house dust mite-induced allergic asthma optimally controlled by pharmacologic treatment and mite-avoidance measures. *Pediatr Allergy Immunol* 2007; 18: 47–57.

159. Compalati E, Passalacqua G, Bonini M, Canonica GW. The efficacy of sublingual immunotherapy for house dust mites respiratory allergy: results of a GA^2LEN meta-analysis. *Allergy* 2009; 64: 1570–1579.

160. Rodriguez Santos O. Sublingual immunotherapy in allergic rhinitis and asthma in 2–5 year-old children sensitized to mites. *Rev Alerg Mex* 2008; 55: 71–75. Spanish.

161. Alche JD, Castro AJ, Jimenez-Lopez JC, Morales S, Zafra A, Hamman-Khalifa AM *et al*. Differential characteristics of olive pollen from different cultivars: biological and clinical implications. *J Investig Allergol Clin Immunol* 2007; 17(Suppl 1): 17–23.

162. Radulovic S, Wilson D, Calderon M, Durham S. Systematic reviews of sublingual immunotherapy (SLIT). *Allergy* 2011; 66: 740–752.

163. Marogna M, Colombo F, Cerra C, Bruno M, Massolo A, Canonica GW *et al*. The clinical efficacy of a sublingual monomeric allergoid at different maintenance doses: a randomized controlled trial. *Int J Immunopathol Pharmacol* 2010; 23: 937–945.

164. Pajno GB, Caminiti L, Vita D, Barberio G, Salzano G, Lombardo F *et al*. Sublingual immunotherapy in mite-sensitized children with atopic dermatitis: a randomized, double-blind, placebo-controlled study. *J Allergy Clin Immunol* 2007; 120: 164–170.

165. Takai T, Ikeda S. Barrier dysfunction caused by environmental proteases in the pathogenesis of allergic diseases. *Allergol Int* 2011; 60: 25–35.

166. Akdis M, Blaser K, Akdis CA. T regulatory cells in allergy: novel concepts in the pathogenesis, prevention, and treatment of allergic diseases. *J Allergy Clin Immunol* 2005; 116: 961–968.

167. Wohlleben G, Erb KJ. Atopic disorders: a vaccine around the corner? *Trends Immunol* 2001; 22: 618–626.

168. Lee CC, Chiang BL. RNA interference: new therapeutics in allergic diseases.*Curr Gene Ther* 2008; 8: 236–246.

169. Wang LC, Lee JH, Yang YH, Lin YT, Chiang BL. New biological approaches in asthma: DNA-based therapy. *Curr Med Chem* 2007; 14: 1607–1618.

170. Chuang YH, Yang YH, Wu SJ, Chiang BL. Gene therapy for allergic diseases.*Curr Gene Ther* 2009; 9: 185–191.

171. Alton EW, Griesenbach U, Geddes DM. Gene therapy for asthma: inspired research or unnecessary effort? *Gene Ther* 1999; 6: 155–156.

172. Casanovas M, Martin R, Jimenez C, Caballero R, Fernandez-Caldas E. Safety of immunotherapy with therapeutic vaccines containing depigmented and polymerized allergen extracts. *Clin Exp Allergy* 2007; 37: 434–440.

173. Valenta R, Niederberger V. Recombinant allergens for immunotherapy. *J Allergy Clin Immunol* 2007; 119: 826–830.

174. Nelson HS. Allergen immunotherapy: where is it now? *J Allergy Clin Immunol* 2007; 119: 769–779.

175. Crameri R, Rhyner C. Novel vaccines and adjuvants for allergen-specific immunotherapy. *Curr Opin Immunol* 2006; 18: 761–768.

176. Wagner A, Forster-Waldl E, Garner-Spitzer E, Schabussova I, Kundi M, Pollak A *et al*. Immunoregulation by *Toxoplasma gondii* infection prevents allergic immune responses in mice. *Int J Parasitol* 2009; 39: 465–472.

177. Taher YA, van Esch BC, Hofman GA, Henricks PA, van Oosterhout AJ. 1alpha,25-dihydroxyvitamin D_3 potentiates the beneficial effects of allergen immunotherapy in a mouse model of allergic asthma: role for IL-10 and TGF-beta. *J Immunol* 2008; 180: 5211–5221.

178. Hsieh KH. Evaluation of efficacy of traditional Chinese medicines in the treatment of childhood bronchial asthma: clinical trial, immunological tests and animal study. Taiwan Asthma Study Group. *Pediatr Allergy Immunol* 1996; 7: 130–140.

179. Xiang YZ, Shang HC, Gao XM, Zhang BL. A comparison of the ancient use of ginseng in traditional Chinese medicine with modern pharmacological experiments and clinical trials. *Phytother Res* 2008; 22: 851–858.

180. Jacobsen L, Niggemann B, Dreborg S, Ferdousi HA, Halken S, Host A *et al*. Specific immunotherapy has long-term preventive effect of seasonal and perennial asthma: 10-year follow-up on the PAT study. *Allergy* 2007; 62: 943–948.

181. Canonica GW, Baena-Cagnani CE, Bousquet J, Bousquet PJ, Lockey RF, Malling HJ *et al*. Recommendations for standardization of clinical trials with Allergen Specific Immunotherapy for respiratory allergy. A statement of a World Allergy Organization (WAO) taskforce. *Allergy* 2007; 62: 317–324.

Chapter 10

TOPICAL APPLICATION OF RECOMBINANT TYPE VII COLLAGEN INCORPORATES INTO THE DERMAL–EPIDERMAL JUNCTION AND PROMOTES WOUND CLOSURE

Xinyi Wang[1], Pedram Ghasri[1], Mahsa Amir[1], Brian Hwang[1], Yingpin Hou[1], Michael Khilili[1], Andrew Lin[1], Douglas Keene[2], Jouni Uitto[3], David T Woodley[1], and Mei Chen[1]

[1]Department of Dermatology, University of Southern California, Los Angeles, California, USA

[2]Department of Molecular and Medical Genetics, Shriners Hospital for Children, Portland, Oregon, USA

[3]Department of Dermatology and Cutaneous Biology, Jefferson Medical College, Philadelphia, Pennsylvania, USA

ABSTRACT

Patients with recessive dystrophic epidermolysis bullosa (RDEB) have incurable skin fragility, blistering, and skin wounds due to mutations in the gene that codes for type VII collagen (C7) that mediates dermal–epidermal adherence in human skin. In this study, we evaluated if topically applied human recombinant C7 (rC7) could restore C7 at the dermal–epidermal junction (DEJ) and enhance wound healing. We found that rC7 applied topically onto murine skin wounds stably incorporated into the newly formed DEJ of healed wounds and accelerated wound closure by increasing re-epithelialization. Topical rC7 decreased the expression of fibrogenic transforming growth factor-β2 (TGF-β2) and increased the expression of anti-fibrogenic TGF-β3. These were accompanied by the reduced expression of connective tissue growth factor, fewer α smooth muscle actin (α-SMA)–positive myofibroblasts, and less deposition of collagen in the healed neodermis, consistent with less scar formation. In addition, using a mouse model in which skin from C7 knock out mice was grafted onto immunodeficient mice, we showed that applying rC7 onto RDEB grafts with wounds restored C7 and anchoring fibrils (AFs) at the

DEJ of the grafts and corrected the dermal–epidermal separation. The topical application of rC7 may be useful for treating patients with RDEB and patients who have chronic skin wounds.

INTRODUCTION

Chronic cutaneous ulcers are an enormous health care problem consuming ~$20 billion per year in the United States.[1] The precise mechanisms involved in the healing of skin wounds are not fully understood. Traditionally, the phases of wound healing have been divided into (i) clot formation and inflammation, (ii) re-epithelialization, (iii) granulation tissue formation by the processes of fibroplasia and neo-angiogenesis, and (iv) tissue remodeling.[2] Except for topical platelet-derived growth factor (GF), there are no biological agents approved by the Federal Drug Administration for the enhancement of skin wound healing. Topical PDGF has been in use for several years, but is indicated only for diabetic skin wounds, and it has proven to have only modest efficacy.[3–5]

Patients with recessive dystrophic epidermolysis bullosa (RDEB) have inherited skin fragility resulting in widespread bullae and open erosive skin wounds that heal with scarring.[6] Scarring results in many complications, such as a fibrotic fusion of the fingers and toes, esophageal stenoses, joint contractures, poor dentition, decreased ability to open the mouth, fixed tongue, and nutritional deficiencies. The chronic cycles of wounding and healing in patients with RDEB, coupled with fibrosis, are thought to be the reason that these patients develop aggressive skin squamous cell carcinomas that often result in premature demise.

RDEB is due to a gene defect in *COL7A1* that codes for type VII collagen α chains.[7–9] Type VII collagen (C7) is the major component of anchoring fibrils (AFs), large 700–900 nm structures that hold the epidermis and dermis of skin together.[10,11] patients with RDEB have a paucity or complete absence of functional C7 and functional AFs in the dermal–epidermal junction (DEJ) of their skin, resulting in poor dermal–epidermal adherence. Structurally, C7 is a composed of three identicalα chains, each consisting of a central collagenous triple-helical domain and two flanking amino- and carboxy-terminal non-collagenous domains NC1 and NC2, respectively.[10–13] Within the extracellular space, C7 molecules form antiparallel dimers, which aggregate laterally to form AFs, wheat-stack shaped structures, that serve to anchor the epidermis onto the underlying dermis.

Like patients with chronic skin wounds, the current treatment of patients with RDEB is only supportive care because there is no consistently effective treatment. Recently, several clinical trials for RDEB have been initiated. Wang et al. demonstrated that the intradermal injection of allogeneic fibroblasts into patients with RDEB would result in increased endogenous mutated C7 in their skin, fewer new skin bullae and improved dermal–epidermal adherence.[14] Wagner and coworkers have shown that bone marrow/stem cell transplantation into patients with RDEB results in increased C7 in the patients' skin and improved clinical outcomes.[15] Although promising, neither therapy is consistently effective or without risk.

Our laboratory has developed *in vitro* and *in vivo* strategies geared to more direct RDEB therapy. We and others showed that C7-deficient RDEB skin cells (keratinocytes and dermal fibroblasts) when gene corrected to synthesize and secrete C7 (via infection with a minimal lentiviral vector or a phi C31 integrase), revert from the abnormal RDEB cellular phenotype and are normalized with regards to cell growth, cell motility, and cell matrix attachment.[16,17] Using a mouse model with human RDEB skin equivalents generated with RDEB fibroblasts and keratinocytes and engrafted onto immunodeficient mice and a C7-knockout mouse model ($Col7a1^{-/-}$, which recapitulates features of RDEB), we intradermally injected: (i) gene-corrected RDEB fibroblasts that could now synthesize and secrete C7,[18] (ii) lentiviral vectors expressing C7,[19] and (iii) C7 protein itself.[20,21] All three strategies resulted in new C7 and AFs in the RDEB skin and a reversal of the poor dermal–epidermal adherence. We recently developed an alternative strategy using an intravenous approach with molecularly engineered RDEB fibroblasts over-expressing human C7.[22] We demonstrated that intravenously injected fibroblasts homed to murine skin wounds, continuously delivered C7, which then incorporated into the healed wound's DEJ, formed stable AFs, and accelerated wound closure.

In the study described herein, we sought to determine if simply applying human recombinant C7 (rC7) topically would be useful in promoting wound healing. Using full-thickness wounds in athymic nude mice, we found that topically applied rC7 stably incorporated into the healing wounds' DEJ. Furthermore, topical application of rC7 to wounds promoted re-epithelialization and dramatically accelerated wound closure. Moreover, in a second animal model in which we grafted RDEB-like, C7-null skin from C7-knockout mice onto athymic nude mice, we found that topically applied rC7 incorporated into the DEJ of the RDEB skin grafts and corrected the grafts' poor dermal–epidermal adherence and AF defect. These studies provide the first evidence for the potential use of topically applied rC7 to improve skin wound healing and reverse the molecular and structural defects in RDEB.

RESULTS

Topical rC7 Incorporated into the Regenerated BMZ

We purified milligram quantities of rC7 from conditioned media of gene-corrected RDEB fibroblasts as described earlier[17,20] and used it to evaluate the feasibility of topical rC7 application for healing of skin wounds and RDEB treatment. We made 1.0 cm² full-thickness excision wounds on the mid-back of athymic nude mice (n = 20) and topically applied 30 µg of rC7 in a 10% carboxymethylcellulose vehicle to the wound site. Skin biopsies were obtained from healed wounds at various time points after the topical application and subjected to immunostaining with an antibody specific for human C7. As shown in **Figure 1a**, the topically applied rC7 incorporated into the DEJ of the healed skin wounds as early as 2 weeks and was persistent for at least 2 months after a single application (panels A and D). In contrast, wounds treated with vehicle (n = 20) lacked human C7 at the DEJ (panel B). Interestingly, wounds treated with the NC1 domain of C7 (the amino-terminal half the C7 molecule) (n = 10 mice) did not have NC1 at their DEJ, suggesting that the ability of rC7 to incorporate into the DEJ requires full-length C7 (panel C). These experiments demonstrated that rC7 applied topically onto full thickness skin wounds was able to incorporate stably into the DEJ of the healed skin.

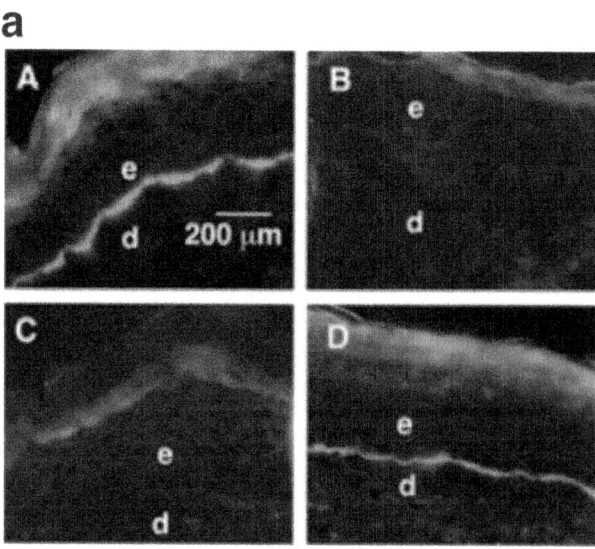

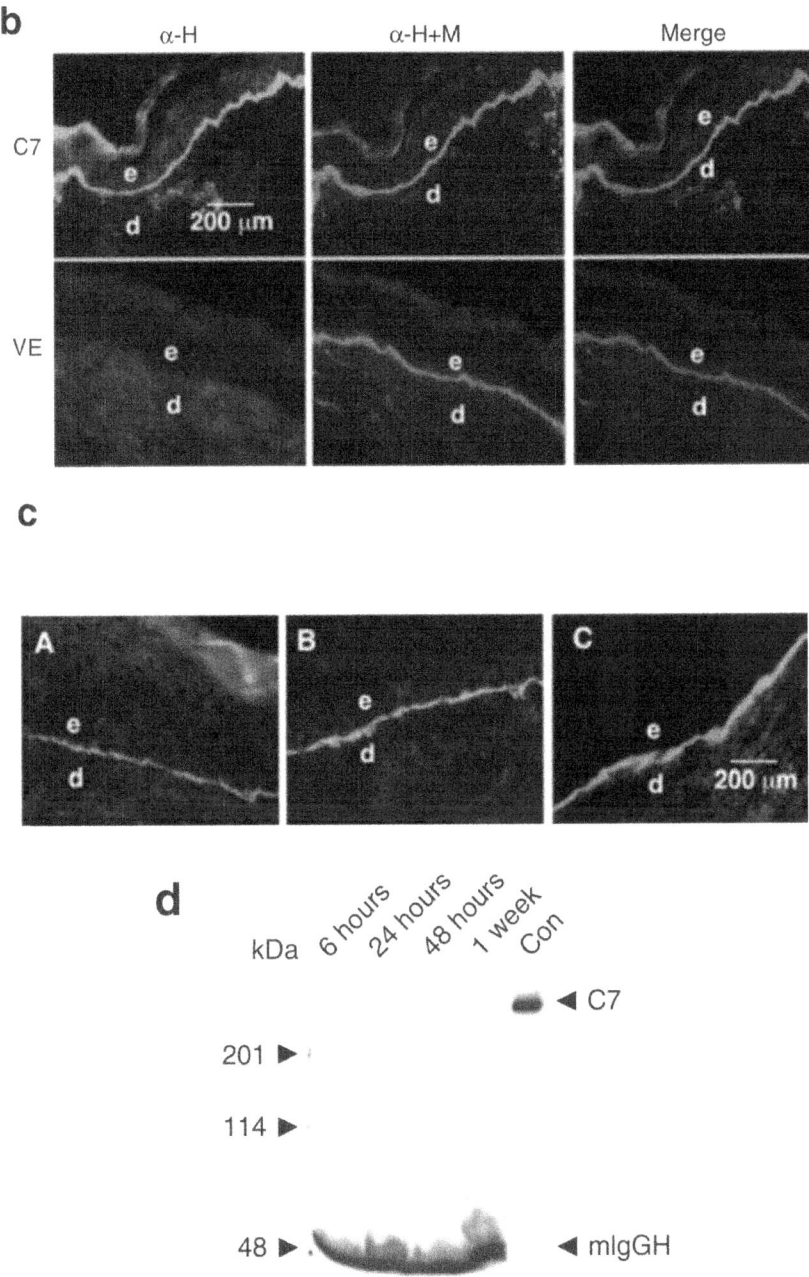

Figure 1. Topically applied rC7 stably incorporated in the regenerated DEJ in the mouse skin. (**a**) Immunofluorescence staining of the mouse skin was performed with

an antibody specific for human C7 at 2 weeks after the topical application of 30 μg of rC7, NC1, or vehicle alone. Note that the healed wounds treated with rC7 ($n = 30$ mice) demonstrated a linear pattern of C7 deposition at the DEJ (panel A). In contrast, no human C7 was detected in mice treated with vehicle alone ($n = 20$ mice) or NC1 ($n = 10$ mice) (panels B and C). Panel D shows the stable incorporation of human rC7 at the mouse DEJ at 8 weeks after the initial topical application of rC7. (**b**) Immunofluorescence staining of mouse skin was performed 2 weeks after the topical application of rC7. The skin sections were labeled with either a monoclonal antibody specific for human C7 (green, panel α-H) or a rabbit polyclonal antibody that recognizes both mouse and human C7 (red, panel α-M+H). Merged images demonstrate colocalization of topically applied human rC7 with endogenous mouse C7 at the mouse DEJ. The lower panel depicts staining of wounds treated with the vehicle (VE) alone. (**c**) Dose-dependent deposition of human rC7 at the mouse DEJ after topical rC7 application. Immunofluorescence staining of biopsy specimens were performed with an antibody specific for human C7 after the animal's wounds were treated with 8 μg (A), 16 μg (B), or 32 μg (C) of topical rC7, respectively. Scale bar: 200 μm. (**d**) Sera were taken from mice at the time indicated after topically applied with 30 μg rC7 and subjected to 4–15% SDS-PAGE followed by immunoblot analysis using an anti-NC1 antibody. Purified rC7 of 10 ng was run as a control (Con). The positions of full-length 290 kDa C7, 50 kDa mouse IgG heavy chain (mIgGH), and molecular weight markers are indicated. d, dermis; e, epidermis.

To confirm that the human C7 was correctly localized within the DEJ of the healing mouse skin, we colabeled the same vertical sections with a polyclonal antibody that recognizes both mouse and human C7 (α-H+M) and our human specific anti-C7 antibody (α-H). As shown in **Figure 1b**, these two antibodies showed perfect colocalization in the mouse's DEJ when the images were merged. Therefore, the topically applied rC7 incorporated to the same location of the DEJ as the mouse's endogenous C7. Note that wounds treated with the vehicle alone (VE in **Figure 1b**) showed the presence of mouse but not human C7.

After demonstrating the ability of the topical rC7 to incorporate into the mouse's DEJ, we wished to determine the minimal concentration of protein required for DEJ deposition. We applied 8, 16, and 32 μg of rC7 topically on the wounds. As shown in **Figure 1c**, there was a dose-dependent increase in the incorporation of rC7 into the mouse's DEJ.

Because rC7 was administered deeply into the wounds, it was important to determine if the topical rC7 was transported into the animals' circulation. To examine this issue, we performed immunoblot analysis on sera obtained from treated mice at 6 hours, 24 hours, 48 hours and 1 week after topical application of rC7. As shown in **Figure 1d**, 10 ng of rC7 running as a control was readily detected by our immunoblot analysis. However, we did not detect any rC7 in the blood stream.

Topical rC7 Promoted Wound Healing via Re-Epithelialization of the Epidermis

After having shown the incorporation of topically applied rC7 into the DEJ of healed wounds, we wished to determine if the topical application of rC7 accelerates the closure of skin wounds. Following full-thickness skin excision, we monitored wound closure over a 14-day period and found that rC7-treated wounds exhibited marked acceleration of wound closure compared with vehicle-treated wounds (**Figure 2a,b**).

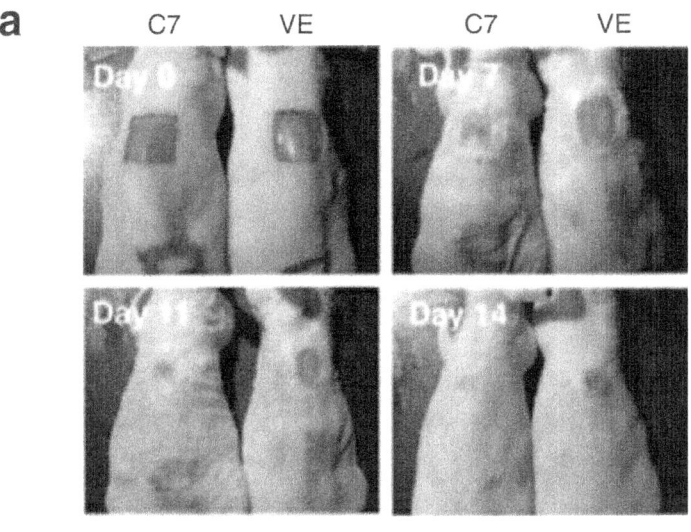

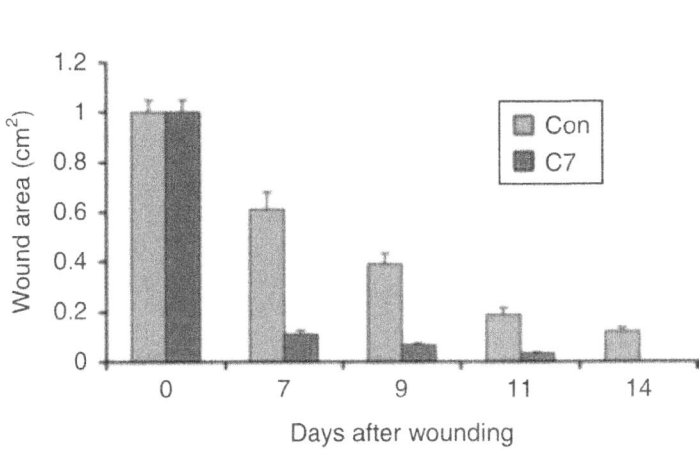

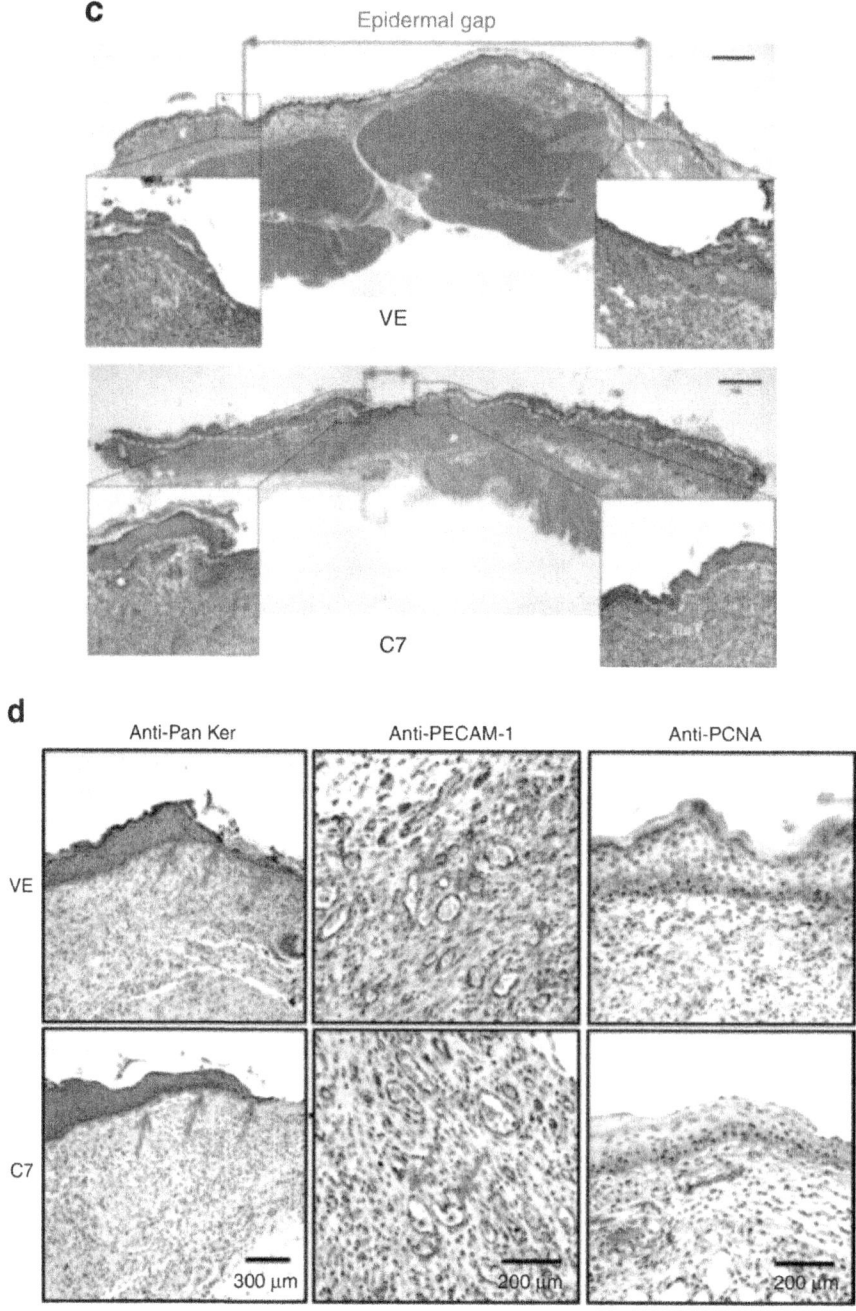

Figure 2. Topical application of rC7 promoted wound healing. A 1.0 cm²(1 × 1 cm) square full-thickness excision wound was made on the mid-back of 8- to 10-week-old

athymic nude mice, and rC7 (30 g) was applied topically once on day 0 ($n = 20$ mice per group). (**a**) Representative days 0, 7, 11, and 14 wounds are shown. (**b**) Mean ± SD open wound area measurements at days 0, 7, 9, 11, and 14 after wounding ($n = 20$ mice for each group). (**c**) On day 7, excisional biopsies of full-thickness wounded skin with a portion of unwounded skin were obtained from wounds treated with topical vehicle (VE) or rC7 (C7). The biopsy specimens were stained by H&E and photographed with a light microscope. Independently, photographed images with identical magnifications were reconstituted to show the unhealed areas of the wounds. Red dotted lines indicate the unhealed wound area. Yellow dotted lines mark the newly re-epithelialized epidermis. The fronts of newly re-epithelialized epidermis were enlarged, as shown in higher magnified images. Scale bars: 0.33 mm. (**d**) Immunohistochemistry analysis of biopsy specimens of day 7 full-thickness wounds treated with either vehicle (VE) or rC7 (C7) with anti-pan keratin (keratinocytes), anti-PECAM-1 (endothelial cells), and anti-PCNA antibodies. Ten randomly selected images per each condition from three independent experiments were analyzed for consensus. Representative images are shown. Scale bars: 0.3 mm (left column); 0.2 mm (middle and right columns). In the left column, arrows point to the keratin-labeled, re-epithelializing tongue (ReT); in the middle column, the arrows point to blood vessels; in the right column, the circles point to proliferating keratinocytes.

Re-epithelialization is a crucial part of wound healing and is primarily mediated by migrating keratinocytes. We previously showed that C7 strongly promotes human keratinocyte migration *in vitro*.[23] To determine whether rC7-driven migration of the keratinocytes across the wound bed contributed to re-epithelialization and enhanced wound healing, as shown in **Figure 2c**, we completely excised the entire wounds plus some normal unwounded surrounding skin 7 days after treatment and subjected the tissue to H&E staining, microscopic analyses, and measurements. The percent of re-epithelialization across each wound was measured by light microscopy using a reticle to measure the proportion of each wound that was covered by a neo-epidermis in relation to the entire original open wound length. As shown in **Figure 2c**, wounds that were treated with rC7, compared with those treated with vehicle alone, exhibited a significant reduction in the epidermal gap, which is the length of the un-re-epithelialized open wound. Conversely, the rC7-treated wounds exhibited significantly more re-epithelialization than vehicle-treated wounds (**Figure 2c**, dotted yellow lines). The re-epithelializing tongue (ReT) can be clearly visualized in the enlarged images on both sides of the wounds.

To confirm the H&E results and also to analyze for the presence of keratinocytes and the presence of blood vessel formation in wounds treated with either rC7 or vehicle, we performed immunohistochemistry on the biopsy specimens with antibodies specific for keratinocytes and endothelial cells. As shown in **Figure 2d**, anti-pan keratin antibody staining clearly labeled the ReT

(**Figure 2d**, red lines and arrows, left panels). There were no differences in the numbers of endothelial cells or blood vessels, as assessed by anti-PECAM-1 antibody staining (**Figure 2d**, arrows, middle panels), between vehicle- and rC7-treated wounds.

Re-epithelialization depends upon keratinocyte migration and cellular division. To evaluate if rC7 affects cellular proliferation, we conducted immunohistochemistry on biopsy specimens with an antibody specific to PCNA, a marker for cell proliferation. As shown in **Figure 2d**, there were no significant differences in the numbers of PCNA positive cells between vehicle- and rC7-treated wounds. We also compared the thickness of the epidermis (measured from basement membrane layer to the stratum corneum of the epidermis) in the healing wounds at days 7 and 19 after wounding (**Table I**). At each time point, no significant differences in epidermal thickness were observed in wounds treated with rC7 or vehicle. Taken together, these data indicate that the acceleration of wound healing by rC7 is likely mediated by promoting re-epithelialization rather than by increasing wound angiogenesis or cellular proliferation.

Table 1. Epidermal thickness of wounds in rC7- and vehicle-treated wounds.

	Epidermis thickness (days after treatment)	
	Day 7	Day 19
C7 ($n = 6$)	61.07 ± 7.02	59.22 ± 6.72
Control ($n = 6$)	68.85 ± 7.82	63.82 ± 7.85

We previously showed that the non-collagenous NC1 domain of C7 failed to promote keratinocyte migration *in vitro*.[23] To see if the ability of C7 to promote skin cell migration is critical for its ability to enhance skin wound closure *in vivo*, we performed wound healing experiments using topically applied NC1 domain at the same molar concentration as full-length rC7. As shown in **Figure 3**, the wounds of mice treated with topical NC1 exhibited similar wound healing rates as wounds treated with vehicle at days 9, 12, and 14 (**Figure 3a,b**). Collectively, these data suggest a direct correlation between the ability of C7 to promote keratinocyte migration and its ability to accelerate wound closure.

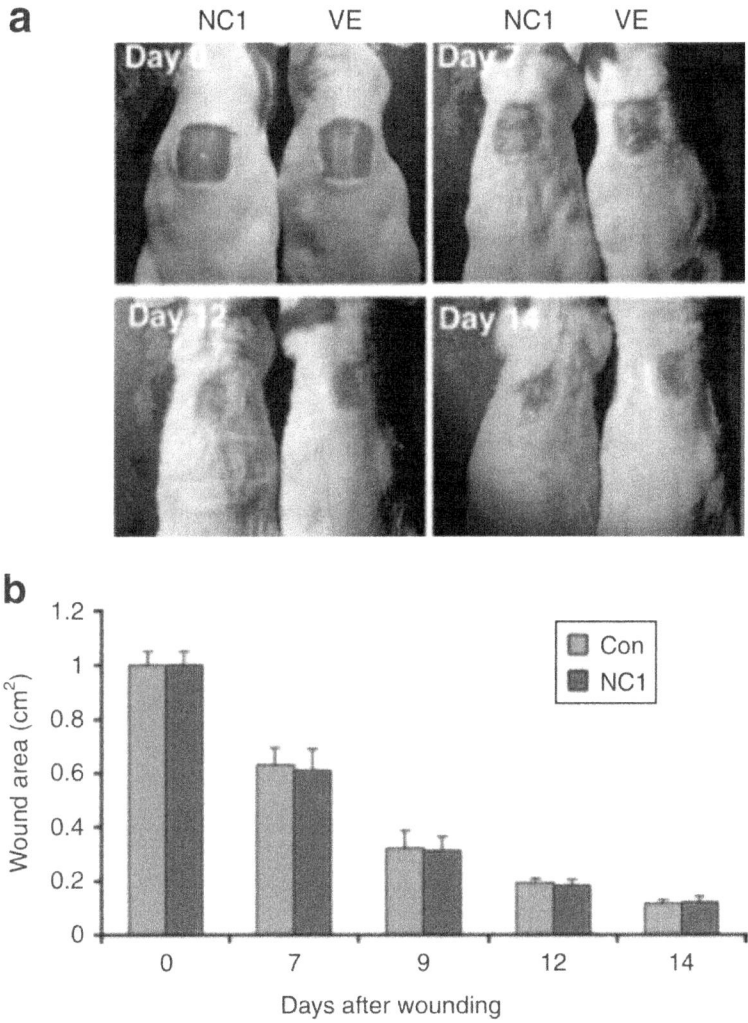

Figure 3. Topical application of NC1 did not promote wound healing. A 1.0 cm²(1 × 1 cm) square full-thickness excision wounds were made on the mid-back of 8- to 10-week-old athymic nude mice and purified NC1 domain (30 g) (n = 10 mice) or vehicle (VE) (n = 10 mice) were applied topically once on day 0. (**a**) Representative days 0, 7, 12, and 14 wounds are shown for wounds treated with NC1 or vehicle alone (VE). (**b**) Mean ± SD wound size measurements at days 0, 7, 9, 12, and 14 after wounding treated with NC1 (n = 10 mice for each group).

Topical rC7 Inhibited Fibrosis of the Wounded Skin

During the course of our experiments, we consistently noticed that rC7-treated wounds appeared less scarred than vehicle-treated wounds. To determine if rC7 has anti-contraction activity, we performed a standard fibroblast-populated collagen I lattice contraction assay.[24] Previously, we have shown that collagen lattices are contracted by human dermal fibroblasts pulling on a specific domain of type I collagen, the slightly unraveled telopeptides.[25] In this experiment, we performed this assay with and without the presence of purified rC7 or its NC1 domain. As expected, maximum contraction occurred in the presence of serum GFs while there was no contraction without them (**Figure 4a**). Interestingly, full-length rC7 inhibited contraction of the collagen lattices, while the NC1 domain did not. As we showed previously that the NC1 domain of C7 was unable to promote skin cell migration *in vitro* or accelerate wound closure *in vivo,* our data suggest that the domain of C7 that promotes cell migration resides within the same domain of C7 that is required for the inhibition of collagen lattice contraction. Furthermore, rC7 inhibited collagen lattice contraction in a dose-dependent manner with maximum levels of inhibition at a concentration of 25 µg/ml (**Figure 4b**).

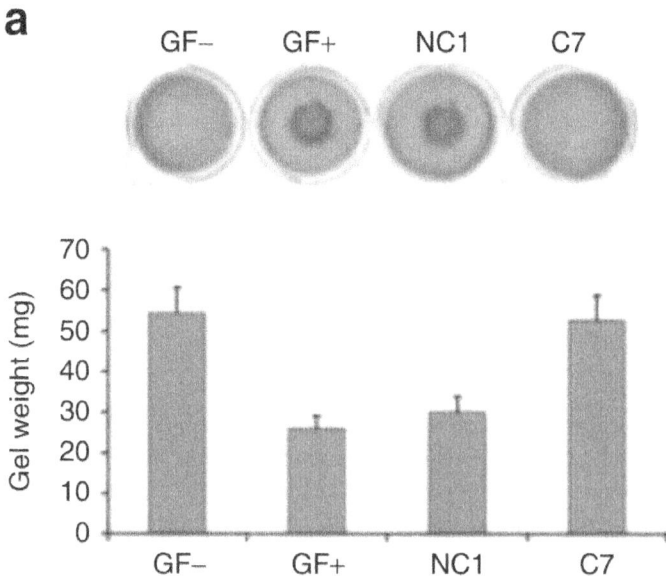

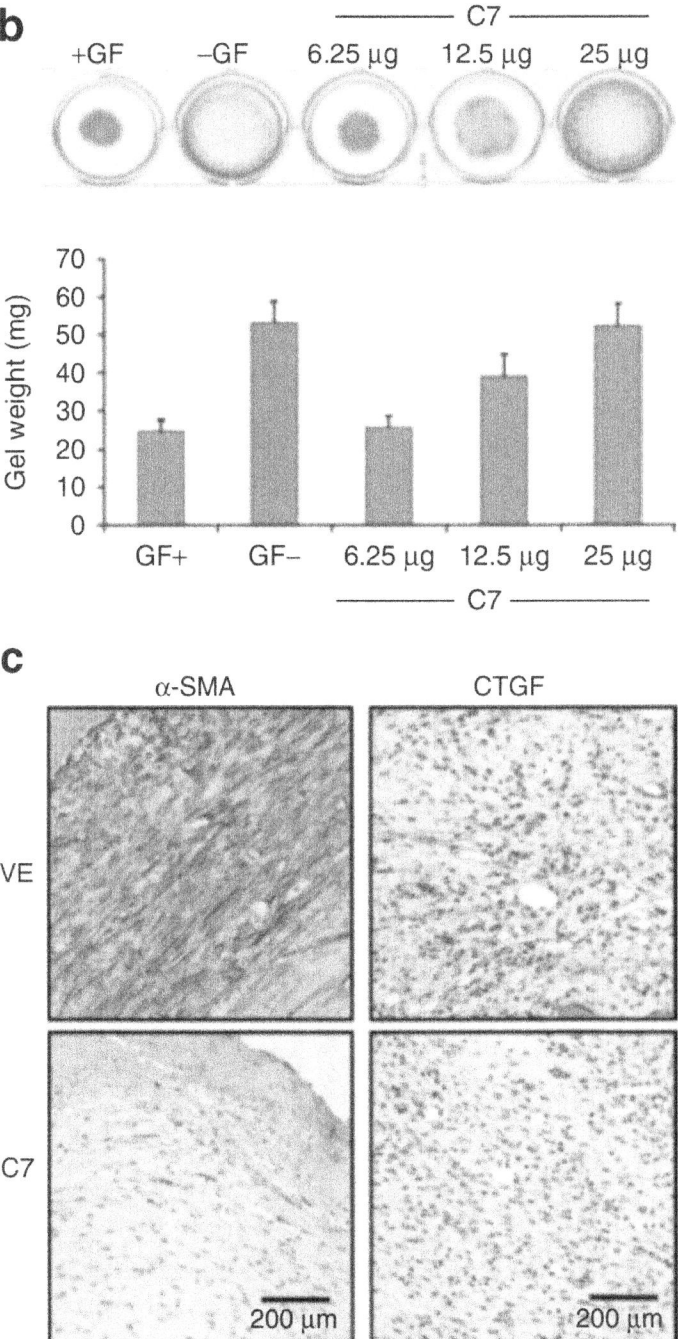

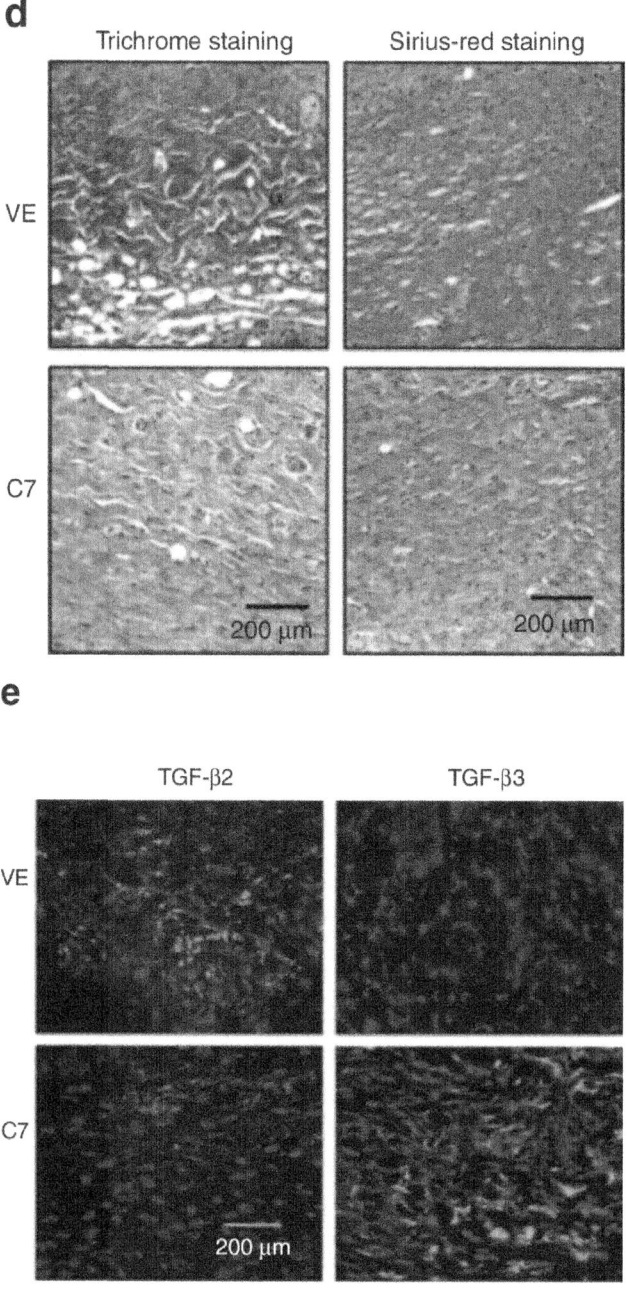

Figure 4. C7 inhibited the contraction of collagen lattices *in vitro* and reduces the presence of myofibroblasts, connective tissue growth factor (CTGF) expression, and collagen deposition *in vivo*. (**a**) Early passage, human dermal fibroblasts were isolated

and mixed with serum and type I collagen. Collagen lattices, with size of 0.6 cm^3 and cell density at 100,000 cells/cm^3, were released and floated in 1 ml DMEM. rC7 (C7) or NC1 of 30 µg were added to the serum containing medium and incubated at 37 °C for 24 hours. Contraction assays were also carried out without serum added (GF−) or with serum added (GF+). Contraction of the lattices was determined by measuring the gel weight. These data represent the mean ± SD of triplicate determinations in one representative experiment. Similar results were obtained in two other independent experiments. (**b**) Dose-dependent inhibition of contraction of collagen lattices by rC7. rC7 at the indicated concentrations were added to the contraction assays. These data represent the mean ± SD of triplicate determinations in one representative experiment. Similar results were obtained in two other independent experiments. (**c**) Immunohistochemistry staining of healed mouse skin was performed with the antibodies specific for αsmooth muscle actin (α-SMA) and connective tissue growth factor (CTGF) at 2 weeks after topical application of vehicle (VE) and rC7 (C7). Ten randomly selected images per condition from three independent experiments were analyzed. The representative images are shown. Scale bars: 0.2 mm. Note that the wounds treated with topical rC7 showed significantly diminished α-SMA–positive fibroblasts (myofibroblasts) as well as decreased CTGF expression. (**d**) Collagen deposition was assessed by Masson's trichrome staining and Picrosirius red staining of healed wounds at 14 days after topical application of rC7 (C7) or vehicle (VE). Ten randomly selected images per condition from three independent experiments were analyzed. The representative images are shown. Scale bars: 0.2 mm. Note that rC7-treated wounds have less collagen deposition in the dermis by both staining methods compared with vehicle-treated wounds which show perturbed fiber architecture. (**e**) Immunofluorescence staining of healed mouse skin was performed with the polyclonal antibodies specific for TGF-β2 and TGF-β3 at 2 weeks after topical application of vehicle (VE) and rC7 (C7). Ten totally randomly selected images per condition from three independent experiments were analyzed. The representative images are shown. Scale bars: 0.2 mm. Note that the wounds treated with topical rC7 display reduced expression of TGF-β2 and increased expression of TGF-β3.

The transformation of fibroblasts to α smooth muscle actin (α-SMA)–positive myofibroblasts is responsible for converting granulation tissue into a permanent scar. Therefore, the upregulation of SMA is associated with fibrosis.[26] To determine if rC7 also inhibits wound fibrosis *in vivo*, we performed immunohistochemistry on healed skin wounds with antibodies specific for α-SMA at 2 weeks after the topical application of rC7 or vehicle. As shown in **Figure 4c**, there was a significant decrease in the quantity of myofibroblasts in wounds treated with topical rC7- versus vehicle-treated wounds. Connective tissue GF (CTGF/CCN2) is required for TGF-β–induced transdifferentiation of fibroblasts to myofibroblasts.[27] As shown in**Figure 4c**, rC7 also reduced the expression of CTGF/CCN2 protein at 2 weeks in comparison with vehicle control. Histological evaluation of healed skin tissues stained with Masson's trichrome and Picrosirius red stain revealed less dermal collagen deposition in rC7-treated wounds compared with vehicle-treated wounds (**Figure 4d**). In addition, rC7-treated wounds revealed collagen fibers organized in a parallel fashion compared with vehicle-treated wounds which showed a haphazard, disarrayed arrangement of collagen fibers. Collectively, the *in vivo* data showed that the topical application of rC7 to skin wounds resulted in a significant reduction in fibroblast-to-myofibroblast differentiation, CTGF expression and collagen deposition, suggesting that C7 may inhibit excessive fibrosis that leads to scar formation.

Transforming GF-β (TGF-β) is a multifunctional GF involved in many aspects of wound healing.[2] However, an excess of TGF-β can lead to pathological scar formation. There are three mammalian TGF-β isoforms: TGF-β1, TGF-β2, and TGF-β3. TGF-β1 and TGF-β2 isoforms exhibit pro-scarring properties, whereas TGF-β3 displays anti-scarring properties.[28] To determine the effects of topical application of rC7 on the expression of TGF isoforms, we performed immunostaining on healed skin wounds with antibodies specific for TGF-β1, TGF-β2, and TGF-β3 at 2 weeks after the topical application of rC7 or vehicle. As shown in **Figure 4e**, the expression of pro-fibrogenic TGF-β2 was significantly reduced while the expression of anti-fibrogenic TGF-β3 was significantly increased in healed wounds treated with topical rC7- versus vehicle-treated wounds. We did not detect any significant difference in the expression of TGF-β1 (data not shown). These data indicate that one mechanism by which C7 affects the fibrotic pathway is by regulating the expression of TGF-β isoforms.

We also assessed the effect of topical rC7 on the healing wounds in human skin grafts. We grafted 1.5 × 1.5 cm pieces of human skin onto the backs of athymic nude mice as previously described.[29] Eight weeks after grafting, the human skin grafts were wounded with an 8 mm punch biopsy tool. We then topically applied rC7 or vehicle onto the wounds. As shown in **Figure 5**, human skin wounds healed significantly faster when treated with topical rC7 compared with vehicle controls. Taken together, these data indicate that topically applied rC7 to healing wounds accelerates the closure of skin wounds in both murine and human skin graft model systems.

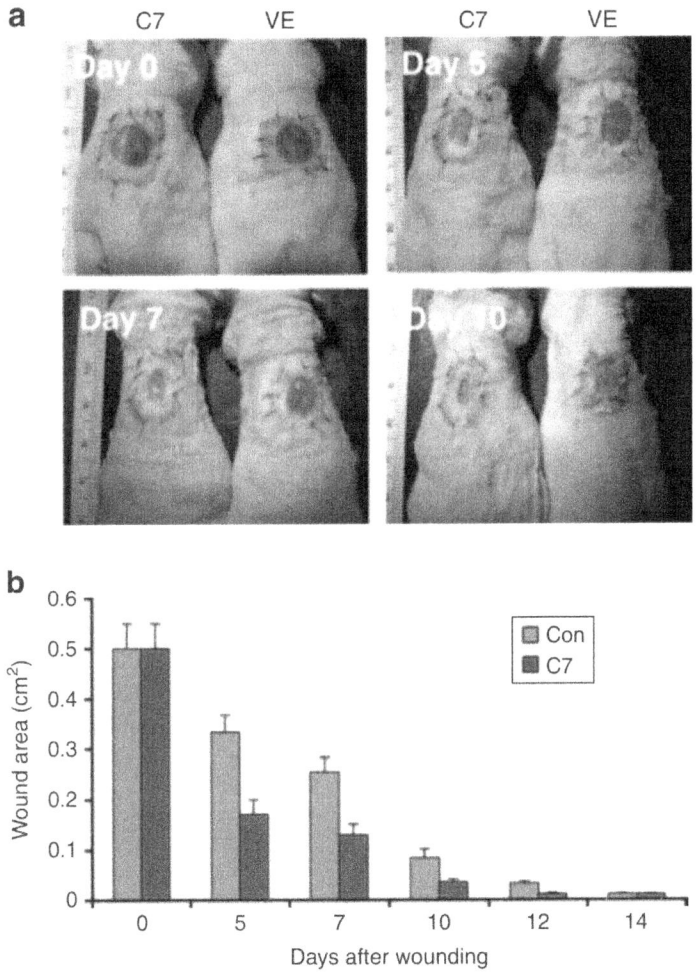

Figure 5. Topically applied rC7 promoted wound closure of human skin. A 0.5 cm^2 (8 mm diameter punch biopsy) full-thickness excision wound was made in human skin grafted onto the mid-back of 8- to 10-week-old athymic nude mice. The wounds were then treated with topically applied 30 μg rC7 (C7) or vehicle (VE) (n = 5 mice per group). (**a**) Representative days 0, 5, 7, and 10 wounds are shown. Wound sizes were reduced in wounds treated with rC7 compared with vehicle control at 5, 7, and 10 days after wounding. (**b**) Mean + SD wound size measurements at days 0, 5, 7, 10, 12, and 14 after wounding (n = 5 mice for each group).

Topical rC7 Restored C7 Expression in Engrafted RDEB Mouse Skin *in Vivo*

Patients with RDEB manifest with widespread open wounds and chronic erosions. To evaluate the efficacy of topical rC7 for a murine model of RDEB wounds, we transplanted skin from C7-knock out mice onto the backs of the athymic nude mice. These *COL7A1*-null mice have no C7 at the DEJ of their skin, and they entirely lack AFs.[30] Clinically, these newborn mice exhibit extensive blisters and die within the first week of life. Thus, these C7-null mice recapitulate many of the clinical, genetic, and ultrastructural features of severe patients with RDEB. Because these RDEB mice are very fragile and small and die within a few days after birth, it was not technically feasible to apply rC7 topically onto these mice. As an alternative, we killed these mice and transplanted their skin onto the backs of adult athymic nude mice. After engraftment of the transplanted RDEB mouse skin, we made a 6 mm punch biopsy wound in the transplanted RDEB graft and topically applied rC7. As shown in **Figure 6a** (panels A and D), RDEB skin grafts before treatment showed histological evidence of dermal–epidermal separation and entirely lacked C7 at the DEJ, characteristic of RDEB-diseased donor skin. Similarly, vehicle-treated wounds displayed dermal–epidermal separation and exhibited no C7 at the DEJ (**Figure 6a**, panels B and E). In contrast, RDEB skin grafts that received topical rC7 exhibited improved dermal–epidermal adherence and had C7 restored at the DEJ in a linear distribution (**Figure 6a**, panels C and F).

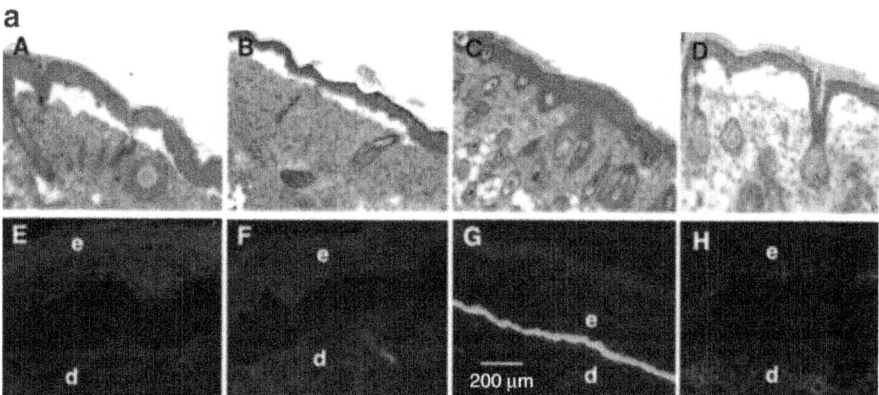

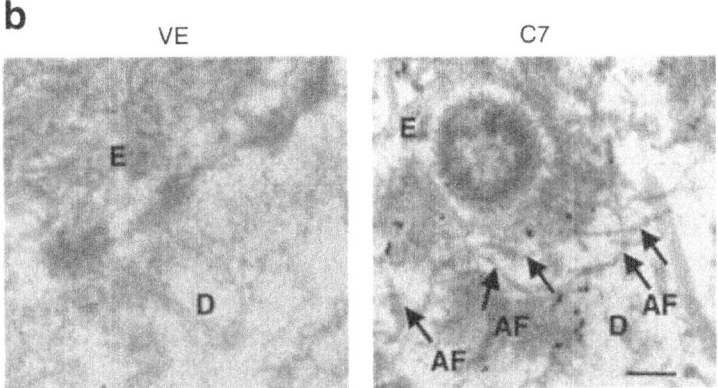

Figure 6. Topical application of rC7 incorporated into the DEJ of RDEB mouse skin grafts and forms AFs *in vivo*. (**a**) Histological appearance (A–D) and immunofluorescence staining (E–H) of engrafted RDEB mouse skin using a polyclonal antibody to the NC1 domain of C7. Panels A and E, 2 weeks after grafting of RDEB mouse skin and before treatment ($n = 30$ mice); panels B and F, wounded and engrafted RDEB skin 2 weeks after topically applied vehicle ($n = 4$ mice); panels C and G, wounded and engrafted RDEB skin 2 weeks after 30 μg of rC7 was applied topically ($n = 15$ mice); panels D and H, unwounded engrafted RDEB skin 2 weeks after 30 μg of rC7 was applied topically ($n = 4$ mice). (**b**) Immunogold labeling of engrafted murine RDEB skin topically applied with 30 μg rC7 (C7) or vehicle (VE) was performed using an anti-NC1 polyclonal antibody and revealed that the topical rC7 incorporated into the RDEB skin grafts and formed AFs. Note restoration of numerous arching AFs depicted with arrows and labeled with gold particles decorating the DEJ in rC7-treated RDEB skin grafts. AF, anchoring fibrils; d, dermis; e, epidermis; Scale bar: 100 nm.

Full-thickness, skin wounds are much different from the superficial bullous wounds of RDEB skin, which occur at the dermal–epidermal interface. Having shown that topical rC7 incorporated into the DEJ of full-thickness, skin wounds, we wished to evaluate if topical rC7 would be able to incorporate into the engrafted RDEB skin wounds by performing experiments without wounding the RDEB skin grafts. As shown in **Figure 6a**, the rC7-treated RDEB skin grafts displayed dermal–epidermal separation and lacked C7 expression at the DEJ (**Figure 6a**, panels D and H). These results indicate that wounding the RDEB skin is a prerequisite for topical rC7 to incorporate into the DEJ.

Immunoelectron microscopy showed that topically applied rC7 restored the formation of AFs in the grafted RDEB skin (**Figure 6b**), suggesting correction of the major ultrastructural disease abnormality. Taken together, these data suggest that protein-based therapy by topical application of rC7 may correct the defects in RDEB skin, namely lack of C7, lack of AFs and compromised dermal–epidermal adherence.

DISCUSSION

In this study, we used a murine wound healing model and a RDEB skin transplant model to evaluate the feasibility of topically applied rC7 to treat RDEB and normal skin wounds. Our studies showed that topical rC7 stably incorporated into the DEJ of healing wounds and accelerated wound closure by increasing re-epithelialization. Furthermore, we observed that C7-mediated acceleration of wound closure was accompanied by decreases in the expression of CTGF and the presence of α-SMA myofibroblasts, markers associated with scar formation. In addition, our results indicated that there is a correlation between the domain of C7 that promotes keratinocyte migration and the domain of C7 that accelerates wound closure. With the RDEB skin transplantation model, we demonstrated that topical rC7 incorporated correctly into the DEJ of RDEB skin, formed AFs and corrected the RDEB clinical phenotype.

Numerous GFs and other agents that could potentially enhance tissue regeneration have been identified, but their therapeutic application has been limited in clinical medicine for several reasons. These include difficulty in maintaining bioactivity of locally applied therapeutic agents in regenerating tissue because of a lack of retention of the agent, poor tissue penetration, and instability of therapeutic proteins in the protease-rich environment of a healing skin wound. It is thought that for these reasons GFs, such as PDGF-BB (Regranex, Healthpoint Biotherapeutics, Fort Worth, TX), have only a modest effect on wound healing and need to be administered daily. C7 is a large, stable collagen molecule with a molecular mass of 900 kDa. It is a stable molecule that is relatively resistant to protease degradation. In fact, we have left purified C7 in neutral buffer at room temperature for up to 3 months or at 4 °C for up to 6 months and found no degradation or loss of biological activity (data not shown). The stability of C7, and its large size may offer advantages over GFs for use as a wound healing agent.

As for treatment of RDEB, our data demonstrate that topical rC7 stably incorporated into the mouse DEJ for up to 2 months. The actual turnover time of C7 in human skin is not known. Our study with human RDEB skin equivalents transplanted onto immunodeficient mice shows that human rC7 administered to skin equivalents endures for at least 3 months.[20] We also know that one does not have to have 100% of the normal complement of AFs to have physiological normal dermal–epidermal adherence, and in fact, only about 35% of the normal AF complement is needed to prevent clinical skin fragility and blistering.³¹ With regards to patients with RDEB, it may be possible to apply topically rC7 and improve the patient's skin fragility and bullae/erosion formation with periodic applications. This will be determined by future clinical trial.

We have not done parallel systematic studies to compare experiments comparing the delivering of rC7 by the three different potential administration routes; topical application, intravenous infusion and intradermal injection. We showed previously that the intradermal injection of 20 μg of rC7 was sufficient to restore C7 and reverse the RDEB phenotype in a square centimeter of RDEB skin equivalent.¹⁹ In a recently published paper using an intravenous injection approach, we showed that functional correction and clinical benefit was achieved using 60 μg of rC7 in ~1 square centimeter of grafted RDEB mouse skin.³² In the topical rC7 study herein, 30 μg of topical rC7 was used to achieve reversal of the RDEB phenotype. It appears that topical and intradermal routes of delivery may require less rC7 in comparison with intravenously administration of rC7. As we have not compared systematically all three approaches at the same time, we cannot extrapolate and say which route will be the most efficient and efficacious in actual patients with RDEB.

We previously showed that the wounds of mice intravenously injected with gene-corrected RDEB fibroblasts over-expressing C7 demonstrated accelerated wound healing compared with control mice injected with uncorrected RDEB cells.²² In the present study, we demonstrated that rC7 topically delivered to skin wounds accelerated their wound closure. How exogenously delivered rC7 promotes healing is unknown. It is known that the process of wound healing involves a series of synchronized events, namely clot formation, inflammation, re-epithelialization, production of granulation tissue, fibroplasia, angiogenesis, and connective tissue remodeling.² The migration of keratinocytes is the first essential step for wound re-epithelialization. When the skin is wounded, the keratinocytes come into contact with collagens and dermal glycoproteins, such as fibronectin, and start to migrate across the wound bed.³³,³⁴ Our previous studies demonstrated that C7 is the most potent pro-motility matrix among other extracellular cellular matrices (including collagens type I and IV, and fibronectin) for driving human keratinocyte migration. However, during the normal wound healing process, C7 was detected very late and appeared at the DEJ on day 7 after wounding.³⁵ It is possible that exogenously delivered rC7 by topical application during early wound healing will provide a matrix that promotes cell motility and allows keratinocytes to migrate and close the wounds. Consistent with this notion, we showed in this study that rC7-treated wounds demonstrated increased re-epithelialization. This notion is further supported by our data showing that the NC1 domain of C7 that was unable to promote keratinocyte migration *in vitro* also failed to accelerate skin wound closure *in vivo*.

Our data in this study showed that rC7 inhibits collagen lattice contraction *in vitro*. In addition, topical rC7 reduces the number of myofibroblasts, the expression of CTGF and disorganized collagen deposition in the neodermis of healed skin wounds. Myofibroblasts and CTGF are key players in fibrotic processes. These observations raise the question of whether exogenous rC7 topically applied to skin wounds could result in wounds with less scarring and fibrosis. There are several lines of evidence to support this notion. First, patients with RDEB who lack functional C7 heal wounds with significant scarring. Second, hypomorphic RDEB mice that only express C7 at 10% of the normal level exhibit enhanced fibrotic processes with excessive contraction and deposition of extracellular matrix.[36] This fibrosis is associated with an upregulation of TGF-β and CTGF, neoexpression of α-SMA–positive cells, and induction of tenascin C. Third, when fetuses *in utero* are subjected to skin wounds within the first trimester, the skin wounds heal without scars, so-called "scarless healing".[37] It is interesting to note that during fetal life, the fetus is bathed in amniotic fluid and surrounded with an amniotic membrane that is very rich in C7 and laminin 332, two large matrix molecules that improve wound healing.[38] In concordance with these observations, amniotic membranes have been used topically on burn wounds, corneal abrasions, and skin wounds to promote re-epithelialization and healing.[39] One could speculate that the molecular basis for these clinical observations is that the C7 within amniotic membranes promotes scarless healing in fetal skin wounds.

Many soluble factors participate in wound repair. One factor that has been implicated in scar formation is TGF-β. The TGF-β family of proteins consists of several structurally related but functionally distinct isoforms. In mammals, three isoforms, TGF-β1, -β2, and -β3, have been identified. These isoforms play distinct roles in the modulation of wound repair.[2] TGF-β1 and TGF-β2 possess pro-fibrogenic activities, while TGF-β3 exhibits anti-fibrogenic activity.[28] In this study, we showed that topical rC7 induced the upregulation of TGF-β3 (anti-fibrogenic isoform) and downregulation of TGF-β2 (the pro-fibrogenic isoform). These data, taken together with the reduction of several scarring parameters (α-SMA expression, collagen deposition and CTGF expression), suggest that rC7 not only accelerates wound closure but also contributes to improve wound healing quality with less scarring by modulating the expression of the TGF-β family.

The mice we used for the wound healing studies are athymic nude mice that are unable to produce T cells, and are therefore immunodeficient. We chose these mice simply because they cannot mount an immune response and produce anti-C7 antibodies after topical application of human C7. In addition, they do not reject grafted human skin or RDEB mouse skin. However, the rC7-mediated accelerated wound closure and reduced expression of α-SMA and CTGF were also observed when wound healing experiments were conducted with immunocompetent mice (data not shown).

Various therapeutic strategies for RDEB are under investigation, including gene, cell and protein therapies, but to date, no consistently low risk, effective treatment is available.[40] Currently, there are several "proof-of-principal" clinical studies that have been planned or initiated to study therapeutic approaches for RDEB. Bone marrow/stem cells have been administered to a limited number of patients with RDEB with some success, but also with considerable mortality.[15] The intradermal injection of C7-producing, allogeneic, human dermal fibroblasts into patients with RDEB has increased the expression of patients' own mutated C7 at their DEJ and improved their dermal–epidermal adherence.[14] However, the injected cells did not persist longer than 2 weeks in the skin of the patients with RDEB'. Using two preclinical animal models, we showed that intradermal injection of C7 expressing fibroblasts, C7 protein and lentivectors expressing C7 into RDEB mice all restored C7 and AFs at the DEJ and significantly prolonged the survival of the RDEB mice.[17–21] In a more recently study, we showed that rC7 administered intravenously to mice homed to engrafted RDEB mouse skin, incorporated into the DEJ of the grafts, and restored C7, AFs and dermal–epidermal adherence.[32]

If topically administered rC7 improves the health of patients with RDEB, it would have many advantages over other treatments such as lack of exposure to live cells, exogenous DNA or RNA or viral vectors and local application. Compared with bone marrow/stem cell transplantation and intravenous or intradermal injection of C7, topically applied rC7 would likely be painless and logistically simple. It would not require sophisticated medical expertise or equipment. Lastly, the topical application of rC7 could promote the closure of the existing wounds of the patient with RDEB. Nevertheless, there are several limitations associated with the topical approach. This includes the inability to treat esophageal lesions that are common in the most severely affected patients with RDEB. In addition, as rC7 does not penetrate intact, unwounded skin and only works locally in existing open wounds, another limitation is that topical rC7 will not prevent the onset of new skin blisters when applied to non-lesional RDEB skin. Lastly, protein replacement therapy requires lifelong repeated administration of rC7 regardless of the mode of administration: intradermal injection, topical application, or intravenous injection.

In summary, our studies demonstrate that rC7 applied topically onto skin wounds stably incorporated into the newly regenerated DEJ of healed wounds and accelerated wound closure. In addition to being a therapy for chronic skin wounds, this strategy may be particularly useful for patients with RDEB who have multiple open wounds and often die from metastatic squamous cell carcinomas arising in chronic skin wounds. The topical application of rC7 may provide a feasible treatment option for not only accelerating wound healing, but also normalizing the devastating cutaneous phenotype of RDEB. These preclinical results are very encouraging though additional studies are needed to determine the safety and efficacy in humans.

MATERIALS AND METHODS

Purification of rC7

Recombinant C7 was purified from serum-free media from RDEB dermal fibroblasts stably transduced with a lentiviral vector coding for full-length C7 as described.[20] Briefly, serum-free media were equilibrated to 5 mmol/l EDTA, 50 µmol/l PMSF, and 50 µmol/l NEM and precipitated with 300 mg/ml ammonium sulfate at 4 °C overnight with constant stirring. Precipitated proteins were collected by centrifuging at 1.2×10^6 g/minute for 1 hour, resuspended, and dialyzed in Buffer A (65 mmol/l NaCl, 25 mmol/l Tris–HCl, pH 7.8). Following dialysis, insoluble material was collected by centrifugation at 8,600g for 20 minute, and the pellet redissolved in Buffer B (50 mmol/l Tris–HCl pH 7.5, 150 mmol/l NaCl, 5 mmol/l EDTA, 2 mmol/l NEM, 2 mmol/l PMSF). The solution was clarified as above, and the supernatant, S1', was passed over a Q-sepharose column (Pharmacia, Piscataway, NJ) equilibrated in the same buffer. Elution was then carried out with a linear gradient from 0.2 to 1.0 mol/l NaCl of appropriate volume size. The rC7 eluted at 0.7–1 mol/l NaCl. The NC1 domain of C7 was purified from 293 cells stably transfected with cDNA encoding for NC1 as described.[41]

Topical Application of rC7 onto Mice

We first made a 1.0×1.0 cm full-thickness excisional wound by lifting the skin with forceps and removing full thickness skin with a scissors on the mid-back of 8- to 10-week-old athymic nude mice (Simonsen Laboratory, Gilroy, CA). Immediately after wounding, 8–32 µg of rC7 or 30 µg NC1 in a 10% carboxymethylcellulose gel were applied to the wound surface. We then covered the wound with a band-aid and a Coban self-adherent wrap, to prevent desiccation. We treated 30 mice with topical rC7, 10 mice with NC1, and 20 mice with the vehicle as a negative control. Two to 8 weeks after topical application, biopsies from the wounds were obtained and subjected to immunostaining using an antibody specific for human C7 (clone LH 7.2;

Sigma, St Louis, MO) or a rabbit polyclonal antibody that recognizes both human and mouse C7.[42] To measure the wound size, standardized digital photographs were taken of the wounds at various days after wounding and open wound areas determined with an image analyzer (AlphaEase FC version 4.1.0; Alpha Innotech, Johannesburg, South Africa) on a personal Macintosh computer. The total pixels that covered the unhealed areas were drawn onto the digital photographs using a pattern overlay in ImageJ (http://rsbweb.nih.gov/ij/). The number of pixels covering an open wound area on a given day was divided by the number of pixels spreading over the initial wound on day 0 to obtain the percentage of closure.

For RDEB mouse skin transplantation studies, 1.2 by 1.2 cm of skin from 2 to 5 days old, newborn RDEB mice was transplanted onto the back of athymic nude mice. At 12–14 days after skin transplantation and engraftment, the RDEB skin grafts were either unwounded or wounded with a 6 mm punch biopsy instrument. We then topically applied 30 μg of rC7 ($n = 15$ mice) or vehicle ($n = 4$) to the wounded RDEB skin or unwounded RDEB skin ($n = 4$) as described above.

For evaluating wound healing of human skin, a 1.5×1.5 cm square of full-thickness human skin was grafted onto athymic nude mice as previously described.[29] Eight weeks after grafting the engrafted human skin was wounded using an 8 mm punch biopsy tool. We then topically applied rC7 (30 μg) or vehicle to the wounds and bandaged as described above. The assessment of wound healing was then performed using area planimetry, as described above for the murine wounds. All animal studies were conducted using protocols approved by the University of Southern California Institutional Animal Use Committee.

Immunofluorescence Staining and Ultrastructural Analysis of Tissue

Five-micrometer thick sections of OCT-embedded frozen tissues were cut on a cryostat, fixed for 5 minutes in cold acetone, and air-dried. Immunolabeling of the tissue was performed using standard immunofluorescence methods as described previously.[17–20] Briefly, for single- and double-immunofluorescence staining, sections were blocked with M.O.M. Mouse IgG Blocking Reagent (Vector Laboratories, Burlingame, CA) for 1 hour at room temperature. Primary antibodies were diluted in phosphate buffered saline with 1% bovine serum albumin. For C7 staining, we used monoclonal antibodies against human C7, clone LH 7.2 (Sigma), or a rabbit polyclonal antibody that recognizes both mouse and human C7.[42] For double-immunofluorescence staining, we incubated the mouse monoclonal anti-human C7 antibody together with a rabbit polyclonal antibody to both human and mouse C7. For TGF-β staining, we used polyclonal antibodies against TGF-β1 (sc-146, Santa Cruz Biotechnology, Santa Cruz, CA), TGF-β2 (sc-90, Santa Cruz Biotechnology), or TGF-β3 (sc-82, Santa Cruz Biotechnology). All primary antibody dilutions were 1:200. After incubation for 1 hour at room temperature, sections were washed in phosphate buffered saline three times and stained for 1 hour with FITC-conjugated goat anti-mouse IgG1 with or without Cy3-conjugated goat anti-rabbit IgG (Sigma) diluted 1:300 in phosphate buffered saline with 1% bovine serum albumin. Slides were mounted with 40% glycerol. Photographs of stained sections were taken using a Zeiss Axioplan fluorescence microscope equipped with a Zeiss Axiocam MRM digital camera system (Carl Zeiss International, Göttingen, Germany).

Immunogold electron microscopy was performed on the engrafted RDEB mouse skin using a standardized method as described previously.[43,44] To assess human AF formation and ultrastructure, 40 micron sections were fixed in 0.1% glutaraldehyde, rinsed in 0.15 mol/l Tris pH 7.5, then incubated in our polyclonal anti-NC1 antibody followed by 5 nm gold secondary antibody and enhancement as described.[43,44]

Histological analysis of tissues. The mice whose wounds were treated with topical rC7 or vehicle were euthanized at 7 or 14 days after treatment. The wounds, together with unwounded skin margins, were excised and put into 10% formaldehyde. H&E staining was carried out as previously described.[45] To show the entire wound, multiple overlapping photographs were taken under a microscope (Nikon, Eclipse TE2000-U, ×4; Nikon, Tokyo, Japan) and used to reconstitute the entire wound. A standard immunohistochemistry staining procedure was carried out as described.[46] We used a mouse monoclonal antibody to pan keratin (Clone 80; Abcam, Cambridge, MA), a rabbit polyclonal antibody to PECAM-1 (Clone M-20; Santa Cruz Biotechnology), a mouse monoclonal antibody to α-SMA (Clone 1A4; Dako Denmark A/S, Glostrup, Denmark), a mouse monoclonal antibody to PCNA (Clone PC10; EMD Millipore, Billerica, MA), and a rabbit polyclonal antibody to CTGF (Abcam). All antibodies were used in 1:100 dilutions.

Collagen Lattice Contraction Assay (FPCL)

Type I collagen (Sigma) lattices were prepared as previously described.[24] Human dermal fibroblasts were cultured in Dulbecco's modified Eagle's medium (Invitrogen, Carlsbad, CA) containing 10% fetal bovine serum with antibiotics. The lattices were prepared with a final fibroblast density of 100,000 cells/ml in a 0.6 ml volume per well in 24-well, non-tissue culture plates (Becton Dickinson, Le Pont De Claix, France). After polymerization of the collagen (VitrogenTM, Cohesion Technologies, Palo Alto, CA), the lattices were incubated in 1.0 ml Dulbecco's modified Eagle's medium with or without fetal bovine serum. Either NC1 or rC7 was added to the individual dishes at concentrations ranging from 6.25–30 µg. The collagen lattices were incubated at 37 °C with 5% CO_2 and the contraction of the lattices was measured by weight.[24] All experiments were carried out in triplicate and repeated three times, and data points and error bars in the figures represent averages and SDs.

ACKNOWLEDGMENTS

This work was supported by grants (NIH RO1 AR47981 to M.C, RO1 AR33625 to M.C. and D.T.W., Sponsored Research Project from Lotus Tissue Repair to M.C. and D.T.W.). However, the content is the sole responsibility of the authors and does not necessarily represent the views of Lotus Tissue Repair or its affiliates. We thank Sara Tufa for technical support of immuno-EM. Microscopy services were provided by the Cell and Tissue Imaging Core of USC Research Center for Liver Diseases (NIH grants No. P30 DK048522 and S10 RR022508). M.C. and D.T.W. are consultants for Lotus Tissue Repair, and hold stock in the company. M.C., D.T.W., and the University of Southern California hold patents for recombinant type VII collagen and have filed a Conflict of Interest Declaration with Randoph W Hall, Vice Provost for Research Advancement at the University of Southern California.

REFERENCES

1. Gordois, A, Posnett, J, Borris, L, Bossuyt, P, Jönsson, B, Levy, E et al. (2003). The cost-effectiveness of fondaparinux compared with enoxaparin as prophylaxis against thromboembolism following major orthopedic surgery. *J Thromb Haemost* 1: 2167–2174.

2. Singer, AJ and Clark, RA (1999). Cutaneous wound healing. *N Engl J Med* 341: 738–746.

3. LeGrand, EK (1998). Preclinical promise of becaplermin (rhPDGF-BB) in wound healing. *Am J Surg* 176(2A Suppl): 48S–54S.

4. Smiell, JM, Wieman, TJ, Steed, DL, Perry, BH, Sampson, AR and Schwab, BH (1999). Efficacy and safety of becaplermin (recombinant human platelet-derived growth factor-BB) in patients with nonhealing, lower extremity diabetic ulcers: a combined analysis of four randomized studies. *Wound Repair Regen* 7: 335–346.

5. Wieman, TJ, Smiell, JM and Su, Y (1998). Efficacy and safety of a topical

gel formulation of recombinant human platelet-derived growth factor-BB (becaplermin) in patients with chronic neuropathic diabetic ulcers. A phase III randomized placebo-controlled double-blind study. *Diabetes Care* 21: 822–827.

6. Lin, AN and Carter, DM (eds) (1992). *Epidermolysis Bullosa: Basic and Clinical Aspects.* Springer-Verlag: New York.

7. Uitto, J and Christiano, AM (1994). Molecular basis for the dystrophic forms of epidermolysis bullosa: mutations in the type VII collagen gene. *Arch Dermatol Res* 287: 16–22.

8. Uitto, J and Christiano, AM (1992). Molecular genetics of the cutaneous basement membrane zone. Perspectives on epidermolysis bullosa and other blistering skin diseases. *J Clin Invest* 90: 687–692.

9. Parente, MG, Chung, LC, Ryynänen, J, Woodley, DT, Wynn, KC, Bauer, EA *et al.* (1991). Human type VII collagen: cDNA cloning and chromosomal mapping of the gene. *Proc Natl Acad Sci USA* 88: 6931–6935.

10. Burgeson, RE (1993). Type VII collagen, anchoring fibrils, and epidermolysis bullosa. *J Invest Dermatol* 101: 252–255.

11. Sakai, LY, Keene, DR, Morris, NP and Burgeson, RE (1986). Type VII collagen is a major structural component of anchoring fibrils. *J Cell Biol* 103: 1577–1586.

12. Morris, NP, Keene, DR, Glanville, RW, Bentz, H and Burgeson, RE (1986). The tissue form of type VII collagen is an antiparallel dimer. *J Biol Chem* 261: 5638–5644.

13. Bruckner-Tuderman, L, Nilssen, O, Zimmermann, DR, Dours-Zimmermann, MT, Kalinke, DU, Gedde-Dahl, T Jr *et al.* (1995). Immunohistochemical and mutation analyses demonstrate that procollagen VII is processed to collagen VII through removal of the NC-2 domain. *J Cell Biol* 131: 551–559.

14. Wong, T, Gammon, L, Liu, L, Mellerio, JE, Dopping-Hepenstal, PJ, Pacy, J *et al.* (2008). Potential of fibroblast cell therapy for recessive dystrophic epidermolysis bullosa. *J Invest Dermatol* 128: 2179–2189.

15. Wagner, JE, Ishida-Yamamoto, A, McGrath, JA, Hordinsky, M, Keene, DR, Woodley, DT *et al.* (2010). Bone marrow transplantation for recessive dystrophic epidermolysis bullosa. *N Engl J Med* 363: 629–639.

16. Ortiz-Urda, S, Thyagarajan, B, Keene, DR, Lin, Q, Fang, M, Calos, MP *et al.* (2002). Stable nonviral genetic correction of inherited human skin disease.*Nat Med* 8: 1166–1170.

17. Chen, M, Kasahara, N, Keene, DR, Chan, L, Hoeffler, WK, Finlay, D *et al*. (2002). Restoration of type VII collagen expression and function in dystrophic epidermolysis bullosa. *Nat Genet* 32: 670–675.

18. Woodley, DT, Krueger, GG, Jorgensen, CM, Fairley, JA, Atha, T, Huang, Y *et al*. (2003). Normal and gene-corrected dystrophic epidermolysis bullosa fibroblasts alone can produce type VII collagen at the basement membrane zone. *J Invest Dermatol* 121: 1021–1028.

19. Woodley, DT, Keene, DR, Atha, T, Huang, Y, Ram, R, Kasahara, N *et al*. (2004). Intradermal injection of lentiviral vectors corrects regenerated human dystrophic epidermolysis bullosa skin tissue in vivo. *Mol Ther* 10: 318–326.

20. Woodley, DT, Keene, DR, Atha, T, Huang, Y, Lipman, K, Li, W *et al*. (2004). Injection of recombinant human type VII collagen restores collagen function in dystrophic epidermolysis bullosa. *Nat Med* 10: 693–695.

21. Remington, J, Wang, X, Hou, Y, Zhou, H, Burnett, J, Muirhead, T *et al*. (2009). Injection of recombinant human type VII collagen corrects the disease phenotype in a murine model of dystrophic epidermolysis bullosa. *Mol Ther* 17: 26–33.

22. Woodley, DT, Remington, J, Huang, Y, Hou, Y, Li, W, Keene, DR *et al*. (2007). Intravenously injected human fibroblasts home to skin wounds, deliver type VII collagen, and promote wound healing. *Mol Ther* 15: 628–635.

23. Woodley, DT, Hou, Y, Martin, S, Li, W and Chen, M (2008). Characterization of molecular mechanisms underlying mutations in dystrophic epidermolysis bullosa using site-directed mutagenesis. *J Biol Chem* 283: 17838–17845.

24. Han, YP, Nien, YD and Garner, WL (2002). Recombinant human platelet-derived growth factor and transforming growth factor-beta mediated contraction of human dermal fibroblast populated lattices is inhibited by Rho/GTPase inhibitor but does not require phosphatidylinositol-3' kinase.*Wound Repair Regen* 10: 169–176.

25. Woodley, DT, Yamauchi, M, Wynn, KC, Mechanic, G and Briggaman, RA (1991). Collagen telopeptides (cross-linking sites) play a role in collagen gel lattice contraction. *J Invest Dermatol* 97: 580–585.

26. Akasaka, Y, Ono, I, Tominaga, A, Ishikawa, Y, Ito, K, Suzuki, T *et al*. (2007). Basic fibroblast growth factor in an artificial dermis promotes apoptosis and inhibits expression of alpha-smooth muscle actin, leading

to reduction of wound contraction. *Wound Repair Regen* 15: 378–389.
27. Grotendorst, GR, Rahmanie, H and Duncan, MR (2004). Combinatorial signaling pathways determine fibroblast proliferation and myofibroblast differentiation. *FASEB J* 18: 469–479.
28. Shah, M, Foreman, DM and Ferguson, MW (1995). Neutralisation of TGF-beta 1 and TGF-beta 2 or exogenous addition of TGF-beta 3 to cutaneous rat wounds reduces scarring. *J Cell Sci* 108 (Pt 3): 985–1002.
29. Kim, YH, Woodley, DT, Wynn, KC, Giomi, W and Bauer, EA (1992). Recessive dystrophic epidermolysis bullosa phenotype is preserved in xenografts using SCID mice: development of an experimental *in vivo* model. *J Invest Dermatol* 98: 191–197.
30. Heinonen, S, Männikkö, M, Klement, JF, Whitaker-Menezes, D, Murphy, GF and Uitto, J (1999). Targeted inactivation of the type VII collagen gene (Col7a1) in mice results in severe blistering phenotype: a model for recessive dystrophic epidermolysis bullosa. *J Cell Sci* 112 (Pt 21): 3641–3648.
31. Kern, JS, Loeckermann, S, Fritsch, A, Hausser, I, Roth, W, Magin, TM *et al.* (2009). Mechanisms of fibroblast cell therapy for dystrophic epidermolysis bullosa: high stability of collagen VII favors long-term skin integrity. *Mol Ther* 17: 1605–1615.
32. Woodley, DT, Wang, X, Amir, M, Hwang, B, Remington, J, Hou, Y *et al.* (2013). Intravenously Injected Recombinant Human Type VII Collagen Homes to Skin Wounds and Restores Skin Integrity of Dystrophic Epidermolysis Bullosa. *J Invest Dermatol.* (e-pub ahead of print)
33. Kirfel, G and Herzog, V (2004). Migration of epidermal keratinocytes: mechanisms, regulation, and biological significance. *Protoplasma* 223: 67–78.
34. Li, W, Fan, J, Chen, M and Woodley, DT (2004). Mechanisms of human skin cell motility. *Histol Histopathol* 19: 1311–1324.
35. Larjava, H, Salo, T, Haapasalmi, K, Kramer, RH and Heino, J (1993). Expression of integrins and basement membrane components by wound keratinocytes. *J Clin Invest* 92: 1425–1435.
36. Fritsch, A, Loeckermann, S, Kern, JS, Braun, A, Bösl, MR, Bley, TA *et al.* (2008). A hypomorphic mouse model of dystrophic epidermolysis bullosa reveals mechanisms of disease and response to fibroblast therapy. *J Clin Invest* 118: 1669–1679.
37. Namazi, MR, Fallahzadeh, MK and Schwartz, RA (2011). Strategies for prevention of scars: what can we learn from fetal skin? *Int J Dermatol* 50: 85–93.

38. Natarajan, E, Omobono, JD 2nd, Guo, Z, Hopkinson, S, Lazar, AJ, Brenn, T et al. (2006). A keratinocyte hypermotility/growth-arrest response involving laminin 5 and p16INK4A activated in wound healing and senescence. *Am J Pathol* 168: 1821–1837.
39. Trelford, JD and Trelford-Sauder, M (1979). The amnion in surgery, past and present. *Am J Obstet Gynecol* 134: 833–845.
40. Uitto, J (2012). Molecular therapeutics for heritable skin diseases. *J Invest Dermatol* 132(E1): E29–E34.
41. Chen, M, Marinkovich, MP, Veis, A, Cai, X, Rao, CN, O'Toole, EA et al. (1997). Interactions of the amino-terminal noncollagenous (NC1) domain of type VII collagen with extracellular matrix components. A potential role in epidermal-dermal adherence in human skin. *J Biol Chem* 272: 14516–14522.
42. Chen, M, Petersen, MJ, Li, HL, Cai, XY, O'Toole, EA and Woodley, DT (1997). Ultraviolet A irradiation upregulates type VII collagen expression in human dermal fibroblasts. *J Invest Dermatol* 108: 125–128.
43. Sakai, LY, Keene, DR, Morris, NP and Burgeson, RE (1986). Type VII collagen is a major structural component of anchoring fibrils. *J Cell Biol* 103: 1577–1586.
44. Sakai, LY and Keene, DR (1994). Fibrillin: monomers and microfibrils. In Ruoslahti, E and Engvall, E (eds). *Methods in Enzymology*, vol. 245. Academic Press: New York. pp. 47–50.
45. van den Dolder, J, Mooren, R, Vloon, AP, Stoelinga, PJ and Jansen, JA (2006). Platelet-rich plasma: quantification of growth factor levels and the effect on growth and differentiation of rat bone marrow cells. *Tissue Eng* 12: 3067–3073.
46. Cheng, CF, Sahu, D, Tsen, F, Zhao, Z, Fan, J, Kim, R et al. (2011). A fragment of secreted Hsp90a carries properties that enable it to accelerate effectively both acute and diabetic wound healing in mice. *J Clin Invest* 121: 4348–4361.

CITATION

CHAPTER 1
Volker F. Wendisch, Steffen N. Lindner and Tobias M. Meiswinkel (2011). Use of Glycerol in Biotechnological Applications, Biodiesel- Quality, Emissions and By-Products, Dr. Gisela Montero (Ed.), ISBN: 978-953-307-784-0, InTech, DOI: 10.5772/25338.

CHAPTER 2
Alban SM, de Moura JF, Thomaz-Soccol V, Sékula SB, Alvarenga LM, Mira MT, et al. (2014) Phage Display and Synthetic Peptides as Promising Biotechnological Tools for the Serological Diagnosis of Leprosy. PLoS ONE 9(8): e106222. doi:10.1371/journal.pone.0106222

CHAPTER 3
Seiya Sato and Hiroaki Itamochi (2015). DNA Repair and Chemotherapy, Advances in DNA Repair, Prof. Clark Chen (Ed.), ISBN: 978-953-51-2209-8, InTech, DOI: 10.5772/59513.

CHAPTER 4
C. Vilos and L. A. Velasquez, "Therapeutic Strategies Based on Polymeric Microparticles," Journal of Biomedicine and Biotechnology, vol. 2012, Article ID 672760, 9 pages, 2012. doi:10.1155/2012/672760

CHAPTER 5

Akinori Nakamura and Shin'ichi Takeda, "Mammalian Models of Duchenne Muscular Dystrophy: Pathological Characteristics and Therapeutic Applications," Journal of Biomedicine and Biotechnology, vol. 2011, Article ID 184393, 8 pages, 2011. doi:10.1155/2011/184393

CHAPTER 6

Zhe Liu, Fabian Kiessling, and Jessica Gätjens, "Advanced Nanomaterials in Multimodal Imaging: Design, Functionalization, and Biomedical Applications," Journal of Nanomaterials, vol. 2010, Article ID 894303, 15 pages, 2010. doi:10.1155/2010/894303

CHAPTER 7

João Conde, João Rosa, João C. Lima, and Pedro V. Baptista, "Nanophotonics for Molecular Diagnostics and Therapy Applications," International Journal of Photoenergy, vol. 2012, Article ID 619530, 11 pages, 2012. doi:10.1155/2012/619530

CHAPTER 8

Simona Casarosa, Yuri Bozzi and Luciano Conti, Neural stem cells: ready for therapeutic applications? Molecular and Cellular Therapies 2014 2:31, DOI: 10.1186/2052-8426-2-31

CHAPTER 9

Yi-Ling Ye, Ya-Hui Chuang, and Bor-Luen Chiang. Strategies of mucosal immunotherapy for allergic diseases, Cellular & Molecular Immunology (2011) 8, 453–461; doi:10.1038/cmi.2011.17

CHAPTER 10

Xinyi Wang, Pedram Ghasri, Mahsa Amir, Brian Hwang, Yingpin Hou, Michael Khilili, Andrew Lin, Douglas Keene, Jouni Uitto, David T Woodley, and Mei Chen ().Topical Application of Recombinant Type VII Collagen Incorporates Into the Dermal–Epidermal Junction and Promotes Wound Closure, Molecular Therapy (2013); 21 7, 1335–1344. doi:10.1038/mt.2013.87

INDEX

A

Adeno-associated virus (AAV) 117, 129
Allergic diseases 234
Alzheimer's disease (AD) 205
Amino acids 22
amyotrophic lateral sclerosis (ALS) 205
anchoring fibrils (AFs) 265, 266
antisense oligonucleotides (AOs) 117
ataxia telangiectasia mutated (ATM) 70

B

Basal Progenitors (BPs) 199
base excision repair (BER) 71
Becker muscular dystrophy (BMD) 117
biodiesel production 1, 2, 3, 19, 28, 33
bioluminescence imaging (BLI) 140
Biosynthesis 12, 24, 25, 31, 47, 49
bovine serum albumin (BSA) 54

C

Cancer chemotherapy 69, 90
cardiomyocytes 101
Central Nervous System (CNS) 198, 209
Citric acid 21, 30, 45
computed tomography (CT) 138
Computed tomography (CT) 140
connective tissue growth factor (CTGF) 278
Corynebacterium glutamicum 2, 11, 22, 30, 31, 33, 34, 35, 36, 37, 38, 39, 40, 41, 42, 43, 44, 45, 46, 47, 49
creatine kinase (CK) 119
cross-linked iron oxide (CLIO) 153
cyclin-dependent kinase (CDK 70

D

dendritic cells (DCs) 236
dermal–epidermal junction (DEJ) 265, 266
Dihydroxyacetone 2, 5, 30, 32, 34, 37, 39, 42, 48
DNA damage response (DDR) 70
double-strand break (DSB) 72
Duchenne muscular dystrophy (DMD) 115, 116
dystrophin-glycoprotein complex (DGC) 116

E

embryoid bodies (EBs) 204
embryonic stem cells (ESCs) 203
endemic controls (EC) 54

epidermal growth factor receptor (EGFR) 176
Ethanol 11, 19, 20, 32, 40

F

Fluorescence imaging (FLI) 140
fluorescence resonance energy transfer (FRET) 154
Food and Drug Administration (FDA) 94
fossil fuels 1

G

Glycerol 1, 2, 3, 9, 13, 19, 21, 24, 26, 27, 28, 30, 32, 33, 34, 37, 38, 39, 40, 42, 44, 45, 46, 47, 48, 49, 50, 297
good medical practice (GMP) 199
growth factor (GF) 266

H

homologous recombination (HR) 72
house dust mite (HDM) 235
household contacts (HC) 54
Huntington's disease (HD) 205

I

induced pluripotent stem cells (iPSCs) 203
inner cell mass (ICM) 203
Iron oxide nanoparticles (IONPs) 146

L

leprosy 53, 54, 55, 56, 59, 60, 61, 63, 64, 65, 66, 67, 68
localized surface plasmon resonance (LSPR) 173

M

magnetic resonance imaging (MRI) 138, 178
Magnetic resonance imaging (MRI) 178
multibacillary (MB) 54

multiple sclerosis (MS) 205
muscle-specific creatine kinase (MCK) 118, 119
Mycobacterium leprae 53, 54, 66, 67, 68
myofibroblasts 265, 278, 279, 284, 286

N

nanomedicine 135, 136, 145, 146, 154, 161, 163, 165
nanoparticles (NPs) 173
Nanophotonics 171, 172, 178, 183, 298
National Nanotechnology Initiative (NNI) 135
near-infrared fluorescence (NIRF) 148
near infrared (NIR) 178
neural stem cell (NSC) 197
new drug application (NDA) 94
non-homologous end joining (NHEJ) 72
nuclear magnetic resonance (NMR) 178
nucleotide excision repair (NER) 71

O

optical coherence tomography (OCT) 178
Optical Coherence Tomography (OCT) 179, 192
optical imaging (OI) 138

P

Parkinson's disease (PD) 205
paucibacillary (PB) 55
personalized medicine 93
phenomena 171
phosphate-buffered saline (PBS) 57
phosphoenolpyruvate 4, 5, 10, 21, 22, 31, 33, 45
photoacoustic imaging (PAI) 178, 179
photodynamic therapy (PDT) 154
plasmon resonant particles (PRPs) 174
polymeric microparticles 93, 94, 95, 96, 100, 101, 102, 103, 106
Positron emission tomography (PET) 141

prostate specific membrane antigen (PSMA) 154
pyrroloquinoline quinone 6, 7

Q

Quantum dots (QDs) 153

R

radial glia (RG) 199
reactive oxygen species (ROS) 73
recessive dystrophic epidermolysis bullosa (RDEB) 265, 266
rostral migratory stream (RMS) 200

S

signal intensity (SI) 152
single-strand breaks (SSBs) 71
smart crosslinked iron oxide 148
specific immunotherapy (SIT) 233, 234
subcutaneous immunotherapy (SCIT) 233
subgranular zone (SGZ) 200
sublingual immunotherapy (SLIT) 233, 244, 262
substantia nigra (SN) 205
subventricular zone (SVZ) 200
Succinic acid 20, 39
surface-enhanced Raman scattering (SERS) 154
Surface enhancement raman spectroscopy (SERS) 175

T

translesion synthesis (TLS) 72
tuberculosis (TB) 56
two-photon luminescence (TPL) 178, 180

W

World Health Organization (WHO) 55